MARINE NAVIGATION AND SAFETY OF SEA TRANSPORTATION

T0132562

Marine Navigation and Safety of Sea Transportation

Advances in Marine Navigation

Editor

Adam Weintrit
Gdynia Maritime University, Gdynia, Poland

CRC Press
Taylor & Francis Group
Boca Raton   London   New York   Leiden

CRC Press is an imprint of the
Taylor & Francis Group, an **informa** business

A BALKEMA BOOK

Published by:
CRC Press/Balkema
P.O. Box 447, 2300 AK Leiden, The Netherlands
e-mail: Pub.NL@taylorandfrancis.com
www.crcpress.com – www.taylorandfrancis.com

First issued in paperback 2020

ISBN 13: 978-0-367-57642-4 (pbk)
ISBN 13: 978-1-138-00106-0 (hbk)

Visit the Taylor & Francis Web site at
http://www.taylorandfrancis.com

and the CRC Press Web site at
http://www.crcpress.com

Typeset by V Publishing Solutions Pvt Ltd., Chennai, India

List of reviewers

Prof. Roland **Akselsson**, Lund University, Sweden
Prof. Yasuo **Arai**, Independent Administrative Institution Marine Technical Education Agency,
Prof. Vidal **Ashkenazi**, Nottingham Scientific Ltd, UK
Prof. Terje **Aven**, University of Stavanger (UiS), Stavanger, Norway
Prof. Michael **Baldauf**, Word Maritime University, Malmö, Sweden
Prof. Andrzej **Banachowicz**, West Pomeranian University of Technology, Szczecin, Poland
Prof. Marcin **Barlik**, Warsaw University of Technology, Poland
Prof. Michael **Barnett**, Southampton Solent University, United Kingdom
Prof. Eugen **Barsan**, Constanta Maritime University, Romania
Prof. Milan **Batista**, University of Ljubljana, Ljubljana, Slovenia
Prof. Angelica **Baylon**, Maritime Academy of Asia & the Pacific, Philippines
Prof. Knud **Benedict**, University of Wismar, University of Technology, Business and Design, Germany
Prof. Christophe **Berenguer**, Grenoble Institute of Technology, Saint Martin d'Heres, France
Prof. Tor Einar **Berg**, Norwegian Marine Technology Research Institute, Trondheim, Norway
Prof. Jarosław **Bosy**, Wroclaw University of Environmental and Life Sciences, Wroclaw, Poland
Sr. Jesus **Carbajosa** Menendez, President of Spanish Institute of Navigation, Spain
Prof. A. Güldem **Cerit**, Dokuz Eylül University, Izmir, Turkey
Prof. Shyy Woei **Chang**, National Kaohsiung Marine University, Taiwan
Prof. Frank **Coolen**, Durham University, UK
Prof. Stephen J. **Cross**, Maritime Institute Willem Barentsz, Leeuwarden, The Netherlands
Prof. Krzysztof **Czaplewski**, Polish Naval Academy, Gdynia, Poland
Prof. Decio Crisol **Donha**, Escola Politécnica Universidade de Sao Paulo, Brazil
Prof. Eamonn **Doyle**, National Maritime College of Ireland, Cork Institute of Technology, Cork, Ireland
Prof. Milan **Džunda**, Technical University of Košice, Slovakia
Prof. Bernd **Eisfeller**, University of FAF, Munich, Germany
Prof. Ahmed **El-Rabbany**, University of New Brunswick; Ryerson University in Toronto, Ontario, Canada
Prof. Alfonso **Farina**, SELEX-Sistemi Integrati, Rome, Italy
Prof. Andrzej **Fellner**, Silesian University of Technology, Katowice, Poland
Prof. Andrzej **Felski**, Polish Naval Academy, Gdynia, Poland
Prof. Włodzimierz **Filipowicz**, Gdynia Maritime University, Poland
Prof. Börje **Forssell**, Norwegian University of Science and Technology, Trondheim, Norway
Prof. Jens **Froese**, Jacobs University Bremen, Germany
Prof. Masao **Furusho**, Kobe University, Japan
Prof. Wiesław **Galor**, Maritime University of Szczecin, Poland
Prof. Stanislaw **Górski**, Gdynia Maritime University, Poland
Prof. Marek **Grzegorzewski**, Polish Air Force Academy, Deblin, Poland
Prof. Lucjan **Gucma**, Maritime University of Szczecin, Poland
Prof. Stanisław **Gucma**, Maritime University of Szczecin, Poland
Prof. Jerzy **Hajduk**, Maritime University of Szczecin, Poland
Prof. Qinyou **Hu**, Shanghai Maritime University, China
Prof. Stojce Dimov **Ilcev**, Durban University of Technology, South Africa
Prof. Toshio **Iseki**, Tokyo University of Marine Science and Technology, Japan,
Prof. Ales **Janota**, University of Žilina, Slovakia
Prof. Jacek **Januszewski**, Gdynia Maritime University, Poland
Prof. Mirosław **Jurdziński**, Gdynia Maritime University, Poland
Prof. John **Kemp**, Royal Institute of Navigation, London, UK
Prof. Lech **Kobyliński**, Polish Academy of Sciences, Gdansk University of Technology, Poland
Prof. Krzysztof **Kołowrocki**, Gdynia Maritime University, Poland
Prof. Serdjo **Kos**, University of Rijeka, Croatia
Prof. Pentti **Kujala**, Helsinki University of Technology, Helsinki, Finland
Prof. Krzysztof **Kulpa**, Warsaw University of Technology, Warsaw, Poland
Prof. Shashi **Kumar**, U.S. Merchant Marine Academy, New York
Prof. Bogumił **Łączyński**, Gdynia Maritime University, Poland

Prof. Andrzej **Lenart**, Gdynia Maritime University, Poland
Prof. Nadav **Levanon**, Tel Aviv University, Tel Aviv, Israel
Prof. Vladimir **Loginovsky**, Admiral Makarov State Maritime Academy, St. Petersburg, Russia
Prof. Mirosław **Luft**, University of Technology and Humanities in Radom, Poland
Prof. Evgeniy **Lushnikov**, Maritime University of Szczecin, Poland
Prof. Jerzy **Mikulski**, Silesian University of Technology, Katowice, Poland
Prof. Wacław **Morgaś**, Polish Naval Academy, Gdynia, Poland
Prof. Reinhard **Mueller**, Chairman of the DGON Maritime Commission, Germany
Prof. Janusz **Narkiewicz**, Warsaw University of Technology, Poland
Prof. Nikitas **Nikitakos**, University of the Aegean, Chios, Greece
Prof. Andy **Norris**, The Royal Institute of Navigation, University of Nottingham, UK
Prof. Stanisław **Oszczak**, University of Warmia and Mazury in Olsztyn, Poland
Prof. Gyei-Kark **Park**, Mokpo National Maritime University, Mokpo, Korea
Mr. David **Patraiko**, The Nautical Institute, UK
Prof. Zbigniew **Pietrzykowski**, Maritime University of Szczecin, Poland
Prof. Boris **Pritchard**, University of Rijeka, Croatia
Prof. Jonas **Ringsberg**, Chalmers University of Technology, Gothenburg, Sweden
Prof. Jerzy B. **Rogowski**, Warsaw University of Technology, Poland
Prof. Hermann **Rohling**, Hamburg University of Technology, Hamburg, Germany
Prof. Władysław **Rymarz**, Gdynia Maritime University, Poland
Prof. Abdul Hamid **Saharuddin**, Universiti Malaysia Terengganu (UMT), Terengganu, Malaysia
Prof. Aydin **Salci**, Istanbul Technical University, Maritime Faculty, ITUMF, Istanbul, Turkey
Prof. Shigeaki **Shiotani**, Kobe University, Japan
Prof. Wojciech **Ślączka**, Maritime University of Szczecin, Poland
Prof. Leszek **Smolarek**, Gdynia Maritime University, Poland
Prof. Jac **Spaans**, Netherlands Institute of Navigation, The Netherlands
Prof. Cezary **Specht**, Polish Naval Academy, Gdynia, Poland
Cmdr. Bengt **Stahl**, Nordic Institute of Navigation, Sweden
Prof. Andrzej **Stateczny**, Maritime University of Szczecin, Poland
Prof. Elżbieta **Szychta**, University of Technology and Humanities in Radom, Poland
Prof. El **Thalassinos**, University of Piraeus, Greece
Prof. Vladimir **Torskiy**, Odessa National Maritime Academy, Ukraine
Prof. Gert F. **Trommer**, Karlsruhe University, Karlsruhe, Germany
Prof. Mykola **Tsymbal**, Odessa National Maritime Academy, Ukraine
Prof. Nguyen **Van Thu**, Ho Chi Minh City University of Transport, Ho Chi Minh City, Vietnam
Prof. František **Vejražka**, Czech Technical University in Prague, Czech
Prof. George Yesu Vedha **Victor**, International Seaport Dredging Limited, Chennai, India
Prof. Peter **Voersmann**, Deutsche Gesellschaft für Ortung und Navigation, Germany
Prof. Vladimir A. **Volkogon**, Baltic Fishing Fleet State Academy, Kaliningrad, Russian Federation
Prof. Ryszard **Wawruch**, Gdynia Maritime University, Poland
Prof. Ruan **Wei**, Shanghai Maritime University, Shanghai, China
Prof. Adam **Weintrit**, Gdynia Maritime University, Poland
Prof. Bernard **Wiśniewski**, Maritime University of Szczecin, Poland
Prof. Krystyna **Wojewódzka-Król**, University of Gdańsk, Poland
Prof. Adam **Wolski**, Maritime University of Szczecin, Poland
Prof. Jia-Jang **Wu**, National Kaohsiung Marine University, Kaohsiung, Taiwan (ROC)
Prof. Hideo **Yabuki**, Tokyo University of Marine Science and Technology, Tokyo, Japan
Prof. Lu **Yilong**, Nanyang Technological University, Singapore
Prof. Homayoun **Yousefi**, Chabahar Maritime University, Iran
Prof. Janusz **Zieliński**, Space Research Centre, Warsaw, Poland

TABLE OF CONTENTS

Advances in Marine Navigation
Introduction

A. Weintrit
Gdynia Maritime University, Gdynia, Poland
Polish Branch of the Nautical Institute

The monograph is addressed to scientists and professionals in order to share their expert knowledge, experience and research results concerning all aspects of navigation, safety at sea and marine transportation.

The contents of the book are partitioned into ten separate chapters: Navigational tools and equipment (covering the subchapters 1.1 through 1.3), ECDIS - Electronic Chart Display and Information Systems (covering the chapters 2.1 through 2.5), e-Navigation concept development (covering the chapters 3.1 through 3.7), Maritime simulators (covering the chapters 4.1 through 4.7), Manoeuvrability (covering the chapters 5.1 through 5.2), GNSS - Global Navigation Satellite Systems (covering the chapters 6.1 through 6.6), Radar, ARPA and anti-collision (covering the chapters 7.1 through 7.7), Watchkeeping and safety at sea (covering the chapters 8.1 through 8.3), Historical aspects of navigation (covering the chapter 9.1) and Safety, reliability and risk assessment (covering the chapters 10.1 through 10.4).

In each of them readers can find a few suchapters. Subchapters collected in the first chapter, titled 'Navigational tools and equipment', concerning the method for compensation of magnetic compass deviation, study on the errors in the free-gyro positioning and directional system and optimal time intervals between observations during heading in data analysis obtained from AIS by ration plotting technique at virtual second observation epoch. Certainly, this subject may be seen from different perspectives.

In the second chapter there are described problems related to ECDIS implementation: inciting the development of engaging screencasts in teaching ECDIS, presentation of magnetic variation on ENC charts according to the rules and good sea practice, implementation of ship collision avoidance supporting system on ECDIS, protection and risks of ENC data regarding safety of navigation, and the fusion of coordinates of ship position and chart features.

Third chapter is about e-Navigation concept development. The readers can find some information about hydrographic data as the basis for integrated e-Navigation data streams, Universal Hydrographic Data Model (UHDM), software quality assurance issues related to e-Navigation, implementation of e-Navigation concept in Turkey, transformations leading to introduction of e-Navigation, e-Voyage Planning and finaly the answer to the question: will satellite-based AIS supersede LRIT?

The fourth chapter deals with marine simulators problems. The contents of the fourth chapter are partitioned into seven subchapters: new concept for maritime safety and security emergency management – simulation based training designed for the safety & security trainer (SST7); advanced ship handling using simulation augmented manoeuvring design and monitoring – a new method for increasing safety & efficiency; simulator programs (2-D and 3-D): influence on learning process of BSMT and BSMAR-E students at Maritime University, Philippines; maritime simulation assessment using behavioral framework; training on simulator for emergency situations in the Black Sea; simulation of BOTAS Ceyhan Marine Terminals; and recommendation for simulator base training programs for ships with azimuthing control devices.

The fifth chapter deals with manoeuvrability. The contents of the fifth chapter are partitioned into two: proposal for global standard maneuvering orders for tugboats, and feasibility analysis of orthogonal anchoring by merchant ships.

In the sixth chapter there are described problems related to GNSS - Global Navigation Satellite Systems: single-frequency horizontal GPS positioning error response to a moderate ionospheric storm over Northern Adriatic, accuracy of the GPS position indicated by different maritime receivers, cost-efficient, subscription-based, and secure transmission of GNSS data for differential augmentation via TV satellite links, method of improving EGNOS service in local conditions, study of the RF front-end of the multi-constellation GNSS receiver, and a new user integrity monitoring for multiple ramp failures.

Seventh chapter is about radar, ARPA and anti-collision. The readers can find some information about improper lights a confusing and fatal factor in the Bluebird/Debonair collision in Dublin Bay, lesson learned during the realization of the automated radar control system for Polish sea-waters, ARPA suitable for automatic assessment of AIS targets, identification of object difficult to detect by synchronous radar observation method, selected methods of UWB radar signal processing, the duty of an anchored vessel to avoid collision and data integration in the Integrated Navigation System (INS) in function of the digital processing algorithms used in the frequency modulated continuous wave (FM-CW) radar.

The eighth chapter deals with watchkeeping and safety at sea. The contents of the eighth chapter are partitioned into three: See more – analysis of possibilities of Implementation AR solutions during bridge watchkeeping, Investigation of officers' navigation and port watches exposed to excessive working hours, and Enhancing of carriers' liabilities in the Rotterdam Rules – too expensive costs for navigational safety?

The ninth chapter deals with historical aspects of navigation. The contents of the ninth chapter concerns the Baltic light vessels in the nineteenth and twentieth centuries.

In the tenth chapter there are described problems related to safety, reliability and risk assessment: Maritime risk assessment: modeling collisions with vessels lying at an anchorage, Reliability of the navigator – navigation complex system, and Fuzzy risk of ship grounding in restricted waters generic competencies for resilient systems.

Each chapter was reviewed at least by three independent reviewers. The Editor would like to express his gratitude to distinguished authors and reviewers of chapters for their great contribution for expected success of the publication. He congratulates the authors for their excellent work.

Chapter 1

Navigational Tools and Equipment

Short Method for Compensation of Magnetic Compass Deviation

E. Lushnikov
Maritime University of Szczecin, Poland

ABSTRACT: It is proposed new method for compensation of magnetic compass deviation at four intermediate course.
It is described the basis of method from point of accuracy and duration of time. It is shown, that new method has absolutely freedom of actions. In generally, the accuracy of new method is better, than classical standard methods. The most important factors of deviation A, B, C, D can be at control by new method.

1 INTRODUCTION

The compensation of magnetic compass deviation require according time, knowledge and skills, means and methods. The improvement of means and methods gives possibility to reach of high accuracy, reduce and simplify the procedural questions.

Still at time of Archibald-Smith made repeated attempts to streamline compensation procedures of deviation, but the most common way now is the way of Erie [1, 5, 8]. You could call it the classical way, with the most widespread in the world.

The coefficients of deviation *A, D* and *E* which are produced by ships soft iron, are usually destroyed at first compensation after the building of the vessel. As a rule this action gives possibility to close this problem [2, 3, 5] for very long time.

The most unstable and most largest coefficients of deviation *B* and *C* shall be destroyed in accordance with the requirements every year. The most common way of compensation today as 100 years ago is the method of Erie. This method is executed on the 4-th main courses N, E, S, W.

2 THE ACCURACY OF ERIE METHOD.

The accuracy of compensation of coefficients *B* and *C* is depending from accuracy of transactions by magnets-compensators at the courses N, E, S, W.

The standard errors of compensation for coefficients B and C are characterized by:

$$m_B = \sqrt{m_E^2 + m_W^2}$$
$$m_C = \sqrt{m_N^2 + m_S^2} \qquad (1)$$

where:

m_B – standard error of compensation for factor *B*;
m_C – standard error of compensation of factor *C*;
m_N – standard error of compensation of deviation to 0^0 at the course *N*;
m_S – standard error of decreasing of deviation in two time at the course of *S*;
m_E – standard error of compensation of deviation to 0^0 at the course *E*;
m_W – standard error of decreasing of deviation in two time at the course of *W*;

The errors of compensation of deviation or his double decreasing is depending from observations of courses and from accuracy of operation by means of magnet-compensators.

The process of observation is very simple and he is not required a special experiences and skills.

The process of installation of magnet-kompensator to according position is not so simple. It is a just what it is the main reason of mistake of compensation the factor *B* and *C*. From this reason the minimization of operations errors by magnet-kompensator is condition of high accuracy of compensation.

The data of observation at the method of Airy do not possibility of preliminary control (before beginning of compensation) of factors *A, B, C, D, E.*

The factors *A* and *E*, which is depending from asymmetrical soft steel *d* and *b* at symmetrical ships is usually [7, 9] at the level $0.2^0 \div 0.6^0$ and is very stable:

$$A = \frac{d-b}{2\lambda}$$
$$E = \frac{d+b}{2\lambda}$$
(2)

The coefficient A may have an additional component of errors due to deflection around the vertical axis of compass binnacle. From this reason, the control of factor A is more necessary than for factor E.

The Coefficient D after his compensation is characterized by high stability [1,2], although originally (before compensation) it reaches 3^0-5^0.

Thus, especially is preferred the monitoring of stability for such factors as D and A but the factor E are usually a little interesting.

3 THE METHOD FOR COMPENSATION OF DEVIATION AT INTERMEDIATE COURSES

To reduce the time of operation, increase accuracy and provide preliminary control coefficients D and A it is proposed the compensation of half-circle deviation at the four intermediate compass courses NE, SE, SW, NW. The monitoring of deviations at these compass courses give possibility to calculate the factors of deviation A, B, C, D by the formulas:

$$A = \frac{\delta_{NE} + \delta_{SE} + \delta_{SW} + \delta_{NW}}{4}$$
$$B = \frac{\delta_{NE} + \delta_{SE} - \delta_{SW} - \delta_{NW}}{2\sqrt{2}}$$
$$C = \frac{\delta_{NE} - \delta_{SE} - \delta_{SW} + \delta_{NW}}{2\sqrt{2}}$$
$$D = \frac{\delta_{NE} - \delta_{SE} + \delta_{SW} - \delta_{NW}}{4}$$
(3)

The accuracy of factors A, B, C, D from the formula (3) is characterized by value:

$$m_A = \frac{m_b}{2};$$
$$m_B = \frac{m_b}{\sqrt{2}};$$
$$m_C = \frac{m_b}{\sqrt{2}};$$
$$m_D = \frac{m_b}{2}$$
(4)

where:
m_A, m_B, m_C, m_D - the standard errors of factors A, B, C, D;
m_b – the standard error of bearing.

If the method of Erie does not allows to have previously information about the value of factors for compensation, the proposed method by the equations (3) gives exhaustive information on this account.

In the most general case, there are still the fifth factor E that, as a rule is very small ($0.2^0 \div 0.6^0$) and stable at the ship of symmetrical design. For this reason, the coefficient of E can be taken from the previous table.

If the coefficients B, or C, or even both of these coefficients are small and there is possibility do not destroy these coefficients, they might not be destroyed. At the way of Erie even detect such situations is not possible in principle, although this situation may actually exist.

Found in the measurements of coefficients A and D allow to check their stability.

Knowing at last compass course $CC = 315^0$ all four coefficient of A, B, C, D we have possibility at the same course to offset any of these coefficient and even all four, if it is necessary.

The coefficient B is destroyed on this course, as is always longitudinal magnets-compensators, coefficient C is destroyed by cross-magnets-compensators, D – by compensatory balls, rods or plates of soft iron, and the coefficient A is compensated by turning of foundation of the compass round the vertical axis.

For compensation of factor B (at last compass course $CC = 315^0$) by means of longitudinal magnet-compensators it is putting the deviation, which must be on this course at condition of $B=0$. On the found factors of deviation A, B, D (the influence of factor E to intermediate courses is zero) using the expression (3) write the deviation in the form:

$$\delta_{NW}^K = A^0 - \frac{\sqrt{2}}{2} \cdot 0^0 + \frac{\sqrt{2}}{2} \cdot C^0 - D^0 + E^0 \cdot 0$$
(5)

Setting by longitudinal magnets-compensators such deviation, the coefficient B is compensated. The magnetic course at this operation becomes equal to:

$$MK = 315^0 + A^0 + 0.7 \cdot C^0 - D^0$$
(6)

For compensation of factor C at proceeding to follow the same course (compass course $CC = 315^0$), by the lateral magnets-compensator set the value of deviation to:

$$\delta_{NW}^K = A^0 - D^0$$
(7)

The magnetic course MC at this situation is equal to:

$$MK = 315^0 + A^0 - D^0$$
(8)

After compensation of coefficient C at necessity of compensation the coefficient D, it is necessary at the same compass course by means of balls or plates of magnetic soft iron to set the deviation:

$$\delta_{NW}^{K} = A^0 \qquad (9)$$

Magnetic course after this operation has the value:

$$MC = 315^0 - A^0 \qquad (10)$$

The coefficient D at this situation *is* compensated.

If it is necessary to compensate for factor A, then it is done without the use of iron by rotating of compasses foundation so that satisfied the condition:

$$\delta_{NW}^{K} = 0^0 \qquad (11)$$

If this condition the magnetic course is equal:

$$MC = 315^0 \qquad (12)$$

Thus, the proposed method, unlike the way of Erie, allows to compensate four factor of deviation A, B, C, D. This is the first and most significant advantage of the proposed method over traditional.

The second advantage is time saving. The compensation of factors B and C is fulfilled by means of two operations instead of four.

The third advantage is increasing of accuracy at compensation of coefficients B and C in $\sqrt{2}$ times because of dual decreasing of number operations by magnets.

The fourth advantage of the method is the fact that the method gives absolute freedom.

If as a result of observations at the four intermediate courses revealed that all coefficients was changed (even significantly), but they is lie at the acceptable boundaries, it is not requiring of compensation, the deviation can be do not destroyed.

It can be calculated the new table of deviation at using of observations information.

Thus, the proposed method gives absolute freedom of action. This fact is the most value at all times and at all peoples.

REFERENCES

[1] M. Jurdziński. Dewiacja i kompensacja morskich kompasów magnetycznych. Gdynia 2000. 182s.
[2] Łusznikow E.M., R.K. Dzikowski. Dewiacja kompasu magnetycznego. Szczecin. 2012.104s.
[3] E.M. Lushnikov. Compensation of magnetic compass deviation at contemporary condition. International scientific conference "Innovation in scientific and education - 2008" Kaliningrad, KGTU, 2008. p. 22-24.
[4] E.M. Lushnikov. The problem of magnetic compass deviation at contemporary condition. International Navigational Symposium "TransNav 09". Gdynia, Maritime University 2009. p. 219-224.
[5] E.M. Lushnikov. About the problem of magnetic compass at contemporary ships. 14th International scientific and technical conference "Marine Traffic Engineering". Szczecin, 2011. p. 293-296.
[6] The problem of Magnetic Compass Deviation at Contemporary Condition International Navigational Symposium "TransNav 09". Gdynia, AM. 2009. p.219-220.
[7] Compensation of Magnetic Compass deviation at single any course. TransNav International Journal of Marine Navigation and Safety of sea Transportation. Volume 5, N3. 2011 pp. 303-307.
[8] Compensation of Magnetic Compass Deviation at any Heading. Scientific Journal of Maritime University of Szczecin. Szczecin. 2010. pp. 46-52.
[9] The problem of magnetic compass at contemporary condition. "X1V International scientific and technical conference on marine traffic engineering". Maritime University of Szczecin. 2011. P.293-296.
[10] Fukuda G. Hajashi S. The Baltic Research for the New Compass System Using Latest MEMS. TransNav International Journal of Marine Navigation and Safety of Sea Transportation. Volume 4. N3. Pp. 317-322.
[11] Е.Л. Смирнов, А.В. Яловенко, В.В. Воронов. Технические средства судовождения. Санкт-Петербург. 1996. 544с.

A Study on the Errors in the Free-Gyro Positioning and Directional System

T.G. Jeong
Korea Maritime University, Busan, Republic of Korea

ABSTRACT: This paper is to develop the position error equations including the attitude errors, the errors of nadir and ship's heading, and the errors of ship's position in the free-gyro positioning and directional system. In doing so, the determination of ship's position by two free gyro vectors was discussed and the algorithmic design of the free-gyro positioning and directional system was introduced briefly. Next, the errors of transformation matrices of the gyro and body frames, i.e., attitude errors, were examined and the attitude equations were also derived. The perturbations of the errors of the nadir angle including ship's heading were investigated in each stage from the sensor of rate of motion of the spin axis to the nadir angle obtained. Finally, the perturbation error equations of ship's position used the nadir angles were derived in the form of a linear error model and the concept of FDOP was also suggested by using covariance of position error.

1 INTRODUCTION

A free–gyro positioning system is to determine the position of a vehicle by using two free gyros. It is an active positioning system like an inertial navigation system (INS) in view of obtaining a position without external source. However, the FPS is to determine its own position by using the nadir angle between the vertical axis of local geodetic frame and the axis of free gyro, while an INS is to do it by measuring its acceleration.

In general the INS comprises a set of inertial measurement units (IMU's), both accelerometers and gyros, the platform on which they are mounted, including the stabilization mechanism if so provided, and the computer that performs the calculations needed to transform sensed accelerations and, in some mechanizations, angles or angular rates into navigationally useful information such as position, velocity and attitude. It is composed of a very complicated structure.

On the other hand, the free-gyro positioning and directional system consists of a set of two sensors of gyro axis motion rate and three sensors of the body frame, two free gyros, and the computer that calculates navigational information, position, etc. It is comparatively simpler than the INS.

Park & Jeong (2004) investigated how to determine the gyro vectors of two free gyros and the position of a vehicle by using the gyros. The errors in the FPS were investigated broadly by Jeong (2005). And the algorithmic design of free gyroscopic compass and positioning was suggested by measuring the earth's rotation rate on the basis of a free gyroscope (Jeong & Park, 2006; Jeong & Park, 2011).

Meanwhile, the free-gyro positioning and directional system is thought to have its own errors. This paper is to analyze such errors theoretically. Firstly, the errors of transformation matrices of the gyro and body frames, i.e., attitude errors, will be examined and the attitude equations be also derived. The perturbations of the errors of the nadir angle including ship's heading will be investigated in each stage from the sensor of rate of motion of the spin axis to the nadir angle obtained. Finally, the perturbation error equations of ship's position used the nadir angles will be derived in the form of a linear error model and the concept of FDOP will be also suggested by using covariance of position error.

Before the errors involved are discussed, the overview of the free-gyro positioning and directional system will be presented.

2 OVERVIEW OF FREE- GYRO POSITIONING AND DIRECTIONAL SYSTEM

2.1 *Determination of ship's position*

The nadir angle, θ, is given by an arbitrary position and gyro vector as shown as equation (1).

$$\cos\theta = -u_x \cos\phi\cos\left(\lambda + \omega_e t\right)$$
$$-u_y \cos\phi\sin\left(\lambda + \omega_e t\right) - u_z \sin\phi \quad (1)$$

Here, ω_e is the (presumably uniform) rate of Earth rotation, λ is the geodetic longitude, ϕ is the geodetic latitude and t denotes time. And u_x, u_y and u_z are the components of the gyro vector, g^i, whose superscript indicates the inertial frame.

If we use two gyro vectors of $g_a^i = \left[u_{ax}, u_{ay}, u_{az}\right]^T$ and $g_b^i = \left[u_{bx}, u_b, u_{bz}\right]^T$ in (1), we can determine the position $\left(\phi, \lambda\right)$ of a vehicle by using two corresponding nadir angles θ_a and θ_b.

Meanwhile the azimuth of the gyro vector from the north, α, is represented by equation (2):

$$\tan\alpha = \frac{E_D}{N_D}, \quad (2)$$

where,

$$N_D = -u_x \sin\phi\cos\left(\lambda + \omega_e t\right)$$
$$-u_y \sin\phi\sin\left(\lambda + \omega_e t\right) + u_z \cos\phi,$$
$$E_D = -u_x \sin\left(\lambda + \omega_e t\right) + u_y \cos\left(\lambda + \omega_e t\right).$$

Once determining the position, we can also obtain the azimuth of a gyro vector by using Eq. (2). Park and Jeong (2004) already suggested the algorithm of how to determine a position.

Fig.1 shows the measurement quantities in the local navigation frame.

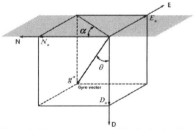

Figure 1. Measurement quantities in the local navigation frame

2.2 *Relation between ship's heading and azimuth of gyro vector*

As Jeong & Park (2006) mentioned, the north component of the earth's rotation rate is $\omega_e \cos\phi$, where ϕ depicts the geodetic latitude of an arbitrary

position. Fig. 2 shows that the angular velocities of the fore-aft and the athwartship components are given by equation (3) (Titterson, et al., 2004), where ψ is ship's heading. And it also shows that ς is the azimuth of a gyro vector from ship's head.

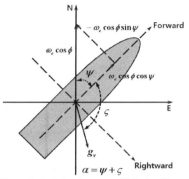

Figure 2. Relation between ship's heading and azimuth of a gyro vector

$$\omega_{Nx} = \omega_e \cos\phi \cos\psi$$
$$\omega_{Ny} = -\omega_e \cos\phi \sin\psi \quad (3)$$

By taking the ratio of the two independent gyroscopic measurement, the heading, ψ, is computed by (4).

$$\psi = -\tan^{-1}\frac{\omega_{Ny}}{\omega_{Nx}} \quad (4)$$

Meanwhile assuming that a gyro vector is ς away from ship's head, its azimuth from North is represented by Eq. (5). Therefore the angular velocity of the horizontal axis of a gyro, ω_H, is given by equation (6) on the navigation frame or local geodetic frame.

$$\alpha = \psi + \zeta \quad (5)$$

$$\omega_H = -\omega_e \cos\phi \sin\alpha \quad (6)$$

Equation (6) shows that if the north component of the earth's rotation rate can be known on the navigation frame, the nadir angle of a gyro vector, θ, is obtained by (7), by integrating Eq. (6) incrementally over a time interval.

$$\theta = \int_{t_1}^{t_2}\omega_H dt \quad (7)$$

2.3 *Algorithmic design of free-gyro positioning and directional system*

Fig. 3 and Fig. 4 show the algorithmic design of free gyros positioning system mechanization. In this mechanization two sensors for sensing the motion rate of the spin axis are mounted in the free gyro. Three sensors for sensing the motion rate of the platform are mounted in orthogonal triad. From the sensors in the gyro frame, the spin motion rate, $\omega_{i/g}^{g}$, is obtained and from the ones in the body frame, $\omega_{i/b}^{b}$, is detected. By using the sum, $\omega_{b/g}^{g}$, of the rates from the free gyro and the ones detected from the body sensors, the transformation matrix, C_g^b, and its inverse are determined. Therefore the spin motion rate, $\omega_{i/g}^{g}$, sensed from the free gyro is transformed into $\omega_{i/g}^{b}$ by using the matrix, C_g^b.

Meanwhile the rate of the earth's rotation, $\omega_{i/e}^{n}$, and the rate of the vehicle movement, $\omega_{e/n}^{n}$, are added to make $\omega_{i/n}^{n}$. Here $\omega_{e/n}^{n}$ is given by:

$$\omega_{e/n}^{n} = \begin{bmatrix} \dot{\lambda}\cos\phi & -\dot{\phi} & -\dot{\lambda}\sin\phi \end{bmatrix}^{T},$$

and $\omega_{i/n}^{n}$ is also expressed as:

$$\omega_{i/n}^{n} = \begin{bmatrix} (\dot{\lambda}+\omega_e)\cos\phi & -\dot{\phi} & -(\dot{\lambda}+\omega_e)\sin\phi \end{bmatrix}^{T}.$$

By using the matrix, C_n^b, it will be transformed into $\omega_{i/n}^{b}$, which is subtracted from the sensed rate from the body, $\omega_{i/b}^{b}$. As a result, $\omega_{n/b}^{b}$ is generated. By using this, the transformation matrix, C_b^n, and the inverse of it, C_n^b, are obtained. And the rate of $\omega_{i/g}^{b}$, is transformed into $\omega_{i/g}^{n}$ by using the matrix of C_b^n.

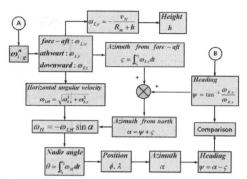

Figure 4. Free gyro positioning and directional system mechanization (2)

For ship's heading, by using the matrix of C_n^b, $\omega_{e/n}^{n}$ is changed into $\omega_{e/n}^{b}$, which is subtracted from $\omega_{i/g}^{b}$, and ω_D is calculated. Finally, using the transformation matrix, C_b^n, we can get the spin motion rate in the NED frame, ω_N, where $\omega_N = \begin{bmatrix} \omega_{Nx} & \omega_{Ny} & \omega_{Nz} \end{bmatrix}^{T}$. As a result, the ship's heading, ψ, is calculated by using the components of the spin motion rate according to equation (4).

Next for ship's position, let's look into the nadir angle (Fig. 4). The motion rate of the spin axis in the navigation frame, $\omega_{i/g}^{n}$, is given by $\omega_L = \begin{bmatrix} \omega_{Lx} & \omega_{Ly} & \omega_{Lz} \end{bmatrix}^{T}$. The azimuth of the gyro vector from the ship's head, ς, can be obtained by integrating ω_{Lz}. Then the azimuth of the gyro vector from the North, α, is obtained.

The horizontal angular velocity, ω_{LH}, and the tilting rate, ω_H, of the spin axis are calculated. And the nadir angle of the gyro vector, θ, can be obtained by integrating ω_H. Finally we can get the ship's position expressed by (ϕ, λ), using equation (1).

Next, using equation (2), we can also obtain the azimuth of the gyro vector, α, and ship's heading, ψ, which are modified by iteration. And ship's position is also corrected. In addition if only we know the northward component of ship's speed, we can also obtain the height, h.

3 PERTURBATION FORM OF ERROR EQUATIONS

This paper derives the error equations of the position and attitude by using linear error model forms (Jekeli, 2001; Roger, 2007).

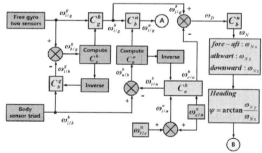

Figure 3. Free gyro positioning and directional system mechanization (1)

3.1 Gyro frame and body frame error equations

Gyro frame error equations can be derived by the transformation matrix, C_g^b. The estimated or computed matrix of it is represented by (8):

$$\bar{C}_g^b = (I - \Xi) C_g^b,$$ (8)

where Ξ is a skew-symmetric matrix, which is equivalent to the vector, $\xi = \begin{bmatrix} \xi_x & \xi_y & \xi_z \end{bmatrix}^T$, and is given by:

$$\Xi = \begin{bmatrix} 0 & -\xi_z & \xi_y \\ \xi_z & 0 & -\xi_x \\ -\xi_y & \xi_x & 0 \end{bmatrix}.$$

The differentiation form of C_g^b can be represented by the following.

$$\dot{C}_g^b = C_g^b \Omega_{b/g}^g$$ (9)

The error matrix of C_g^b is also given by:

$$\delta C_g^b = \bar{C}_g^b - C_g^b = -\Xi C_g^b$$ (10)

Taking the derivative of (10) yields

$$\delta \dot{C}_g^b = -\dot{\Xi} C_g^b - \Xi \dot{C}_g^b = -\dot{\Xi} C_g^b - \Xi C_g^b \Omega_{b/g}^g$$ (11)

The error equation of equation (9) is given by:

$$\delta \dot{C}_g^b = \delta \left(C_g^b \Omega_{b/g}^g \right) = \delta C_g^b \Omega_{b/g}^g + C_g^b \delta \Omega_{b/g}^g$$ (12)

where the perturbation in angular rate, $\delta \Omega_{b/g}^g$, denotes the error in the computed value, $\bar{\Omega}_{b/g}^g$, and is expressed as:

$$\delta \Omega_g^b = \bar{\Omega}_{b/g}^g - \Omega_{b/g}^g.$$ (13)

Substituting (13) into (12) and equating (12) with (11), we can get

$$-\dot{\Xi} C_g^b - \Xi C_g^b \Omega_{b/g}^g = \delta C_g^b \Omega_{b/g}^g + C_g^b \delta \Omega_{b/g}^g.$$ (14)

Substituting (10) into (14) and arranging it yields:

$$\dot{\Xi} = -C_g^b \delta \Omega_{b/g}^g C_b^g.$$ (15)

This is equivalent to the vector form given by:

$$\dot{\xi} = -C_g^b \delta \omega_{b/g}^g,$$ (16)

where $\delta \omega_{b/g}^g$ is the error in the rotation rate of the gyro frame relative to the body frame. This is separated by:

$$\omega_{b/g}^g = \omega_{i/g}^g - C_b^g \omega_{i/b}^b$$ (17)

An error of this equation will be represented by:

$$\delta \omega_{b/g}^g = \delta \omega_{i/g}^g - \left(C_b^g \Xi \omega_{i/b}^b + C_b^g \delta \omega_{i/b}^b \right).$$ (18)

Substituting (18) into (16) and rearranging it gives

$$\dot{\xi} = -\omega_{i/b}^b \times \xi - C_g^b \delta \omega_{i/g}^g + \delta \omega_{i/b}^b.$$ (19)

Equation (19) shows the error dynamics of the gyro frame attitude.

Similarly, we can get the error equations of body frame attitude as the following (20).

$$\dot{\gamma} = -\omega_{i/n}^n \times \gamma - C_b^n \delta \omega_{i/b}^b + \delta \omega_{i/n}^n$$ (20)

Here γ is the error angle of body frame attitude given by $\gamma = \begin{bmatrix} \gamma_x & \gamma_y & \gamma_z \end{bmatrix}^T$. This can be represented equivalently by the skew-symmetric matrix, Γ.

$$\Gamma = \begin{bmatrix} 0 & -\gamma_z & \gamma_y \\ \gamma_z & 0 & -\gamma_x \\ -\gamma_y & \gamma_x & 0 \end{bmatrix}$$ (21)

In addition, the error of the rate of motion of the navigation frame, $\delta \omega_{i/n}^n$, is expressed as:

$$\delta \omega_{i/n}^n = \begin{bmatrix} \delta \dot{\lambda} \cos \phi - (\dot{\lambda} + \omega_e) \delta \phi \sin \phi \\ -\delta \dot{\phi} \\ \delta \dot{\lambda} \sin \phi - (\dot{\lambda} + \omega_e) \delta \phi \cos \phi \end{bmatrix}.$$

3.2 Perturbations of error equations for nadir angle and ship's heading

In Fig. (3), the rate of motion of the spin axis in the navigation frame, $\omega_{i/g}^n$ is given by;

$$\omega_{i/g}^n = C_b^n C_g^b \omega_{i/g}^g.$$

We can get the error of this equation by using the differential operator, δ.

$$\delta \omega_{i/g}^n = -\left(\Gamma C_b^n + C_b^n \Xi \right) C_g^b \omega_{i/g}^g + C_b^n C_g^b \delta \omega_{i/g}^g$$ (22)

Here $\delta \omega_{i/g}^n = \begin{bmatrix} \delta \omega_{Lx} & \delta \omega_{Ly} & \delta \omega_{Lz} \end{bmatrix}^T$.

The error of the azimuth of the gyro vector from ship's head, $\delta \zeta$, and that of the north, $\delta \alpha$, are represented by:

$$\delta \zeta = \int_{t_1}^{t_2} \delta \omega_{Lz} dt,$$

$$\delta \alpha = \delta \zeta + \delta \psi.$$ (23)

The error of the horizontal rate of the spin axis, $\delta \omega_{LH}$, is expressed as:

$$\delta\omega_{LH} = \frac{\omega_{Lx}\delta\omega_{Lx} + \omega_{Ly}\delta\omega_{Ly}}{\sqrt{\omega_{Lx}^2 + \omega_{Ly}^2}} . \tag{24}$$

And also the error of the tilting rate of the spin axis, $\delta\omega_H$, is obtained as:

$$\delta\omega_H = -\delta\omega_{LH}\cdot\sin\alpha - \omega_{LH}\cos\alpha\cdot\delta\alpha . \tag{25}$$

Finally, the error of the nadir angle of the gyro vector, $\delta\theta$, is given by:

$$\delta\theta = \int_{t_1}^{t_2}\delta\omega_H dt . \tag{26}$$

Meanwhile the error of ship's heading is obtained as the following.

The error of the rate of the motion of the spin axis in the body frame, $\delta\omega_D$, is expressed as:

$$\delta\omega_D = -\boldsymbol{\Xi}\boldsymbol{C}_g^b\boldsymbol{\omega}_{i/g}^g + \boldsymbol{C}_g^b\delta\boldsymbol{\omega}_{i/g}^g - \boldsymbol{C}_n^b\left(\boldsymbol{\Gamma}\boldsymbol{\omega}_{e/n}^n + \delta\boldsymbol{\omega}_{e/n}^n\right) \tag{27}$$

And the error of the rate of the motion of the spin axis in the navigation frame, $\delta\omega_N$, is given by:

$$\delta\omega_N = -\boldsymbol{\Gamma}\boldsymbol{C}_b^n\boldsymbol{\omega}_D - \left(\boldsymbol{\Gamma}\boldsymbol{\omega}_{e/n}^n + \delta\boldsymbol{\omega}_{e/n}^n\right) + \boldsymbol{C}_b^n\left(\boldsymbol{C}_g^b\delta\boldsymbol{\omega}_{i/g}^g - \boldsymbol{\Xi}\boldsymbol{C}_g^b\boldsymbol{\omega}_{i/g}^g\right), \tag{28}$$

where $\delta\omega_N = \begin{bmatrix} \delta\omega_{Nx} & \delta\omega_{Ny} & \delta\omega_{Nz} \end{bmatrix}^T$. Therefore the error of ship's heading, $\delta\psi$, is represented by:

$$\delta\psi = \frac{\omega_{Nx}\delta\omega_{Ny} - \omega_{Ny}\delta\omega_{Nx}}{\omega_{Nx}^2\sec^2\psi} . \tag{29}$$

Fig.5 shows the error dynamics in the navigation frame we discussed so far.

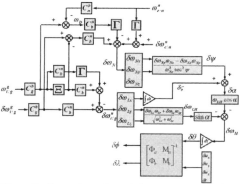

Figure 5. Error dynamics of free gyro positioning and directional system

3.3 Error equations of ship's position

For a development of the position error equation, we differentially perturb eqn. (1), assuming that the earth rate is a constant, and then derive the perturbations of errors resulting in a linear model.

The differential perturbation of eqn. (1) is given by eqn. (30):

$$\cos\phi\cos\left(\lambda + \omega_e t\right)\delta u_x + \cos\phi\sin\left(\lambda + \omega_e t\right)\delta u_y$$
$$+\sin\phi\delta u_z - \sin\theta\delta\theta$$
$$+\cos\phi\left[u_y\cos\left(\lambda + \omega_e t\right) - u_x\sin\left(\lambda + \omega_e t\right)\right]\omega_e\delta t$$
$$=\left[u_x\sin\phi\cos\left(\lambda + \omega_e t\right) + u_y\sin\phi\sin\left(\lambda + \omega_e t\right) - u_z\cos\phi\right]\delta\phi$$
$$+\left[u_x\cos\phi\sin\left(\lambda + \omega_e t\right) - u_y\cos\phi\cos\left(\lambda + \omega_e t\right)\right]\delta\lambda \tag{30}$$

Now, let two gyro-vectors be $\boldsymbol{g}_a^i = \begin{bmatrix} u_{ax}, u_{ay}, u_{az} \end{bmatrix}^T$ and $\boldsymbol{g}_b^i = \begin{bmatrix} u_{bx}, u_b, u_{bz} \end{bmatrix}^T$. And the corresponding nadir angles of θ_a and θ_b are given. The perturbation matrix is arranged by eqn. (31).

$$\begin{bmatrix} Z_a \\ Z_b \end{bmatrix} = \begin{bmatrix} \Phi_a & M_a \\ \Phi_b & M_b \end{bmatrix}\begin{bmatrix} \delta\phi \\ \delta\lambda \end{bmatrix} \tag{31}$$

Here,

$$Z_a = \cos\phi\cos\left(\lambda + \omega_e t\right)\delta u_{ax} + \cos\phi\sin\left(\lambda + \omega_e t\right)\delta u_{ay}$$
$$+\sin\phi\delta u_{az} - \sin\theta_a\delta\theta_a$$
$$+\cos\phi\left[u_{ay}\cos\left(\lambda + \omega_e t\right) - u_{ax}\sin\left(\lambda + \omega_e t\right)\right]\omega_e\delta t ,$$

$$Z_b = \cos\phi\cos\left(\lambda + \omega_e t\right)\delta u_{bx} + \cos\phi\sin\left(\lambda + \omega_e t\right)\delta u_{by}$$
$$+\sin\phi\delta u_{bz} - \sin\theta_b\delta\theta_b$$
$$+\cos\phi\left[u_{by}\cos\left(\lambda + \omega_e t\right) - u_{bx}\sin\left(\lambda + \omega_e t\right)\right]\omega_e\delta t ,$$

$$\Phi_a = u_{ax}\sin\phi\cos\left(\lambda + \omega_e t\right)$$
$$+u_{ay}\sin\phi\sin\left(\lambda + \omega_e t\right) - u_{az}\cos\phi ,$$

$$\Phi_b = u_{bx}\sin\phi\cos\left(\lambda + \omega_e t\right)$$
$$+u_{by}\sin\phi\sin\left(\lambda + \omega_e t\right) - u_{bz}\cos\phi ,$$
$$M_a = u_{ax}\cos\phi\sin\left(\lambda + \omega_e t\right)$$
$$-u_{ay}\cos\phi\cos\left(\lambda + \omega_e t\right) ,$$

$$M_b = u_{bx}\cos\phi\sin\left(\lambda + \omega_e t\right)$$
$$-u_{by}\cos\phi\cos\left(\lambda + \omega_e t\right) .$$

This matrix can be written in the matrix-vector form,

$$\Delta\boldsymbol{Z} = \boldsymbol{\Psi}\Delta\boldsymbol{x} \Longleftrightarrow \Delta\boldsymbol{x} = \boldsymbol{\Psi}^{-1}\Delta\boldsymbol{Z} , \tag{32}$$

where,

$$\Delta\boldsymbol{x} = \begin{bmatrix} \delta\phi \\ \delta\lambda \end{bmatrix}, \boldsymbol{\Psi}^{-1} = \begin{bmatrix} \Phi_a & M_a \\ \Phi_b & M_b \end{bmatrix}^{-1}, \text{ and } \Delta\boldsymbol{Z} = \begin{bmatrix} Z_a \\ Z_b \end{bmatrix}$$

have been introduced. The covariance matrix for eqn. (32) is given by:

$$\text{cov}(\Delta x) = \boldsymbol{\Psi}^{-1}\text{cov}(\Delta Z)(\boldsymbol{\Psi}^{-1})^T . \qquad (33)$$

The assumption may be made that the sensor and time errors show a random behavior resulting in a normal distribution with expectation value zero and variance, σ_s^2. Therefore, measured sensor and time values are linearly independent or uncorrelated. The $\text{cov}(\Delta Z)$ is represented by:

$$\text{cov}(\Delta Z) = \sigma_s^2 \boldsymbol{I} , \qquad (34)$$

where $\boldsymbol{I}$ is the unit matrix.

Substituting eqn. (34) into eqn. (33) yields

$$\text{cov}(\Delta x) = \boldsymbol{\Psi}^{-1}\sigma_s^2 \boldsymbol{I}(\boldsymbol{\Psi}^{-1})^T = \sigma_s^2 \boldsymbol{\Psi}^{-1}(\boldsymbol{\Psi}^{-1})^T$$
$$= \sigma_s^2 (\boldsymbol{\Psi}^T\boldsymbol{\Psi})^{-1} = \sigma_s^2 \boldsymbol{Q}_x \qquad (35)$$

where $\boldsymbol{Q}_x = (\boldsymbol{\Psi}^T\boldsymbol{\Psi})^{-1}$. The cofactor matrix, $\boldsymbol{Q}_x$ (Hofmann-Wellenhof B. et al, 2001) is a 2 x 2 matrix where two components are contributed by the gyro vectors of $\boldsymbol{g}_a^i$ and $\boldsymbol{g}_b^i$. The elements of the cofactor matrix are denoted as:

$$\boldsymbol{Q}_x = \begin{bmatrix} q_{\phi\phi} & q_{\phi\lambda} \\ q_{\phi\lambda} & q_{\lambda\lambda} \end{bmatrix}. \qquad (36)$$

In the cofactor matrix the diagonal elements are used for *FDOP* which is the geometry of two free gyros.

$$FDOP = \sqrt{q_{\phi\phi}^2 + q_{\lambda\lambda}^2} . \qquad (37)$$

Therefore the error of a position in the free-gyro positioning and directional system, fd_{rms}, is given by:

$$fd_{rms} = FDOP\sigma_s . \qquad (38)$$

From eqn. (38) the fd_{rms} can be computed easily. If in this case two-dimensional error distribution is close to being circular, the probability is about 0.63(Kaplan , 2006).

4 CONCLUSIONS

First of all, this paper dealt with the determination of ship's position by free gyros and its algorithmic design briefly. Next, the errors of transformation matrices of the gyro and body frames, i.e., attitude errors, were examined and the attitude equations were also derived. The perturbations of the errors of the nadir angle including ship's heading were investigated in turn. Finally, the perturbation error equations of ship's position used the nadir angles were derived in the form of a linear error model and the concept of FDOP was also suggested by using covariance of position error.

However, the free-gyro positioning and directional system has still many problems to be solved. First the error of a position was experimentally verified. Especially the sensor errors will have to be investigated. In addition, the additional drift needs to be investigated, which occurs when a free gyro is suppressed by the measures which prevent gimbals lock and tumbling. The alignment in the system also needs to be examined.

All these will be dealt with in the next papers.

REFERENCES

[1] Hofmann-Wellenhof B, Lichtenegger H and Collins J(2001), "GPS, Theory and Practice", 5th edition, SpringerWienNewYork, p. 272.
[2] Jekeli, C.(2001), Inertial Navigation Systems with Geodetic Applications, Walter de Gruyter, p.147-157.
[3] Jeong, T.G.(2005), "A Study on the Errors in the Free-Gyro Positioning System (I)", International Journal of Navigation and Port Research, Vol. 29, No. 7, pp. 611-614.
[4] Jeong, T.G. and Park, S.C.(2006), "A Theoretical Study on Free Gyroscopic Compass", International Journal of Navigation and Port Research, Vol. 30, No. 9, pp. 729-734.
[5] Jeong, T. G. and Park, S. C.(2011), "An Algorithmic Study on Positioning and Directional System by Free Gyros", TransNav-International Journal on Marine Navigation and Safety of Sea Transportation, Vol. 5, No.3, pp.297-302.
[6] Kaplan, E. D. et al(2006), Understanding GPS Principles and Application, 2nd edition, ARTECH House, p.331.
[7] Park, S.C. and Jeong, T.G.(2004), "A Basic Study on Position Fixing by Free Gyros", Journal of Korean Navigation and Port Research, Vol.28, No.8, pp. 653-657.
[8] Rogers, R.M(2007), Applied Mathematics in Integrated Navigation Systems, 3rd ed., AIAA, pp.106-111.
[9] Titterton, D.H. and Weston, J.L.(2004), Strapdown Inertial Navigation Technology, 2nd ed., AIAA, p. 287.

Determining of the Distance of the Nearest Approach of Vessels with Using Information from AIS Method of Relative Ship' Plotting with a Virtual Second Observation Epoch

V.M. Bukaty & S.U. Morozova
Baltic State Academy of Fishing Fleet, Kaliningrad, Russia

ABSTRACT: the present article gives grounds to the closest distance of vessels approach determination based on data from AIS by means of the ration plotting technique. The peculiarity of this technique is made up by preliminary calculation of bearing and the distance to the target ship at the second epoch of observation upon the vessels' positions and parameters of their motion at the first epoch of observation. The technique in question does not require time consumption for waiting until the second epoch of observation comes and this builds up its advantage comparing to classical ration plotting technique. Also the article contains estimation of the accuracy for closest distance of vessels approach calculation and optimal time interval in terms of the mean square error minimal criterion for preliminary bearing and distance calculation at the second epoch of observation. Though the technique accuracy is one and half times less than the accuracy of classical ration plotting technique however the described technique is worth considering on the ground of its immediate and time-saving data obtaining.

1 INTRODUCTION

Until recently the minimal closest point of approach of sea-going vessels encounter distance determination has been exclusively figured out based on results of bearing measured by radar and distance to the target vessel at one time moment and repeated bearing and distance measure at the second time moment which is fixed usually 6 minutes later. Then in accordance with the ration plotting technique the closest distance of approach was calculated based on the data obtained as above described with the help of formula known from specified literature [1]:

$$D_{min} = \frac{D_1 D_2 \sin \Delta B}{\sqrt{D_1^2 + D_2^2 - 2 D_1 D_2 \cos \Delta B}}, \qquad (1)$$

as per this formula D_1 and D_2 mean distances to the target vessel at the first and the second epoch of observation, ΔB means target vessel bearings difference between epochs of observation.

But when vessels were equipped with Automated Identification Systems (AIS) the solution of this task became possible by means of the same ration plotting technique based on the data of vessel observer's position and target vessel transmitted by AIS radio channel.

Studies in the Reference source [1-3] present mathematical rationale of the closest approach point of heading in based on data from AIS by ration plotting technique and provide estimation of its calculation possible errors. In virtue of practically equal bearings and distances measurements at both epochs of their measurements mean square error $m_{D min}$ of closest approach is able to be calculated adequately by the following formula:

$$m_{D\,min} = D_{min} \sqrt{2\left(\frac{m_{D1}}{\Delta D}\right)^2 + 2\left(\frac{m_{B1}}{\Delta B}\right)^2}, \qquad (2)$$

where m_D and m_B are mean square errors of distance and bearings measurements respecttively; ΔD is distances differences between epochs of observation.

In the Reference source [1] the strong evidence demonstrates that AIS data use assist calculation of distances to the target vessel and her bearings by high-accuracy ships' positions obtained from SNS and the closest approach point is found far more accurate compared to its calculation based on radar observations results.

2 FUNDAMENTALS OF METHOD

However despite this advantage the ration plotting technique based on use of data from AIS like the use of radar measurements have a considerable defect that highly valuable time at such tasks is spent for expecting the second epoch of observation. Meanwhile the data on target vessel movement parameters obtained from the AIS radio channel as a whole with the data on movement parameters of observer vessel together with bearing B_1 calculation and distance D_1 to target vessel for the first epoch of observation as per formulae

$$D_1 = \sqrt{\Delta\varphi_1^2 + \Delta\lambda_1^2 \cos^2 \varphi_n} , \qquad (3)$$

$$B_1 = \frac{arctg\Delta\lambda_1 \cos\varphi_n}{\Delta\varphi_1} \qquad (4)$$

makes practically simultaneous figuring out (foretell) bearing B_2 and distance D_2 to the target vessel for the second epoch of the observation (virtual one) as per formulae [4],

$$D_2 = \sqrt{\begin{array}{l}[\Delta\varphi_1 + (v_t \cos q_t - v_o \cos q_o)t]^2 + \\ + [\Delta\lambda_1 \cos\varphi_n + (v_t \sin q_t - v \sin q_o)t]^2\end{array}} , \qquad (5)$$

$$B_2 = \frac{arctg\Delta\lambda_1 \cos\phi_n + (v_t \sin q_t - v_o \sin q_o)t}{\Delta\phi_1 + (v_t \cos q_t - v_o \cos q_o)t} , \qquad (6)$$

where

$\Delta_{\varphi1}$ $\Delta_{\lambda1}$ mean the difference of latitudes and difference of longitudes between vessels for the first epoch;

φ_n is a certain intermediate latitude, which can be allowed as arithmetic mean latitude at the low-latitude areas [4];

v_o and v_t mean vessel observer's speed over ground and speed over ground of the target vessel respectively;

q_o and q_t mean course over ground angle of vessel observer and course over ground angle of target vessel respectively;

t is the time interval till the second (virtual) epoch of observation.

With the help of simultaneously calculated pairs of distances and bearings D_1, B_1, and D_2, B_2 by technique of ration plotting the closest distance of vessels' approach is determined graphically or calculated by the same formulae [1].

Our study [5] demonstrates that due to the fact that distance and bearing obtained from AIS for the second epoch of observing have much more considerable errors than distance and bearing for the first epoch of observing (due to the effect caused to them by errors in courses over ground and vessels' speeds) obtained by simultaneous pairs of distances

and bearings, so the closest distance of approach will be obtained with less accuracy. This is apparently a drawback of ration plotting technique with virtual second epoch of observation. But at the same time the obvious advantage of this technique is practically immediate obtaining the information about the distance of the closest approach without loss of time.

As in case of $D_{\min}$ defining by means of ration plotting with virtual second epoch of observations the measurements of distances and especially of bearings are of unequal accuracy, so the mean square error of the closest approach distance should be calculated from the following expression [1]:

$$m_{D\min} = \sqrt{\left[\frac{D_2 \cdot m_{D1}}{\Delta D \cdot D_1}\right]^2 + \left[\frac{D_1 \cdot m_{D2}}{\Delta D \cdot D_2}\right]^2 + \frac{m_{B1}^2 + m_{B2}^2}{\Delta B^2}} \qquad (7)$$

where m_{D1} and m_{B1}, m_{D2} and m_{B2} are mean square errors of distances and bearings for the first and second (virtual) epochs of observations.

To obtain mean square errors m_{D1} and m_{B1}, m_{D2} and m_{B2} of distances and bearings for the first and second (virtual) epochs of observation we have developed the following expressions:

$$m_{D1} = \frac{m_{\varphi\lambda}}{D_1} \sqrt{2(\Delta\varphi_1^2 + \Delta\lambda_1^2 \cos^4 \varphi_n)}, \qquad (8)$$

$$m_{B1} = \frac{m_{\varphi\lambda} \cos\varphi_n}{D_1^2} \sqrt{(\Delta\varphi_1^2 + \Delta\lambda_1^2)}, \qquad (9)$$

$$m_{D2} = \frac{1}{D_2} \sqrt{\begin{array}{l}(\Delta\varphi_1 + v_t \cos q_t - v_o \cos q_o)^2 \times \\ \times \left\{2m_{\varphi\lambda}^2 + \left[\begin{array}{l}m_v^2(\cos^2 q_t + \cos^2 q_o) + \\ + m_q^2(v_t^2 \sin^2 q_t + v_o^2 \sin^2 q_o)\end{array}\right]t^2\right\} + \\ [\Delta\lambda_1 + (v_t \sin q_t - v_o \sin q_o)\sec\varphi_n]^2 \times \\ \times \left\{2m_{\varphi\lambda}^2 + \left[\begin{array}{l}m_v^2(\sin^2 q_t + \sin^2 q_o) + \\ + m_q^2(v_t^2 \cos^2 q_t + v_o^2 \cos^2 q_o)\end{array}\right]t^2 \sec^2 \varphi_n\right\}\end{array}} \qquad (10)$$

$$m_{B2} = \frac{\cos\varphi_n}{D_2^2} \sqrt{\begin{array}{l}(\Delta\varphi_1 + v_t \cos q_t - v_o \cos q_o)^2 \times \\ \times \{2m_{\varphi\lambda}^2 + [m_v^2(\cos^2 q_t + \cos^2 q_o) + m_q^2(v_t^2 \sin^2 q_o + v_o^2 \sin^2 q_o)]t^2\} + \\ + [\Delta\lambda_1 + (v_t \sin q_t - v_o \sin q_o)\sec\varphi_n]^2 \times \\ \times \left\{2m_{\varphi\lambda}^2 + \left[\begin{array}{l}m_v^2(\sin^2 q_t + \sin^2 q_o) + \\ + m_q^2(v_t^2 \cos^2 q_t + v_o^2 \cos^2 q_o)\end{array}\right]t^2 \sec^2 \varphi_n\right\}\end{array}} \qquad (11)$$

in which

$m_{\varphi\lambda}$ is a mean square error of the vessels' positions (for the SNS it is admitted equal to 25 m = =0.0135 miles);

m_v and m_q are mean square errors of the vessels' speeds and courses over ground (they are admitted for calculations as m_v=0.1 knot and m_q=0.5^0=0.0087 radian);

φ_n is intermediate latitude admitted for calculation as equal to 60^0.

3 OPTIMAL TIME INTERVALS UNTIL THE VIRTUAL SECOND EPOCH OF OBSERVATION

As we can see the mean square errors m_{D2} and m_{B2} of distances and bearing for virtual second epoch of observation apart others are also dependent on time interval t taken for calculation of D_2 and B_2. It should seem judging from expressions (10) and (11) the less time interval t is taken the less are errors in distances and bearings for the virtual second epoch of observations and the more accurate the closest distance of approach is obtained. But together with the decreasing of the time interval between the observations the less differences of distances and differences of bearings will become. The decrease of differences of distances and differences of bearings, on the contrary, will result in the increase of the mean square error in the closest distance of approach. Such influence of argument on the behavior of function usually causes the appearance of the function extremum.

Thus in the version studied herewith about use of AIS data we can suppose that the mean square error of the closest distance of approach (7) as the function of the time interval t should possess unimodal extremum, that is minimum. Having taken into consideration formulae from (8) to (11) and dependences ΔD and ΔB from the time interval t we conclude that the expression (7) is an extremely complicated case preventing to find its extremum by the analytical way. That is why we are to analyze the function behavior (7) as the function of the time interval t by numerical (graphical) analysis by calculating all its values, for example the situation of vessels heading in presented at the Figure 1.

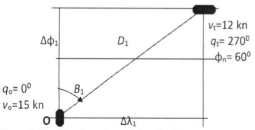

Figure 1. The Situation of vessels heading in.

Here we can see dependence diagrams m_{Dkp} from t for υ_t=12 knots, υ_0=15 knots, q_t=270^0, q_0=0^0 for three versions of the vessels' starting positions

Version 1: $\Delta\varphi1$=4.243′ $\Delta\lambda_1$=8.486′, D_I=6 miles, B_I=45^0, D_{min}=0.600 miles;

Version 2: $\Delta\varphi1$=8.485′ $\Delta\lambda_1$=16.972′, D_I=12 miles, B_I=45^0, D_{min}=1.330 miles;

Version 3: $\Delta\varphi1$=12.728′ $\Delta\lambda_1$=25.456′, D_I=18 miles, B_I=45^0, D_{min}=1.640 miles are pictured at Figure 2.

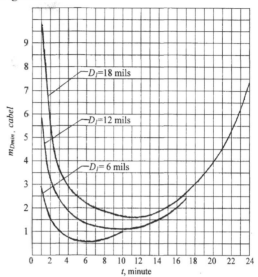

Figure 2. Dependence m_{Dmin} from time interval until the second (virtual) epoch of observations.

As it is evident at the Figure 2 the expression (7) indeed has though not too vividly expressed but nevertheless minimum which at distances in the area of final situation finding of vessels heading in (D1=6 miles) is put for the time interval of 6 minutes, as at the area of preliminary situation finding of vessels heading in (D_I=12 miles and D_I=18 miles) the time intervals become about 9 and 12 minutes respectively.

Thus while use of AIS data for vessels heading in situation finding by means of ration plotting with virtual second epoch of observation for calculating of the closest distance of approach with maximal accuracy the time interval until virtual second epoch of observation should be fixed equal to 6 minutes for the distance of 6 miles between vessels, about 9 minutes for the distance between vessels of 12 miles and about 12 minutes for the distance between vessels approximately 18 miles.

4 CONCLUSION

We bring to the note that if the vessels heading in situation finding were made by ration plotting technique by different time pairs of bearings and distances, according to the formula (6) mean square errors of distance defining of closest approach would make up for Version 1 m_{Dmin}=0.05 miles, for Version 2 m_{Dmin}=0.10 miles and for Version 3

m_{Dmin}=0.15 miles. It is evident that there is one and half times higher level of accuracy of defining D_{min} on heterogeneous at time pairs of bearings and distances than the level of accuracy of its defining on simultaneous pairs of bearings and distances (with the virtual second epoch of observations). But this technique peculiarity is hardly more convincing at practice than the capability of practically immediate obtaining of information about the closest approach distance which is inherent to the analyzed here technique of ration plotting with heterogeneous in time pairs of bearings and distances (with the virtual second epoch of observation), moreover that in case of the virtual second epoch of observation the accuracy of obtaining D_{min} is sufficiently high for a convinced judgment about dangerous and safe vessels heading in.

REFERENCE

[1] Lushnikov E.M. The Navigational Safety of Sailing. Kaliningrad, Publishing Department of the Baltic State Fishing Fleet Academy of the Russian Federation, 2007, 250 pp.

[2] Mikhailov S.A., Orlov E.O. Vessels Heading in Parameters Calculation Methods by means of AIS Data. Odessa, ONMU, Navigation, issue 17, 2010, pp. 113-122.

[3] Mikhailov S.A., Orlov E.O. Errors Estimation of Vessels Heading in Parameters Calculation by AIS Data, Studies at the X-th Scientific and Practical Conference "Practical Problems in Radio Communication and Radio Navigation at GMZLB in systems of AIS, CYPC and PIC", Odessa, ONMA, 2009, pp. 45-47.

[4] Dmitriev V.I., Grigorian V.L., Katenin V.A. Navigation and Pilot-Book. Moscow, Akademkniga Publishing House, 2004. 471 pp.

[5] Bukaty V.M., Morozova S.Yu. Problem of Accuracy in Finding the Closest Vessels Approach Distance Measured by AIS. Studies at the IX-th Inter-universities Scientific and Technical Conference of Postgraduates, Doctor Degree Candidates and Students "Scientific and Technical Developments in Fishing Fleet Problems Solutions and Staff Training Issues", 2008. Kaliningrad, Publishing Department of the Baltic State Fishing Fleet Academy of the Russian Federation, 2009, pp. 6-17.

Chapter 2

Electronic Chart Display and Information Systems (ECDIS)

Inciting the Development of Engaging Screencasts in Teaching ECDIS

S. Bauk
University of Montenegro, Montenegro

R. Radlinger
University of Graz, Austria

ABSTRACT: This paper is intended to motivate teachers of maritime schools and colleges to create interesting and engaging screencasts for teaching students (future seafarers) ECDIS basis, by using Camtasia Studio program. It is a multi-media, user-friendly environment, providing the customers with the variety of possibilities for editing PowerPoint presentations with the introduction of audio, video and different animations, in order to make teaching/learning content more interesting and to point out the most important issues. This complex applied software allows PC screen video capturing, and adding of audio and numerous animated effects to it, as well. Even though the Camtasia Studio possesses a broad palette of advanced features for recording and editing of "lively" and edifying screencasts, in this article only brief descriptions of "reviving" classical PowerPoint presentations, taking the screen captures, and their editing, shall be presented. Several examples will be given, while the Transas demo ECDIS platform will be used as the base for screenshots.

1 INTRODUCTION

Along with the process of developing information - communication technology (ICT) in the domain of marine navigation, computers and following equipment have become the essential part of working environment onboard a ship. New seamen generations are being taught how to use computers and how to understand their operations in principle. They are being introduced to the possibilities of the computer use in the administration on a ship, and especially in conducting as much precise and safer navigation as possible. All information and communication (IC) systems (or, tools) that are being used nowadays in steering the ship and determining its position are near perfection. However, in the actual era of cloud computing, or pervasive computing, navigators essentially have less and less insight of all that computers do, and how they actually operate, since they are simply built into the "things", i.e. navigational devices. Such is the case with modern command bridges of the ships. This is IT break through the maritime navigation. Although taking the technology for absolutely perfect one would be seriously wrong. Ignoring human link in the form of action-reaction chain in the electronic navigation would be unacceptable [13]. At last, captain of the ship is fully responsible for its safety and safety of the crew, passengers, and human lives in general. In addition, first officer is in charge of getting the ship safely to the destination port, i.e. to the port of arrival. Besides, navigation principles have been constant for centuries as well as the parameters linked to the navigation. Though IC systems (tools) are only a component more, some kind of science quality seal in navigating a ship on the primer base of the navigational skills that are to be developed in different manners [6;7]. One of them is, or will be soon, based on more intensive employing up-to-date computer mediated learning and training approaches.

2 BRIEFLY ON ECDIS: AS A PERPETUATOR OF ELECTRONIC NAVIGATION

The ECDIS is an electronic navigation system that integrates real-time navigational data from ship sensors (GPS, Radar, AIS, etc.) and electronic navigational charts (ENCs - Electronic Nautical Charts) [17]. In its very nature, it is a centralizing instrument with the unique function of integrating many aspects of navigation [15]. More explicitly, it allows the integration of numerous operational data, such as ship's course and speed, depth soundings, and radar data into the display. Furthermore, it allows automation of alarm systems to alert the navigator of potentially dangerous situations, and gives him/her a complete picture of the instantaneous situation of the vessel and all charted dangers in the area [9]. ECDIS has been conceived in such a way to support and enforce the transition to the eNavigation concept [17].

Although the International Maritime Organization (IMO) officially approved it as the equivalent to the classical paper charts in November 1995 [9], the transition to its full usage in practical maritime navigation is still slow. The causes are the lack of the official ENCs, the high cost of ECDIS, and a dose of skepticism in accepting this new technology by the traditional marine community. However, ECDIS has benefits in terms of time saving in route planning and monitoring, preventing accidents and thus protecting the ship and marine environment. About 80% of reported accidents at sea are caused by human errors, which are consequences of fatigue, extensive work in navigation and others. Number of accidents at sea could be reduced by the introduction of electronic navigation systems, since they would help in mitigating the navigator fatigue and stress. The ECDIS is, therefore, advanced IT tool that allows navigation during which it monitors the position, and collision/grounding avoidance for 24 hours. These ECDIS functions can be used effectively in restrictive waterway areas, especially during periods of poor visibility, i.e. under conditions of mist and during the night. The DGPS (differential GPS) is crucial in a way that it complements the ECDIS through the provision of continuous and accurate positioning of the ship [8;16]. One of the major problems that prevent wider use of ECDIS is the lack of the official ENC covering the main routes and ports in the world. Otherwise, reasons for concern when using non-official electronic charts are the quality of the source material, data and their updates. Some hydrographical organizations have problems in creating ENC that are consistent with the IHO (International Hydrographic Organization) S-57 standard, and newer ones [19], due to the lack of experts' knowledge in this area, as well as the lack of finances. Though, some HOs expressed concern about the slow development of mechanisms for creating, distributing and updating the ENCs. And, ultimately, with the development and implementation of ECDIS, one of the main goals is to improve communication between the HOs, ECDIS manufacturers and end users, primarily seafarers. The basic components of ECDIS display, i.e. most of the visualized commands of ECDIS (on the exemplar of Navi-Trainer Professional NTPro 4000 nautical simulator manufactured by the Transas Marine) have been described in detail within some previously published papers by the author in this field [1-5]. Also, the basic and some advance features of ECDIS have been covered by numerous referential literature resources, like [9-

17;19-21;24-30]. Though, this paper will not be focused on these issues, but, through introducing readers with Camtasia Studio application software, it will be more focused on stimulating usage of modern multi media tools and technologies in offering students, future seafarers, exciting, engaging, and ultimately, edifying presentations on the ECDIS basis, display elements, visualized commands, its different functions, and relevance toward providing safe navigation in any conditions at the sea. Accordingly, the following parts of the paper will be devoted mostly to the Camtasia Studio as a software tool which allows recoding, editing and sharing presentations/videos on ECDIS through different popular media formats, making students of maritime schools and collages more familiar with this important topic. Of course, it is to be noted that nowadays there is a quite large offer of different application software which can be used for producing:

1 **audio** (Audacity, NCH Wave Pod, Adobe Audition, Cubase Steinberg, Logic Studio, Kristal Audio Engine, etc);
2 **video** (Windows Movie Maker, Adobe Premiere, Avidemux, Magix Video, Video Spin, AVIedit, etc); and,
3 **sreencapturing** (Adobe Capticate, Capture Fox, Camtasia Studio, Jing, ActivePresenter, BB Flashback, BB Flashback Express, ScreenPresso, VirtualDub, etc).

Web can be used as a resource for further search [31;32].

Some of these software tools are proprietary commercial, while some are freeware. And it is difficult to give the recommendation which one to use. Exploring *pros* and *cons* of these and numerous other software is beyond the scope of this article. However, at this moment of the authors' work in this field, the most appropriate *seems* here employed and briefly presented software. However, this does not mean that the teachers/educators at METs should not experiment with other software tools, and that the authors will not do so, what should create new opportunities for exchanging and mutual enriching experiences in this MET domain in the future.

3 BRIEFLY ON CAMTASIA STUDIO: AS A TOOL OF KNOWLEDGE TRANSFER

Camtasia Studio and Camtasia for Mac, are software applications for creating professional-looking presentations, video tutorials and/or screen captures, published by TechSmith [22]. The PowerPoint presentation recordings along with a variety of animated effects, the narrator's voice, background sounds (music), and web camera recordings of the presenter are enabled by this software. Additionally, the whole screen, or the exact pre-specified screen area (of any PC program, or, here the ECDIS Transas demo version) can be captured, and audio may be recorded simultaneously, or embedded latter, from any standard input source device. During the content production the presenter is able to jump from one application to another without interrupting the recording process. The presenter is able to stop recording with a hotkey combination at any time, at which point the software renders the input that has been captured, and applies user-defined settings. After the presentation had been captured, it is possible to revise it by cutting and/or pasting different parts, as needed.

The presenter is also able to overlay the voice sequences, sound effects or music onto the presentation, if it is needed. Camtasia allows audio recording while screen-capturing is in progress, so the presenter can narrate the demonstration as it is carried out. Most presenters, however, prefer to wait until they have finished the screen-capture, and then record the narration from a script as the application is playing back the recorded capture. The program allows files to be stored in its own proprietary format, which is only readable by Camtasia itself; this format allows fairly small file sizes as well as longer presentations [22;23].

The completed video recordings can be also output to several different, popular common (video) file formats, such as AVI, Flash, Quick Time, RealMedia, etc., which can be easily read by most computers. Camtasia Studio can be used for quickly recording, editing and submitting variety of contents in variety of manners, but within this context of learning ECDIS fundamentals, and probably some of its advanced functions in the perspective – *the knowledge transfer is its most basic!*

4 CAMTASIA STUDIO: THE APPLICATIONS IN TEACHING/LEARNING ECDIS

Aiming to incite the creating of engaging screencasts in teaching students ECDIS, some PowerPoint presentations on ECDIS fundamentals concerning historical facts, law and standardization requirements, navigational chart types, and their features, updates, training standards/seafarers, ECDIS as a safety tool for navigation, etc., are narrated, and produced in the appropriate video format for sharing it on the Web. Additionally, several short video tutorials on ECDIS screen elements, visualized commands, and some basic functions (route planning, AIS targets acquisition, Navtex messages overlaying, etc.) are recorded, as well, and produced in order to be shared latter among students and colleagues.

Below are given some *mixed* explanations on both Camtasia Studio (ver. 7.1.1) recording and editing techniques, and presentations on ECDIS

content by means of the new media application in knowledge transfer.

4.1 *Recording PowerPoint presentations on ECDIS*

In general, recording PowerPoint presentations should be realized in two ways: recordings can be done directly from PowerPoint by using Camtasia Studio PowerPoint Add-in tool, or by saving each PowerPoint presentation slide in JPEG format, and importing them into the Camtasia Studio Clip Bin, and latter on, transferring them sequentially to the Timeline. Then, the JPEG files can be edited by associating them with voice narration, web camera recordings, different animated and transitioning effects, etc. The detail description on both procedures can be found in [22]. Here, in introducing students with the ECDIS basis: historical facts, standards, types of navigational charts, performances, educational-training requirements, etc., both ways of PowerPoint recording have been applied.

A screencast which presents the process of editing the PowerPoint recording being made in the second explained manner, i.e. by creating JPEG files of each slide and inserting them latter into Camtasia Studio for creating the unique audio/video record, is given below (Fig. 1). Though, within the PowerPoint presentation, which editing in Camtasia Studio is shown in Fig. 1, the following issues have been considered briefly:
- some historical facts on electronic charts and ECDIS;
- legal aspects and requirements;
- performance standards including: IMO, IHO, and IEC ones;
- ECDIS approvals; seafarers training and familiarization requirements due to the 1.27 IMO model course;
- the problem of some navigators' overconfidence on ECDIS, etc.

Due to the voice narration and web camera recordings, along with the different animations being realized in Camtasia Studio, above listed topics became undoubtedly more interesting; firstly, in terms of keeping up students' attention and most probably, making them curios to learn more and more on this topic in the perspective.

4.2 *Recording the screen captures over EDCIS Transas demo software*

In the process of recording screen captures on ECDIS, the Transas demo version 2.00.012 (2010) has been used as a base upon which the recordings are done. The whole screen is being recorded, along with the presenter narration, and after the recording had been finished, the capture is imported to the Camtasia Studio and edited. Different animated effects (callouts, captions, smart-focus tools: zoom, pan, etc.) are added, in order to make the captures more interesting, and ultimately more edifying to students. Although, all necessary details on screen recording, audio adding, and editing the recordings can be found in [22] – it is on a presenter, here teacher, to optimally allocate the place and duration of each animated effect within the presentation, aiming to make engaging and really worth audio/video record, prepared to be shared among students, colleges, and/or wider, e.g. Web audience.

Some screencasts which present the process of capturing the screen and editing the screen captures taken over ECDIS demo version software are given below (Fig. 2-6). In Fig. 2 the process of explaining the visualized command buttons in the control panel by using Camtasia Studio effects is shown. Here, the true motion mode has been explained [1;4;5]. Though, by using different animations (callouts), it becomes possible to visualize the most important facts on ECDIS operating and to make them more understandable to students.

Figure 1. Editing PowerPoint presentation on ECDIS fundamentals in Camtasia Studio.

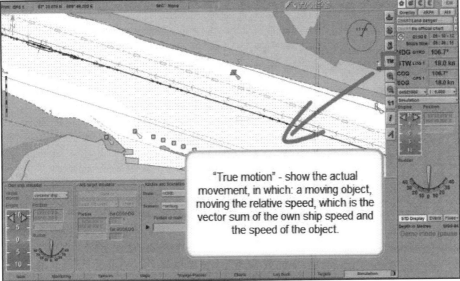

Figure 2. Using Camtasia Studio callouts for explaining ECDIS true motion mode.

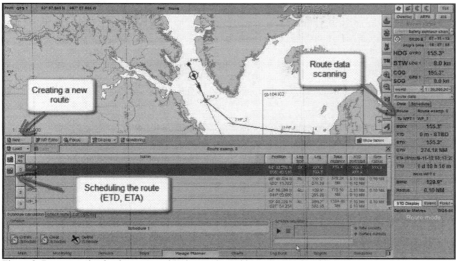

Figure 3. Route creating by the voyage planning mode, and its parameters tracking.

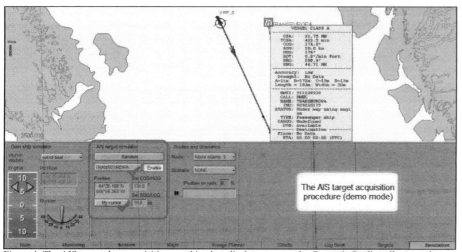

Figure 4. The AIS target data acquisition: marking key display segments by Camtasia Studio callouts, and zoom and pan effects.

The main object of the screen shot shown in Fig. 3, along with the voice narration of the presenter, was the route creating graphically, and scheduling it by entering ETD (Estimated Time of Departure) and ETA (Estimated Time of Arrival). The process of route saving (for later reference and potential output to the autopilot), along with the possibility of deleting some of its segments, or inserting new ones has been presented. The possibility of waypoints' parameter tracking in the control panel from the route data sub-window has been explained, as well, and it is marked on the screen (Fig. 3) as an important segment of ECDIS route monitoring. Within this context of route planning it is to be pointed that the operator should control the route parameters related to the alarms and indicators, like [9]:

– *Cross-track error*: set the distance to either side of the track the vessel can stay before an alarm sounds. This will depend on the phase of navigation, weather and traffic;
– *Safety contour*: set the depth contour line which will alert the navigator that the vessel is approaching shallow water;
– *Course deviation*: set the number of degrees off course the vessel's heading should be allowed to stray before an alarm sounds;
– *Critical point approach*: set the distance before approaching each waypoint or other critical point that an alarm will sound;
– *Datum*: set the datum of the positioning system to the datum of the chart, if different, etc.

Because of the demo version of ECDIS by means of which some Camtasia Studio presentation features have been applied in this work, there are certain limitations in setting on the critical values of the above listed parameters by the user, though for the purpose of continuing to meet the students with the functions of ECDIS, the *real* ECDIS simulator should be necessarily used (e.g. Navi-Trainer Professional NTPro 4000 nautical simulator, or some others, advanced ones versions).

Furthermore, the process of acquisition of the AIS (Automatic Identification System) target data (in manual, not in random mode in here employed ECDIS demo version) has been also shown in the short video presentation (Fig 4). For the purpose of making AIS targets visible and selecting one of them, the AIS overlay command button must be pressed in the command panel in the upper right corner of the display. In the simulation panel the random button has to be switched off and certain available AIS target is to be selected and enabled. Its position can be controlled by inserting manually its coordinates and course, or by cursor, i.e. by positioning it directly at the proper place, along with the direction onto the chart panel. These options are zoom in by zoom and pad (zoom-n-pan) Camtasia Studio tool, and marked in red by the callouts in Fig. 4.

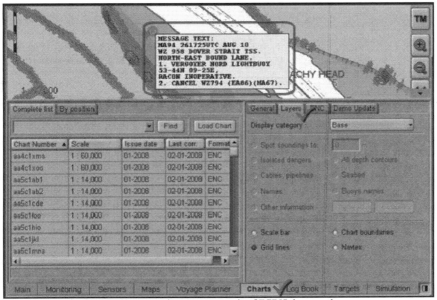

Figure 5. Capturing NAVTEX message in a preset scenario of ECDIS demo mode.

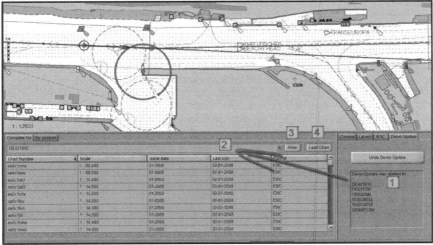

Figure 6. Chart (demo) updating: following the procedure by the Camtasia Studio callouts.

The Fig. 5 presents capturing NAVTEX (Navigational Telex) message in the ECDIS demo version. The Chart functional panel is to be opened, then Layers page within this panel is to be opened, and finally NAVTEX display category is to be turned on. This is very important possibility, especially in the restrictive sailing zones, or those with potential mostly different meteorological dangerous. However, since the ECDIS Transas demo version is used in this paper as a platform for Camtasia Studio recordings, editing, and post-production of the captures, a numerous shortages are present. E.g. acquisition of NAVTEX messages is available only within some preset scenarios, but it is not available in the free, or manually route creating mode. The similar situation is with showing ARPA overlay which is visible on the screenshot (Fig. 5), and by means of which the clear tactical situation with respect to the other vessels or referential objects on the chart can be captured [9]. Namely, in the case when trainers/trainees create the route on their own, this possibility of ARPA overlay is unfortunately unavailable.

In the Fig. 6 the imitation of the chart update procedure is outlined. The ECDIS operator has to find the available update of certain chart in Chart functional panel and to upload it into the system, i.e. to replace the old chart by the new one. The updates are marked in orange (in here used demo ECDIS version) in the new chart version, and the renewed data can be seen for each marked object in the updated chart, simply, by clicking the info button. It is to be mentioned that each vessel should have up-to-date charts for safe navigation. That is the requirement of SOLAS Convention regulation V/27. Updates can be manual or official (automatic or semiautomatic). The manual update is used for navigational warnings sent as MSI (Maritime Safety Information) by NAVTEX, or EGC (Exchange Group Call). Official updates are distributed by RENCs (Regional Electronic Navigational Chart Coordinating Center) throughout the update discs [17].

It is to be pointed out once again that by using the available preset scenarios in ECDIS Transas demo version 2.00.012 (2010) the ARPA overlay and the NAVTEX messages observing are available, as well as the possibility of imitating chart updating procedure. However, since the demo version of ECDIS is in matter, these options are available only for some preset route scenarios, but not in the free route planning mode. Of course, these and others, rather numerous restrictions, as those related to some relevant route parameters tracking (cross-track error, safety contour, course deviation, critical point approach, etc.) should be overcome by using *real* ECDIS simulator, or through underway exercise sequences on real ECDIS [14;15], as the sound and confident platforms for recording, editing, and post-

producing educational/training videos by new media equipment and software tools devoted to providing more efficient knowledge transfer in this domain.

Though, such approach might be a challenge for forthcoming, more extensive and rigorous investigation work. Also, instead of Camtasia Studio, some other applied software can be applied, e.g. Adobe Premiere, since it offers some advanced possibilities of recording, editing and post-producing educational materials.

5 RECOMMENDATIONS: HOW TO RECORD ENGAGING ECDIS LEARNING CAPTURES?

Now, try to answer the key question: What should be the general recommendation for recording and editing engaging ECDIS learning captures? - Regardless of the content of the presentation, the answer is almost the same [18]. First of all, the presenter must have a good knowledge of the area which he/she presents. Additionally, he/she should be well prepared in a sense of having very clear idea about what, in what extent, and in which order it is to be said. The presentation should be clear and concise. And, the presenter should not be "in rush", at all, during the narration/explanation phase(s). Leaving some *free* or *silence* sequences is recommended, as well. The following animated effects should be of the appropriate length, and given in the appropriate amount. Students should be allowed to hear and understand what the presentation is about.

Since the ECDIS is a very important issue, which directly touches the integration of almost all vital navigational equipment and acquisition of the information that they provide, which ultimately implies the safety of navigation – these particularly reinforce previously given suggestions.

6 CONCLUSIONS

The paper contains short description of ECDIS and its importance to the safe navigation, mostly in terms of recalling the author's previous published papers on this topic [1-5] and some sound references in this field [9-17;19-21;24-30]. An emphasis is on introducing new methods and techniques to the process of learning students of maritime schools and colleges ECDIS principles and operational basis.

Accordingly, Camtasia Studio application software has been briefly presented in order to draw the attention of teachers and instructors in a manner how to make their lectures more interesting and engaging for students. The engagement is of crucial importance of the appropriate acquiring of the knowledge. Besides Camtasia Studio, many new technologies are available today for educators to create alternatively learning environment in which students learning should be expanded and reinforced [18]. Learning ECDIS by involving new media tools, such as Camtasia Studio, should be undoubtedly an interesting base for developing a more stimulating learning environment and a new - active knowledge transfer channel between educators/trainers and students and/or trainees in the field.

ACKNOWLEDGEMENT

This work has been supported by JoinEU SEE Erasmus Mundus program and it has been realized at the Academy of New Media & Knowledge Transfer, University of Graz (Austria).

REFERENCES

[1] Bauk S., Dlabač T., "Information-communication Tools in Seamen Training at the Examplar of Navi-Trainer Simulator", *XVI IT Conference*, 22nd -26th February, Žabljak, Montenegro, 2011. (CD issue)

[2] Bauk S., Dlabač T., Džankić R., Radulović V., "On some NTPro Nautical Simulator IT Functions and POB maneuvers", *Proc. of the 15th International Conference on Transport Science*, 27th May, Portorož, Slovenia, 2012. (CD issue)

[3] Bauk S., Dlabač T., Pekić Ž., "Implementing E-learning Modes to the Students and Seafarers Education: Faculty of Maritime Studies of Kotor case study, *Proc. of the 14th International Maritime Science Conference*, 16th-17th June, Split, Croatia, 2012. (CD issue)

[4] Bauk S., Dlabač T., Radulović V., "On Some IT Functions of Navi-Trainer Nautical Simulator", *XVII IT Conference*, 27th February – 2nd March, Žabljak, Montenegro, 2012. (CD issue)

[5] Bauk S., Džankić R., Bridge Management: Upon Some Navigational ICT Tools in Seamen Education and Training, *Proc. of the 14th International Conference on Transport Science*, 27th May, Portorož, Slovenia, 2011. (CD issue)

[6] Barsan E., Muntean C., Project PRACNAV for a Better on Board Training Curricula, *TransNav – International Journal on Marine Navigation and Safety of Sea Transport*, Vol. 4, No. 3, September 2010, pp. 351-355

[7] Barsan E., Surugiu F., Dragomir C., Factors of Human Resources Competitiveness in Maritime Transport, *TransNav – International Journal on Marine Navigation and Safety of Sea Transport*, Vol. 6, No. 1, March 2012, pp. 89-92

[8] Benedict K., Fischer S., Gluch M., et al., "Fast Time Simulation Technology for Investigation of Person over Board Manoeuvres for Improved Training and Support for Application Onboard", *International Conference IMLA 19*, Opatija, Croatia, 2011, pp. 71-84

[9] Bowditch N., The American Practical Navigation, Pub. No. 9, Edited by National Imagery and Mapping Agency (NIMA), Bethesda, Maryland, pp. 199-215, 2002.

[10] Bošnjak R., Vidan P., Belamarić G., "The Term and Development of E-navigation", *Proc.of the 14th International Maritime Science Conference*, 16th-17th June, Split, Croatia, 2012. (CD issue)

[11] Graff J., E-Maritime: An Enabling Framework for Knowledge Transfer and Innovative Information Services Development Across the Waterborne Transport Sector, *TransNav – International Journal on Marine Navigation and Safety of Sea Transport*, Vol. 3, No. 2, June 2009, pp. 213-217

[12] Edmonds D., "10 Things They Should Told You About ECDIS", Paper given at *TransNav*, Gdynia, 2007.

[13] Hech H., et al., *The Electronic Chart, Fundamentals, Functions, Data, and other Essentials*, Geomares Publishing, 2011.

[14] Hempstead C., "Assessing Competence in ECDIS Navigation", *Proc. of the 16th International Navigation Simulation Lecturers' Conference*, 12-16th July, Dalian, China, 2012., pp. 59-74

[15] Hempstead C., "Teaching ECDIS at the U.S. Merchant Marine Academy", PowerPoint presentation, *Internet resource*, download: November, 2012.

[16] International Maritime Organization, "Development of Model Procedure for Executing Shipboard Emergency Measures", STW 41/12/3, London, 2009.

[17] Mulloth A., *Safe Navigation with Electronic Chart Display and Information System (ECDIS) Handbook*, Cengage Larning Asia, 2012.

[18] Patel R., Fension C., "Using PHStat and Camtasia Studio 2 in Teaching Business Statistics", Journal of College Teaching & Learning, Vol. 2, No. 9, pp. 53-57, 2005.

[19] Powell J., The New Electronic Chart Product Specification S-101: An Overview, *TransNav – International Journal on Marine Navigation and Safety of Sea Transport*, Vol. 5, No. 2, June 2011, pp. 167-171

[20] Seefeldt D., Enhance Berth to Berth Navigation Requires High Quality ENC's – The Port ENC – a Proposal for a New Port Related ENC Standard, *TransNav – International Journal on Marine Navigation and Safety of Sea Transport*, Vol. 5, No. 2, June 2011, pp. 163-166

[21] Simović A., *Terestrička navigacija*, 6. izdanje, Školska knjiga, Zagreb, 2001.

[22] TechSmithCoorporation, *Camtasia Studio - Help File Document*, Release 8.0, June 2012, pp. 39-47 (Web resource: http://assets.techsmith.com-/Docs/pdf-camtasiaStudio/Camtasia_Studio_8_Help_File.pdf; downloaded: December, 2013)

[23] TechSmithCoorporation, *Camtasia Studio 8 – Create Engaging Screencasts*, Release 8.0, June 2012. (Web resource: http://assets.techsmith.com/Docs/pdf-camtasiaStudio/Create_Engaging_Screencasts(1).pdf; downloaded: December, 2012)

[24] Tetley L., Calcutt D., *Electronic Aids to Navigation*, 2nd Edition, Elsevier, Lightning Source UK, Ltd., 1988.

[25] Tetley L., Calcutt D., *Electronic Navigation Systems*, 3rd Edition, Elsevier, Lightning Source UK, Ltd., 2004.

[26] Transas, *Navi – Trainer 4000 (ver. 4.50), Navigational Bridge*, Transas Marine GB, Ltd., 2004.

[27] Transas, *Navi – Trainer 4000 (ver. 6.42), Instructor Manual*, Transas Marine GB, Ltd., 2007.

[28] Transas, *Navi – Trainer 4000 (ver. 6.42), Navigational Bridge*, Transas Marine GB, Ltd., 2007.

[29] Transas, *Navi – Trainer 4000 (ver. 6.42), Technical Description and Installation Manual*, Transas Marine GB, Ltd., 2007.

[30] Weintrit A., The Electronic Chart Display and Information System (ECDIS). An Operational Handbook. A Balkema Book. CRC Press, Taylor & Francis Group, Boca Raton - London - New York - Leiden, 2009.

[31] Web resource: http://www.techsupportalert.com/best-free-audio-editing-software.htm

[32] Web resource: http://webseasoning.com/technology/best-free-windows-video-editing-software/1079/#.UP_kph1bbUd

Electronic Chart Display and Information Systems (ECDIS)
Advances in Marine Navigation – Marine Navigation and Safety of Sea Transportation – Weintrit (ed.)

Presentation of Magnetic Variation on ENC Charts According to the Rules and Good Sea Practice

K. Pleskacz
Maritime University of Szczecin, Poland

ABSTRACT: The ECDIS system plays an increasingly important function on vessels nowadays. This author presents requirements for equipment and course indicator control on ships in connection with an ENC, navigator's common tool, critically assessing the method of presenting magnetic variation data and suggesting solutions improving the safety of navigation.

1 INTRODUCTION

The 21st century has witnessed a rapid, almost revolutionary advancement in ship equipment that go in line with development of modern navigational systems as well as broadly understood control systems. However, this author claims that despite technological progress, the compass still remains the most important element of shipboard equipment. It is the compass that can only be relied on during difficult manoeuvres. No other tool exists that the helmsman could depend on while trying to keep the ship on course in accordance with captain's changing orders.

The development of automation may be very beneficial as it relieves seafarers from certain duties. On the other hand, one disadvantage is increased risk of errors made by operators who are very often unaware of all system characteristics and restrictions. In this connection, the author intends to discuss the method of magnetic variation presentation on ENC charts and consequent problems with course indicators control.

The Electronic Chart Display and Information System (ECDIS), already carried by many vessels, will soon be installed on a majority of ships, is far from perfect in terms of navigation safety. Misquoting a verse from Shakespeare's Hamlet we could say 'Something is rotten in the state of ECDIS'. The second meeting of IHO-HSSC (The Harmonized System of Survey and Certification) on 26 – 29 October 2010 in Rostock adopted the document titled *Operating Anomalies Identified in Some ECDIS*, where attention is drawn to insufficient requirements ENC charts have to satisfy.

As a result, situations occur where shallow water is not identified in a standard display and an alarm is not activated when preset depth limits are exceeded. We should note the fact, emphasized in the document, that the above anomaly was <u>accidentally</u> found.

2 A REVIEW OF REGULATIONS AND RELATED DOCUMENTS REQUIRED FOR THE USE OF COMPASSES ON MODERN SHIPS

Chapter V, Regulation 19, par. 2.1.4 of the SOLAS Convention provides requirement that all ships regardless of size should carry navigational charts and nautical publications for planning and presenting voyage routes and for plotting and checking ship's position. At the same time, the same paragraph provides that ECDIS systems may be recognized as satisfying the requirement of that paragraph concerning possession of navigational charts on board. All requirements concerning the ECDIS system are contained in the IMO Resolution A.817 (19), as amended by MSC.64 (67) and MSC.86 (70), Performance Standards for Electronic Chart Display and Information Systems (ECDIS)). One of the standards reads that only an ENC (Electronic Navigational Chart) may be used with an ECDIS system.

ENC is a database, with standardized contents, structure and format, created in government-approved hydrographic offices. ENC contains all cartographic information necessary for safe navigation. To get a certificate of an ENC chart, an electronic chart has to meet S-57 standards (IHO S-57, "IHO Transfer Standard for Digital Hydrographic Data"), issued by the International Hydrographic Organization (IHO). The publication comprises encoding instructions and catalogues for uniform interpretation by hydrographic offices, describes cartographic objects and their characteristics, concepts of base cells and corrections.

Both magnetic compasses and gyrocompasses are subject to regulations of the SOLAS Convention. Chapter V of the Convention includes requirements for ships' navigational systems and equipment.

All ships regardless of their size should carry:
- properly adjusted standard magnetic compass, or other means, independent of any power supply, to determine the ship's heading and display the reading at the main steering position [Reg. 19, par. 2.1.1];
- means of correcting heading and bearings to true at all times [Reg. 19, par 2.1.3];
- All ships of 500 gross tonnage and upwards shall have a gyrocompass, or other means, to determine and display their heading by shipborne non-magnetic means and to transmit heading information for input to radar, AIS or ARPA equipment [Reg. 19, par 2.5.1].

The STCW Convention, in turn, sets forth requirements concerning personnel training and watch keeping. Section A-II/1 defines minimum requirements for watch officers on ships of 500 gross tonnage and upwards. Such officers must know principles of operation of magnetic compasses and gyrocompasses and be able to determine errors of such compasses using both astronomical and terrestrial means. Watch officers must have skills to allow for such errors.

Chapter VIII of the STCW Convention contains standards for watch keeping at sea. Its section 4 – *Watch keeping at sea*, par. 22, includes items that the officer taking over the watch should check personally. These include, among others, operating state of all navigational and safety equipment that may be used during a watch, [par 22.5.1] and possible errors of gyrocompasses and magnetic compasses [22.5.2].

Par. 34 includes requirements on checks that an officer of the watch should regularly make to find out, at least once in a watch, whether there is a specific error of the main magnetic compass and, if possible, after each major course alteration. OOW should ensure that the main compass and gyrocompass are being compared, and repeaters are synchronized with the main compass [par. 34.2].

One necessary element for correct control of magnetic compasses is the knowledge of current variation established for ship's position. This follows from formulas used during a check of a magnetic compass:

$$(\pm\text{Dev.}) = TC - CC - (\pm\text{Var.})$$
or
$$(\pm\text{Dev.}) = TB - CB - (\pm\text{Var.}) \tag{1}$$

where
- Dev. – magnetic compass deviation
- Var. – magnetic compass variation
- TC, TB – true course or bearing
- CC, CB – compass course or bearing

3 SPECIFIC REGULATIONS REFERRING TO THE DISPLAY OF MAGNETIC VARIATION ON ENC CHARTS

Principles of encoding and presenting magnetic variation on ENC charts are included in the S–57 standard. This document contains a requirement that, until a world magnetic model is used in ECDIS, values of magnetic variation should be encoded using the object class MAGVAR. As a minimum, updates should be supplied every five years.

The mandatory attributes are:
- RYRMGV- reference year for magnetic variation;
- VALACM- value of annual change in magnetic variation;
- VALLMA - value of local magnetic anomaly;
- VALMAG - value of magnetic variation.

Values of magnetic variation can be encoded in ENC as an point, line or area object. Details contained in the S–52 document 'Specification for chart content and display aspects of ECDIS', issued by IHO, refer to values of magnetic variation and areas of local magnetic anomalies.

Requirements for graphic presentation of a magnetic variation symbol on an ENC are given in Figure 1.

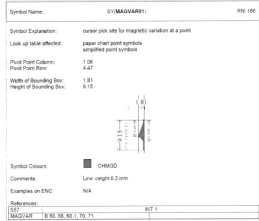

Figure 1. A point symbol presenting magnetic variation [source: S–52, IHO 2008]

The symbol RN: 166, if the symbol is cursor picked, the value of magnetic variation will be displayed.

Another solution is to present variation in a given area. In that case the symbol RN:167 is presented, which does not require to be precisely picked by the cursor that can only be placed in its vicinity.

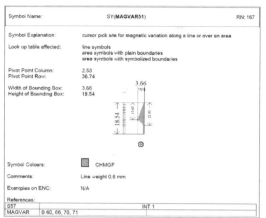

Figure 2. An area symbol presenting magnetic variation [source: S–52, IHO 2008]

Local magnetic variations are represented by the same rules, except that the symbols are fully colorful or have color contours only.

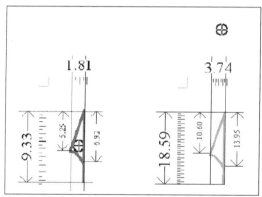

Figure 3. Symbols of local magnetic anomaly represented as a point object and area object [source: S–52, IHO 2008]

Figure 4. Value of variation presented on an ENC [source: simulator TRANSAS Navi-Sailor 3000 ECDIS-i]

After cursor picking of a given symbol and pressing *Enter*, we get information on values of particular attributes.

4 AN ANALYSIS OF PRESENTATION AND IMPLEMENTATION OF REGULATIONS ON MAGNETIC VARIATION BY HYDROGRAPHIC OFFICES

In paragraph 3.7 of Appendix B to S–57 standard legislators used a vague imperative referring to requirements of presenting information on magnetic variation on ENC charts: ... 'The ENC *may* contain information about magnetic variation, ...', which can be understood as a permission rather than a definite data presentation requirement that has to be met.

Similarly, paragraph 6.5, referring to the contents of a presented chart legend uses a 'soft' modal *should*, not *must*: "This legend *should* contain at minimum: .... magnetic variation"... .

As interpretations may vary in reference to international regulations on magnetic variation information presentation, there are hydrographic offices (ENC producers) that do not encode that information setting forth an argument that at the age of GPS is unnecessary. Hydrographic offices that put such information do it in a variety of ways, choosing encoding in the form of a point, line or area or combine them. Some put the information in all usage bands, others only in general and coastal ones.

The confusion that has emerged in this respect is an important factor reducing the safety of navigation.

One of the offices failing to provide information on magnetic variation on their ENCs is the Canadian Hydrographic Service. That office does not even place any such data for regions where magnetic variation amounts to -21°. Thus, any magnetic compass check, and use, is impossible. As international regulations provide that an ENC is equivalent to paper charts and the latter do not have to be carried on board ships, the question to answer is: How can the navigator meet requirements concerning magnetic compass control?

Below is a section of an ENC chart No 176030 and a corresponding raster chart extract. The information on the ENC does not contain any data on magnetic variation. A point symbol to present magnetic variation is not there, either. The ENC chart presentation has a proper scale and display layer that should include all available information about the chart.

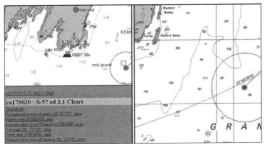

Figure 5. Canadian ENC and raster chart sections [source: simulator TRANSAS Navi-Sailor 3000 ECDIS-i]

The method of encoding magnetic variation as point objects is used on some Polish charts, for instance, a chart of the approach to the port of Szczecin. Along the whole route from Szczecin to the Świnoujście heads there is only one symbol, which cursor picked, reveals the value of magnetic variation. It is located near *Brama Torowa* (Fairway Gate) No 2.

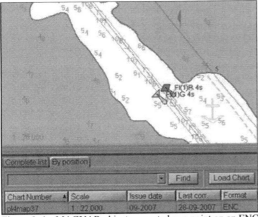

Figure 6. An MAGVAR object presented as a point on an ENC – pl4map37 [source: simulator TRANSAS Navi-Sailor 3000 ECDIS-i]

Where only one piece of information is included for an entire relatively long and narrow fairway, it will not be looked for. The navigator, not knowing where to find it, will probably give up searching along the whole route and the compass will not be checked. The chart scale is large and scanning it visually all over will draw too much of navigator's attention that has to be focused on ship's position. In such a narrow channel the navigator will definitely choose navigating with caution instead of meeting a regulation. The Figure below shows a fragment of the same chart ENC-plmap37, in the area of leading marks, the place most convenient for compass indication control. In that area, in spite of correct scale and chart display layer, information on magnetic variation is not available.

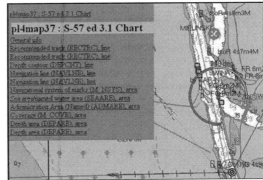

Figure 7. A section of ENC – pl4map37 [source: simulator TRANSAS Navi-Sailor 3000 ECDIS-i]

From the navigator's perspective, area encoding is the most convenient method, used by the Hydrographic Department of the Finnish Maritime Administration.

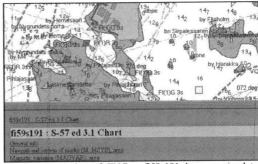

Figure 8. A section of ENC – fi59s191 [source: simulator TRANSAS Navi-Sailor 3000 ECDIS-i]

After a cursor pick of any point on the chart free from other navigational information in the *Info* option, the operator obtains data on all attributes of magnetic variation.

5 PRESENTATION OF MAGNETIC VARIATION ATTRIBUTES IN AN ECDIS SYSTEM

The form of presenting attributes is a certain inconvenience while we are using magnetic variation information contained in ECDIS.

Regardless of the encoding as a point or area object, the present value of variation is not given, only archived data with a trend of changes. This requires calculations each time and carries a risk of human error affecting the system. This is unnecessary complication that should be avoided by presenting an updated variation for a given position. There are free programs for computing variation for a given moment and position. They could be adapted to an ECDIS system that continuously generates necessary input parameters: ship's time and position.

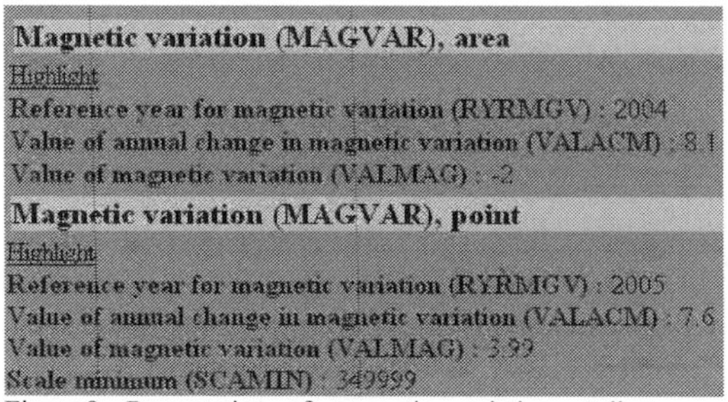

Magnetic variation (MAGVAR), area
Highlight
Reference year for magnetic variation (RYRMGV) : 2004
Value of annual change in magnetic variation (VALACM) : 8.1
Value of magnetic variation (VALMAG) : -2
Magnetic variation (MAGVAR), point
Highlight
Reference year for magnetic variation (RYRMGV) : 2005
Value of annual change in magnetic variation (VALACM) : 7.6
Value of magnetic variation (VALMAG) : 3.99
Scale minimum (SCAMIN) : 349999

Figure 9. Presentation of magnetic variation attributes on ENCs
[source: simulator TRANSAS Navi-Sailor 3000 ECDIS-i]

6 SUMMARY

The role of gyrocompass and magnetic compass during daily operation of the ship cannot be overestimated. Although the gyrocompass is used during most manoeuvres, and the magnetic compass is nowadays regarded as an emergency device, its presence, skillful use and knowledge of data reliability make up an important factor for the safety of navigation. Many a navigator on the bridge has experienced problems with a gyrocompass. Additionally, it should be noted that the magnetic compass is the only course indicator independent of external energy supply.

If we sum up the above remarks and requirements imposed by relevant regulations in force, it seems unquestionable that these course indicators should be regularly controlled. Students of navigation learn about magnetic variation and the role of magnetic compass in the first year of their studies. Bearing this in mind, this author is surprised by the approach to these issues by those both legislators and ENC producers. It should be emphasized that, despite technological progress, the simplest solutions on the ship are still justified. The issue of magnetic variation on ENCs that soon will completely replace paper charts should be appropriately regulated.

Undoubtedly, experienced navigators should be consulted on the issue, an action both necessary and beneficial. Besides, research should be conducted to identify problems that might result from improper or incorrect design, selection, installation and use of automated systems in marine navigation.

Let us hope that in the near future the IHO's Transfer Standards Maintenance and Applications Development Working Group will work out guidelines for uniform principles of encoding magnetic variation on electronic navigational charts that will satisfy good sea practices.

REFERENCES

[1] Bąk A., Dzikowski R., Grodzicki P., Grzeszak J., Pleskacz K., Wielgosz M.: Przewodnik operatora systemu ECDIS Navi-Sailor 3000 ECDIS-i, Akademia Morska, Szczecin 2009
[2] IHO Publication S-52 Appendix 2: Addendum to Annex A, Part 1, Edition 3.4, 2008.
[3] IMO MSC 82/15/2, Research into interaction with automate systems, IMO 2006
[4] IMO MSC/Circ.1091, Issues to be considered when introducing new technology on board ship, IMO 2003
[5] IMO MSC/Circ.1091, Issues to be considered when introducing new technology on board ship, IMO 2003
[6] IMO MSC/Circ/1061, Guidance for the operational use of integrated bridge system (IBS), IMO 2003
[7] International Convention for the Safety of Life at Sea (SOLAS), IMO, 2009, (Consolidated Edition)
[8] NAV: Report to the Maritime Safety Committee. NAV/54/2514, August 2008, IMO, London 2008.
[9] Resolution A.817 (19) with amendments, IMO, London 1996
[10] Simulator TRANSAS Navi-Sailor ECDIS 3000 ECDIS-i

Implementation of Ship Collision Avoidance Supporting System on Electronic Chart Display and Information System

K.S. Ahn & M.S. Hwang
Hyundai Heavy Industries Co. Ltd., Ulsan, Republic of Korea

Y.W. Kim & B.J. Kim
Hyundai e-Marine Co. Ltd., Ulsan, Republic of Korea

ABSTRACT: ECDIS system functioned as a collision avoidance supporting system is now being developed. This system deals with the stationary targets from ENC of ECDIS as well as the moving targets from AIS and ARPA while the own ship is moving. The system consists of three sub systems: the automatic target detection from AIS, ARPA and ENC, the decision support from collision avoidance algorithm and the result display on ECDIS including signal generation for autopilot system. The application domain is limited to the open sea including some ENC objects for real-time onboard calculation, but the extension to coastal waters is now going on. This system has been successfully confirmed by using various simulations and the verification through sea trial is now prepared.

AIS: Automatic Identification System   ARPA: Automatic Radar Plotting Aid
ECDIS: Electronic Chart Display and Information System ENC: Electronic Navigational Chart

1 INTRODUCTION

For the navigational safety, a lot of systems supporting pilot's decision have been developed and the development of new products continues now. However, despites these efforts, collision accidents are still not decreasing.

Collision avoidance system was mentioned so many times as one of the potential applications while introducing e-Navigation strategy since 2007 [1]. However, this kind system still doesn't come to attention even if some systems were already developed [2].

ECDIS (Electronic Chart Display and Information System) for commercial vessels has been mandatory nowadays and become a main navigational system.

This study shows the results of the development of ECDIS system functioned as a collision avoidance supporting system so far. This system deals with both of the stationary targets from ENC of ECDIS and the moving targets from AIS and ARPA while the own ship is moving.

As a result, we could develop a prototype and confirm the applicability of this system.

2 SYSTEM STRUCTURE

Developed collision avoidance supporting system consists of 3 subsystems: the automatic target detection, decision support and result display shown in Figure 1.

Figure 1. Collision avoidance supporting system

2.1 *Automatic target detection*

2.1.1 *AIS and ARPA simulator*

All targets from AIS and ARPA are automatically gathered to ECDIS. Among these data, targets within a predefined range are transferred to decision support part. The data are updated in real time, but transferred to the decision support part in regular intervals.

In this study, these data were virtually simulated. However, these data can be replaced without any modifications by real data of ship in the commercialization stage.

2.1.2 *ENC object*

In the deep and open sea area, ENC objects are not important, thus many researchers generally don't concern these objects. However, ENC objects become important if ship moves to shallow water area because collision to geographical features also becomes critical issue. In this study, point objects such as lights and sunken ship etc. and line objects such as land and safety depth within a range of interest are also considered.

In the shallow waters, too many objects can be detected. For real-time onboard calculation, the search region should be set like Figure 2. In this study, circle having a limited radius is used for the region. Retrieved data are substituted into virtual ships and transferred to decision support part

Figure 2. Detected ENC Objects

The data are updated and transferred to the decision supporting part in regular intervals to save time.

2.1.3 *Coordinate Transform*

All detected data have the latitude and longitude points on the surface of the earth, but decision support part wants the rectangular coordinates of x, y. Therefore, the transformation algorithm was neccessary to solve this problem. For more accurate calculation, the Vincenty formula within 0.5mm of accuracy was used, but it can be changed to Haversine method within 0.3% of accuracy if the additional time saving is needed [3].

2.2 *Decision support*

Algorithm for this system was based on the collision avoidance algorithm of Seoul National University [4] and partially modified in order to be suitable for ECDIS system. Basic algorithm and modifications according to improvement requirements are explained briefly as below.

2.2.1 *Algorithm*

Collision risk was inferred by using fuzzy theory based on the two non-dimensional parameters of TCPA and DCPA shown in Figure 3. TCPA and DCPA mean the time to of closest point of approach

and the distance of closest point of approach, respectively. Hasegawa suggested this concept [5].

The detected targets are assumed to be moving with constant speed and direction at each update.

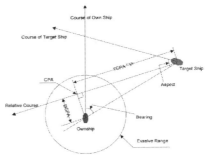

Figure 3. DCPA and TCPA

The defuzzification equation is as follows.

$$Collision\, Risk = \frac{\sum_{i=1}^{n} CR_i \cdot \alpha_i}{\sum_{i=1}^{n} \alpha_i}$$

where,
n = number of reasoning rule
CR_i = singleton value of conclusion part of ith rule
α_i = contribution factor of conditional part of ith rule

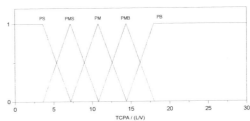

Figure 4. Membership function of TCPA

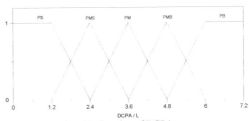

Figure 5. Membership function of DCPA

Minimum operator is used as in the conditional part of ith rule. Parameters of conditional part have membership function of triangular type as shown in Figure 4 and Figure 5.

Table 1 shows the conclusion part having reasoning rule of singleton type which was used for saving computation time. The negative value of TCPA is meaningless because the target was already passed.

Table 1. Reasoning rules of collision risk

		TCPA							
		NB	NM	NS	PS	PMS	PM	PMB	PB
D	PS	-0.2	-0.6	-1.0	1.0	0.8	0.6	0.4	0.2
C	PMS	-0.2	-0.2	-0.6	0.8	0.6	0.4	0.2	0.2
P	PM	-0.2	-0.2	-0.2	0.6	0.4	0.2	0.2	0.2
A	PMB	-0.2	-0.2	-0.2	0.4	0.2	0.2	0.2	0.2
	PB	-0.2	-0.2	-0.2	0.2	0.2	0.2	0.2	0.2

When collision risk exceeds the predetermined threshold, this system produces the recommended paths through the expert system based on CLIPS. This module determines each target as either privileged or burdened and recommends the safest actions according to the action plan. Figure 6 shows the course alteration diagram suggested by the Royal Institute of Navigation in 1970 [6].

Figure 6. Course Alteration Diagram

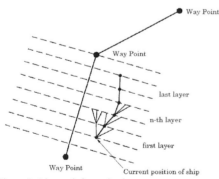

Figure 7. A* search for optimal route

The chain of the safest actions are determined by using A* search method shown in Figure 7.

A* algorithm finds the optimal path which minimizing the cost function as below.

Estimated total cost
= Cost from start to current node
+ Estimated cost from current node to final node

The example cost function between two nodes is as follows.

Cost from node A to B
*= Collision risk at A * Time from A to B*

Various routes when using the different cost functions and restrictions can be recommended. Current system shows various different routes contingent to the membership functions, path restriction and cost function.

2.2.2 *Modification according to Improvement Requirements*

Original program changes membership functions of conditional part by using genetic algorithm. However, in this study, constant membership functions based on previous simulations was used because that is the time consuming process.

It has been required that the system to issue an earlier avoidance order in order to prevent the drastic rudder change which possibly causes dramatic ship's roll motion which can cause the passengers to feel uncomfortable and cause damage to the cargo. To fulfill this requirement, membership function of TCPA was changed like Figure 8 which notices danger early on. The results of before and after of modification are shown in Figure 9.

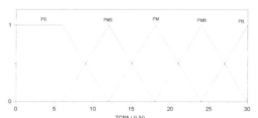

Figure 8. Modified Membership Function of TCPA

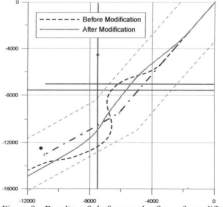

Figure 9. Results of before and after of modification of Membership Function of TCPA

Another requirement was that a ship should not off course too much from the original course. To satisfy this requirement, the concept of cross track distance (XTD) generally used in the track control system was applied [7]. This is illustrated in Figure 10. The region can be optionally selected by users from 0.5 nm to 2 nm.

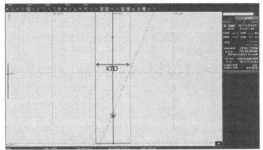

Figure 10. Cross track distance (XTD)

For smoother path, the concept of line of sight guidance principle, where the ship switches the way point to the next point when it came in the circle of acceptance, was applied in the final simulation [8].

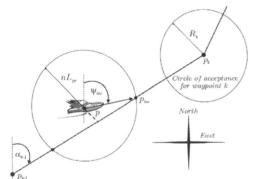

Figure 11. Line of Sight guidance principle

2.3 *Result display*

In this study, ECDIS system of Hyundai e-MARINE was used.

2.3.1 *Why displayed on ECDIS*

All the results should be displayed on ECDIS which pilots always keep an eye on. If the collision avoidance supporting system had own display out of pilots sight, collision information might be unnoticed to the pilots. Furthermore, ECDIS is not only the window for representing ENC, but also all navigational data including AIS, ARPA and anemometer etc. are gathered without any exertion. Following Figure 12 shows general external inputs to ECDIS.

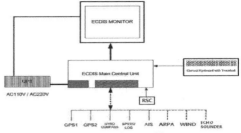

Figure 12. ECDIS system and external inputs

And ECDIS can control the autopilot system through the special communication protocol. As a result, the selected results can be directly transferred to autopilot system when pilots determine the recommended route as appropriate.

2.3.2 *Displayed items*

This system shows the recommended routes with respect to the given cost function or restriction, current value of collision risk and the upcoming object and most dangerous object in the region.

2.3.3 *Region of Display*

This system can show the results within just a predefined detecting region because the results exceeding the region are meaningless.

2.3.4 *Updates*

This system is updated at regular intervals such as every 3 or 5 minutes, otherwise this system cannot consider the unexpected situations such that new targets suddenly appeared. And if the own ship exceeds the given region, the update must be carried out.

3 RESULTS

3.1 *Collision Avoidance in open sea with ship*

The simple collision situation was assumed like Figure 13.

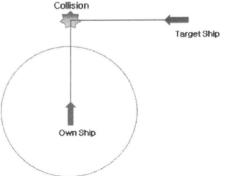

Figure 13. Collision Situation 1

If target are came within a predefined, warning message and recommended route are shown on the ECDIS display like Figure 14.

Figure 14. Collision Avoidance with ship

3.2 Collision Avoidance in open sea with ships and ENC objects

Nowadays, the multi-ship encounter situations are also regarded [9]. Under the circumstances, we set the complicated scenario like Figure 15. On the given course, there are several ENC object like a wreck and several ships are crossing each other.

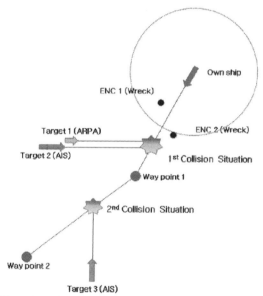

Figure 15. Collision Situation 2

Objects within predefined area (e.g. circle with radius of 5 nm) are considered to the collision avoidance calculation. Thus, ENC1, ENC 2, target 1, target 2 and target 3 are recognized to own ship consecutively. If the collision risk is larger than the predefined value, the alarm message is displayed on the ECDIS display like Figure 16. Several routes can be recommended depending on their purpose.

Figure 16. Collision Avoidance with respect to Target 2

After selecting one of the routes, user can determine to use the automatic track control or to steer the ship manually refering the route, or can cancel all the routes. Figure 17 and Figure 18 shows the trajectories of each ships as time goes by.

Figure 17. Collision Avoidance with respect to Target 1

Figure 18. Collision Avoidance with respect to Target 3

3.3 Application to coastal waters

Developed system only focuses on the application to ship-ship collision in open sea but having some ENC objects. However, we have a plan extending application domain to the more shallow waters. For this purpose, the preliminary research was carried out and some problem-solving schemes to lessen the calculation load could be suggested as follows. During this research, only ENC data were considered.

In the case of coastal waters, although the search region is set to 5 nm, there were many detected objects like Figure 19.

Figure 19. Search region for collision avoidance

It is found that the filtering scheme of the intersection within the cross track distance is needed like Figure 20. As a result, many of them can be filtered out and the collision avoidance simulation can be carried out smoothly without a time lag.

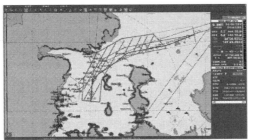
Figure 20. Intersection of detection region and cross track distance

And dectection range should be adjusted to smaller value in case of coastal waters.

These schemes were successfully applied like Figure 21.

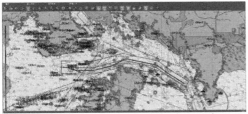
Figure 21. Collision avoidance simulation in coastal waters

The changes of search region can be automatically performed through numbering ENC objects around ship.

Despite this effort, this system showed a time lag in some cases, so improvement work is now carrying out.

4 CONCLUSION AND FUTURE WORKS

ECDIS system functioned as a collision avoidance supporting system is now being developed. For this purpose, the collision avoidance algorithm of Seoul National University and the ECDIS system of Hyundai e-Marine was used. The prototype was developed and the system validation was finished.

Developed system consists of 3 sub-systems: the automatic target detection, decision support and result display. Target detection part includes not only the target ships from AIS and ARPA but also ENC objects such as the point objects representing lights and sunken ships etc and line objects representing land within a range. Decision support part has been improved to communicate with the ECDIS and satisfy the requirements. Result display part is in charge of displaying the recommended route through the decision support part and performing the virtual simulation.

During its development, the various requirements of users and schemes for real-time onboard calculation could be reflected in the system. The applicability of this system could be confirmed through various simulations.

As a next step toward commercialization, the verification through sea trial is now planned. And in order for fast commercialization, the application domain was primarily limited to the open sea including some ENC objects, but the extension to more shallow waters is also going on.

REFERENCES

[1] Ward, N. and Leighton, S. "Collision Avoidance in the e-Navigation Environment", Proc. 17th IALA Conference, pp.4-10, 2010.
[2] Totem Plus Ltd. "COLREGS Adviser [online] Available at : http://www.totemplus.com/colregs.html.
[3] Vincenty, T. "Direct and Inverse Solutions of Geodesics on the Ellipsoid with application of nested equations", Survey Review, v. XXII, no 176, 1975
[4] Lee, H.J., Rhee K.P. "Development of Collision Avoidance System by using Expert System and Search Algorithm", International Shipbuilding Progress, v. 48(3), pp.197-212., 2001.
[5] Hasegawa, K. and Kouzuki, A. "Automatic Collision Avoidance System for Ships Using Fuzzy Control (in Japanese)", Journal of the Kansai Society of Naval Architects, v.205, pp.1-10, 1987.
[6] Cockcroft, A.N. and Lameijer, J.N.F. "A Guide to the Collision Avoidance Rules", Butterworth-Heinemann Ltd., 2009.
[7] ECDIS Ltd. "ECDIS Ltd Articles and Press Releases", [online] Available at : http://www.ecdis.org/media/?p=334.
[8] Fossen, T.I. "Guidance and Control of Ocean Vehicles", John Wiley and Sons Ltd., 1994.
[9] Szłapczynski, R. "Evolutionary Sets of Cooperating Trajectories in Multi-Ship Encounter Situations - Use Cases", International Journal on Marine Navigation and Safety of Sea Transportation, v. 4(2), pp.191-196., 2010.

Protection and Risks of ENC Data Regarding Safety of Navigation

S. Kos & D. Brčić
Faculty of Maritime Studies, University of Rijeka, Croatia

D. Pušić
Croatian Hydrographic Institute, Split, Croatia

ABSTRACT: In accordance with SOLAS Convention amendments concerning navigational equipment (Chapter V: Safety of Navigation; Regulation 19: Carriage Requirements for Shipborne Navigational Systems and Equipment), ECDIS (Electronic Chart Display and Information System) and ENCs (Electronic Navigational Chart) should be implemented on existing and new built cargo and passenger vessels. This process started with 1st of July 2012. In accordance with amendments, the number of vessels which should and/or must carry ECDIS equipment onboard increases significantly. It applies both on national and world fleet vessels, depending from their size and type. With the objective to protect the official electronic navigational charts, the International Hydrographic Organization adopted the security plan and standards of data protection have been accepted (IHO Data Protection Scheme: Standard S-63, edition 1.1.1). With present, increasing expansion of internet-based communications, computer users (that could be unqualified and, even more important, unauthorized persons) are allowed to access ENC software and official electronic charts in various ways, thereby opening the possibility/opportunity of abusive actions. The proposed paper structurally analyzes the usage and security implications of mentioned, illegal software. Illegal ENC security issues are discussed in the context of safety-of-navigation related possible – real scenarios using unproven and non-validated electronic navigational charts.

1 INTRODUCTION

Onboard vessels, paper charts are increasingly replaced by electronic database maps. Today, on merchant and passenger ships, Electronic Chart Display and Information Systems (ECDIS) and Electronic Navigational Charts (ENCs) are already widely used without the obligation to hold classic navigational paper charts. In the very meaning, the real ECDIS implementation process is still to come, as well as the time when one will look at the paper charts as a matter of the past, even nostalgically. However, this transition should be perceived considering both its benefits and disadvantages.

It is estimated that between 4000 and 7000 approved ECDIS systems onboard vessels and ashore are already implemented (MCA 2012). Regarding SOLAS Regulations, this number is still rising, and it will reach its peak at the final deadline of the implementation process, what will be presented in further text. Considering the increased development of new techniques and technologies, particularly information (and related) technologies, and including development of ever-advancing computing systems, it is expected that ECDIS systems will be implemented on all kind of vessels increasingly. Regardless of obligation of approved ENC/ECDIS possession, there appears a possibility of non-official electronic charts and software usage, which are their unlicensed copies. This illegal software can be already downloaded from particular Internet sites, as well as corresponding electronic charts. Unofficial ENC data and unofficial software, once used for navigation purposes, may pose great risk to the safety of navigation. Whether accidental or intentional, the usage of illegal data may result in errors in conducting navigation, which can lead to catastrophic consequences.

The proposed paper describes the process of protection and encryption of ENC data in accordance with International Hydrographic Organisation (IHO) Standards. Here, unconditional reliability of ENC data and ECDIS software is emphasized. Disadvantages of used encryption are listed, as well as methods in which encryptions can be broken, thus enabling the application access without authorization. In the following chapters security threats to illegal software are described. By using dedicated programming tools, authors have performed and presented possible scenarios of electronic charts and software abuse.

The paper concludes with summarized results of the research, emphasizing risks and dangers of using unofficial data. Moreover, attention is drawn to intentional editing which is, in the light of growing ECDIS usage, increasingly harder to control.

2 BACKGROUND

As in IMO *ECDIS Performance Standards*, ECDIS equipment is defined as a *"mean of navigation information system which, with adequate back-up arrangement can be accepted as complying with the up-to-date chart required by regulation V/19 and V/27 of the 1974 SOLAS Convention, by displaying selected information from a system electronic navigational chart (SENC) with positional information from navigation sensors to assist the mariner in route planning and route monitoring, and by displaying additional navigation-related information if required"* (IMO MSC.232(82) 2006, Weintrit 2009).

Implementation of ECDIS/ENC onboard vessels has officially started with 1st of July, 2012. The transitional period will last until 1st of July, 2018, its features depending on particular vessel's age, type and size (IMO MSC.282(86) 2010). On Figure 1 ECDIS implementation time schedule is illustrated. According to Convention (SOLAS 2011) every ship carrying two independent ECDIS systems can be considered as paperless. Each independent system includes set of appropriate electronic navigational charts, independent set of mandatory sensors and unintended power supply (IMO MSC.232(82) 2006).

Here, it should be stressed out the increased use of electronic charts onboard vessels which, according to SOLAS Convention, do not have to be equipped with official ECDIS. It refers primarily to yachts, mega-yachts and other leisure vessels. Those vessels today are using Electronic Chart Systems (ECSs). *ECS is a navigation information system that electronically displays vessel position and relevant nautical chart data and information from an ECS Database on a display screen, but does not meet all the IMO requirements for ECDIS and is not intended to satisfy the SOLAS Chapter V requirements to carry a navigational chart* (ISO-19379 2003). Globally, due to its cost, ECDIS is still relatively difficult to apply, although the ECDIS implementation onboard non-SOLAS vessels is noticed.

ECDIS is highly complex-to-make and hard-to-develop product. As such, it is still inaccessible to the majority of end-users. Becoming the subject of increased interest, there appears a tendency to purchase and to use illegal and unlicensed computer software and electronic charts copies. Small CD&DVD media prices and easy-to-access Web are resulting in simple software and data illegal reproduction and distribution. For the end-user, the usage of unlicensed data delivers legal consequences, however there are much harder and more-direct impacts that can occur in the form of unreliability of the system due to usage of unlicensed software and data.

Recent studies (BSA 2011) have shown that at the global level, more than 57% of all PC users admit they are using illegal software. The studies were conducted on 15000 respondents from 33 countries. It is estimated that the overall harm committed by using illegal software exceeds US$ 63 billion.

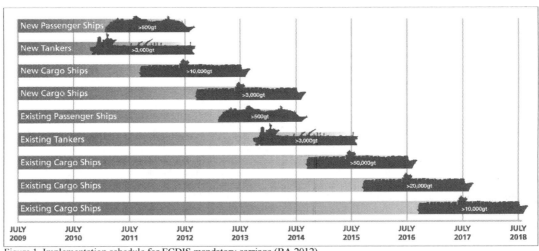

Figure 1. Implementation schedule for ECDIS mandatory carriage (BA 2012).

Analysis performed by authors showed that various ECDIS software is available online. According to research indicators, the easiest way to reach this unlicensed and illegal software is through *torrents*. However, the purchase through various websites is not negligible. Particular issue are representing forums closed for public, where logged users share illegal software. Currently, a dozens of websites offering illegal and unlicensed software are available. With the possibility to access the majority of web resources easily, illegal software is simple to find and download. When considering illegal software ease of access together with the ECDIS or ECS software high market price, it is to expect the rise of non-approved, potentially dangerous systems usage.

3 ENC DATA PROTECTION AND ENCRYPTION

3.1 *ECDIS and ENC*

In the strictest sense, ECDIS equipment consists of hardware, software and data. However, for the complete ECDIS functionality inputs from various sensors/devices are provided, ranging from sensors essential for the sole performance of the system (e.g. devices providing position – GPS/GNSS, speed – log, and bearings - gyrocompass) to less-required ones (e.g. water temperature sensors)[1]. In this paper, software and data components are specifically concerned.

According to IHO Standards, Electronic Navigational Chart (ENC) is the latest version of official data onboard ships intended for navigation (IHO S-57 2000, Duplančić Leder 2006). In order to facilitate processing of ENC data, geospatial grip is divided in *cells*. Electronic Navigation Cells are geographical areas which contain ENC data (Duplančić Leder 2006). Each cell must be contained in physically separate, uniquely designated file on removable media, known as Data set file (IHO S-63 2012). EN cell's geospatial size is chosen by ENC manufacturer in a way that the file does not exceed 5 Mb. The size of cell has to be large enough in order to prevent creation of excessive number of cells. Cells must be rectangular (two meridians and two parallels). Overlapping is allowed between cells of same navigational purpose only - i.e. *usage bands* (IHO S-57 2000, Vanem et al. 2007) - however data within cells must not overlap (Duplančić Leder 2006).

The name of each ENC is classified according to S-57 standard (IHO Transfer Standard for Digital Hydrographic Data), as: CCPXXXXXXXX:EEE, where mark CC denotes the manufacturer, mark P denotes navigational purpose, XXXX is the mark of the chart's serial number and EEE denotes the extension of the EN cell (Duplančić Leder 2006).

3.2 *ENC encryption*

Official ENC data (IHO S-63 2012) are encrypted by means of digital signature that is addition of electronic security mark to data files. With file's digital signature, verification publisher identity is rendered possible, as well as insight/check of possible data file alterations. If the file does not contain valid digital signature, there is no certainty in specified source's reliability, and there appears the possibility of malicious modification of the file.

Digital signature technique used by Primar (North European Region Electronic Navigational Chart Center) and in S-63 scheme involves *asymmetric encryption methodology*, or *public-key cryptography*.

This encryption algorithm relies on the fact that is possible to edit and *digitally sign* data file (cell) with one key (*secret private key*) on behalf of one organisation, and that other organisation can, with digital signature by different, *public key*, confirm, unlock and decrypt the same ENC data file. Interconnection of public and private keys is called the *key pair* (Scheiner 1996). ENC data encryption process is schematically presented on Figure 2.

In ENC data encryption, only one algorithm is used. Files are encrypted within ENC base cell or within the updated cell within S-57. Images and text are not encrypted. Each EN cell data file is encrypted with a unique key. The same key is used for the update of the cell. Encryption is performed with a 40-bit key. During the encryption, a 64-bit *block cipher* is used, known as Blowfish algorithm. All operations are based on the XOR (exclusive disjunction code) mathematical operation, including additional operations over 32-bit words.

In Blowfish algorithm variable length keys (up to 448 bits) are used, and it is specially designed for 32-bit computers (Scheiner 1996).

In the described way ENC data are ensured, and the integrity of ENC services is maintained, providing more data services and covering larger customer base. Purpose of data protection is threefold:
- Anti-Piracy: By ENC data encryption, unauthorized use of data is prevented;
- Selective Access: ENC data are limited for client's licensed cells only;
- Authentication: Provision of ENC data from approved sources is guaranteed.

---

[1] ECDIS sensors are classified as (listed in order of importance): must-have; need-to-have, good-to-have and nice-to-have. Acting together, they enable unique and full integration of the Electronic Chart Display and Information System.

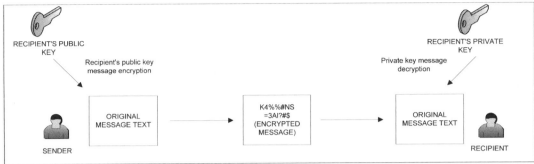

Figure 2. Public key cryptography process (Scheiner 1996).

Piracy protection and selective access are achieved by encrypting the ENC information and providing cell permits to decrypt them. (IHO S-63 2012). Data from server (*data servers*) will encrypt ENC information provided by national manufacturers before the same information are retrieved by the customer (*data client*). Authentication is provided through digital signature within data. This method allows mass distribution of encrypted ENCs on the media (e.g. CD/DVD) and it can be used by all customers and users which are holding a valid license (IHO S-63 2012).

4 ENC AND ECDIS SOFTWARE RELIABILITY

Reliability of navigational charts and publications implies user's trust in their information content in a specified time (Kasum et al. 2005). In its previous, paper format, the navigational chart was considered as a main navigation tool as well, leaving no space for errors or misinterpretation.

One major drawback/disadvantage of encryption using Blowfish algorithm can be noticed at once; there has to be an agreement regarding communication private key between the sender and the recipient.

Agreement process represents special problem, given that implies unsecure communication (through e-mail, telephone, etc.), which can compromise the privacy of the key.

The next theoretically possible method of software cracking is that the forger succeeds to reconstruct whole one-way function (algorithm). However, it is very difficult to achieve due to enormous computing resources which would then have to be used, even for the simplest algorithm implementations of modern public keys. For the forger, somewhat simpler way is breaking of one coded message, but even this method demands for virtually immeasurable time period.

In theory, every symmetric algorithm is subject to *brutal force* breakage that is consecutive testing of all possible secret key values. However, if the set of values is large, and the encryption algorithm is complex, this process is practically impossible.

Besides the installation drive, for the ECDIS software installation *USB dongle key lock* is required. It has to be connected to computer's serial or parallel communication port in order to run certain application. On *dongle* license data are stored, and it acts as a *security token* for the software authentication check, with the aim to prevent unauthorized usage. The level of security that can be achieved is usually determined by the performance of the application, and not the one by the dongle hardware key. However, the dongle validity check code within the application can be encrypted, including all communication between application *and* the dongle.

By imitating the performance of USB dongle key is possible to forge the license, or copy the contents from the dongle, and to further distribute the content. These actions are rendered possible with various *emulating* software. Such license counterfeiting method is very popular, and it is used in nearly all unlicensed ECDIS software available online.

4.1 *Illegal software security threats*

In order to protect their products as efficiently as possible, software manufacturers are using methods of USB dongle protection or programming method of protection within the software itself. In all methods, the software executable code is subject of encryption, thus disabling unauthorized access and alterations of the code, however to the *certain extent*.

The Reverse Engineering (RE) within the software itself is the most common security threat. RE is an disassembling process for detailed examination and analysis of a product or device to discover the concepts involved in manufacture, usually in order to produce similar data (Eliam & Chikofsky 2007). The code can be altered with online available tools and software, most of which are free. This software is as listed below (Raymond et al. 2004):

- Dissasemblers; backtracking assembler operations
- Debuggers; program testing and troubleshooting
- Decompilers; retrieving the software source code
- Hexadecimal editors; binary files manipulation.

Since the central component of each ECDIS/ECS is actually an configured computer, with its software installed as on every other computing system, it can be stated that each of these systems can be subject of attempt to unauthorized access, opening the possibility for intentional (or due to ignorance) data alterations and software modifications. Considering ECDIS as primary navigation system onboard ships, dangers arising from illegal software versions and data usage should be stressed out. The following chapters are elaborating the abuse of ECDIS software and data.

4.2 Patching: executable code modifications, an example of GPS coordinates spoofing

With *debugging software* (application for tracking the software executable code), it is possible to modify the code of the program. Under the scenario of unauthorized code changes, GPS signal reception code can be edited.

Among other parameters, GPS receiver displays the values of geographic coordinates in specific moment. The receiver is detached from the ECDIS console, representing, as every other sensor, a separate device. It is connected with ECDIS through (mostly) NMEA 0183 communication protocol. By knowing standardized protocols, that are communication modes between GPS and ECDIS (or between ECDIS and *any other* sensor) it is (easily) possible to find the information on the value of the coordinates in the software's source code. By programming the specific loop or code, it is possible to generate an artificial GPS signal error at a certain time, all performing in the executable code of the software. With the generated code activation, a part of in-advance-written additional code will be activated as well, thus shifting the received GPS position for a specific amount.

Execution of the above code represents the generic method for error modification, or intentional causing of vessel's position error.

Figure 3 represents the snapshot of the executable code editing process. On Figure 4 vessel's original GPS position is displayed on the electronic chart, while spoofed, shifted GPS position of the vessel is shown on Figure 5.

Figure 3. Example of MaxSea© software executable code editing using Ollydb© debugging free software: Programming the new subroutine. Made by the authors.

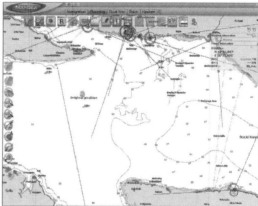

Figure 4. MaxSea© electronic chart display showing vessel's original GPS position. Made by the authors.

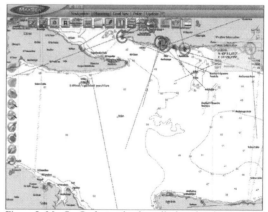

Figure 5. MaxSea© electronic chart display showing vessel's edited/spoofed GPS position. Made by the authors.

4.3 *Electronic chart content modification*

Another possible, although more demanding way to adversely affect the ECDIS software is to change the sole content of the electronic chart. If an unauthorized person can reach the code and monitor the program execution, it will be able to identify, *understand*, and crack the software protection method. According to web research results, among currently "cracked" software electronic charts are found as well (Figure 6).

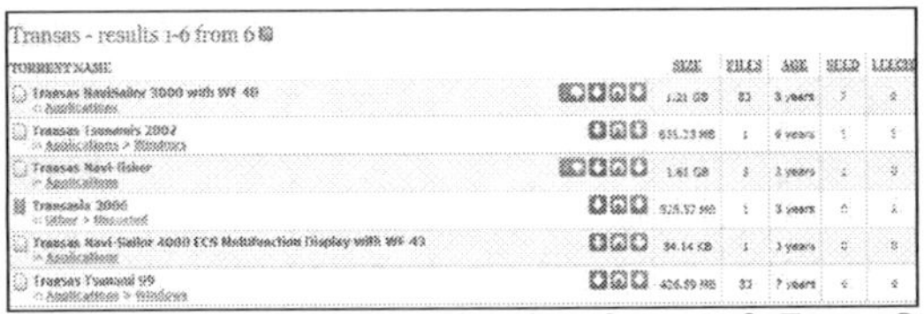

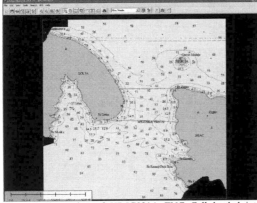

Figure 6. Internet-available illegal software of Transas© manufacturer (KAT, 2012).

Using editing tools, the chart content can be simply modified. The procedure of electronic chart content modification is graphically demonstrated on following figures. Modifications were performed on Electronic Navigational Chart ENC HR5C021A: 'Splitska vrata', area of Middle Adriatic, East Coast (HHI 2008). Manual removal of certain information/data (isobaths, depths, toponyms, etc.) is shown, including deletion of an entire geographical object – the island. Such processed electronic chart can be posted on Web, ready for further publishing and distribution (Figure 9).

Figure 7. Screenshot of HR5C021A ENC Cell loaded into Global Mapper© Software. Made by the authors.

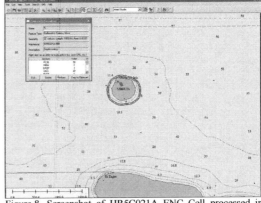

Figure 8. Screenshot of HR5C021A ENC Cell processed in Global Mapper© Software: Deletion of isobath, depths, toponyms and symbols. Made by the authors.

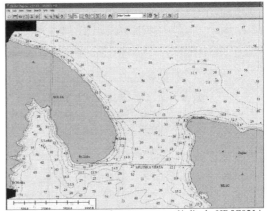

Figure 9. Screenshot of post-processed/edited HR5C021A ENC Cell. The figure reveals the absence of (existing) island. Made by the authors.

5 DISCUSSION

Looking at the example of GPS coordinates modification, it is needless to point out the safety of navigation endangering. ECDIS operator/officer on watch does not have to know for the modified, malicious code execution, neither for the size of the generated GPS position shift. This would consequently result in navigation conducting error, opening possibilities for further dangerous risks. Any unauthorized chart content modification would lead to incalculable consequences when using charts for navigational purposes. For example, if only one depth shoal, depth symbol on the chart and shoal symbol on the waterway would be removed or modified, it is to expect that, by using such deliberately altered chart, vessel would run aground in due time.

In the previous chapter some of the possible program modifications were described. However, the system may be affected in many other ways. Given that the malicious actions are targeting the executable code directly, every data entering the system can become subject of abuse. Outcome of any of these scenarios could lead to unwanted and deleterious consequences. Furthermore, malicious software (*viruses*) can pose a serious threat to such illegally modified systems as well; however, they were not analyzed in the paper.

The reasons for these intrusions may be different. It is much more important that they are possible. With near one hundred thousand SOLAS vessels worldwide (Lloyd's 2009) and the obligation of ECDIS implementation, great risks can be identified. This problem should be approached systematically from several standpoints, particularly legislative and educational.

The elaborated issue is very dependent on the profile of the end-user. It seems unlikely that such illegal software could find their place on SOLAS vessels. However, the issue is more pronounced on non-SOLAS, ECS vessels, where the crew does not have to be necessarily trained appropriately, thus unaware of the problems which could arise. Considering good seamanship, every sailor should verify the accuracy of the data *on master devices*, aware that data on ECDIS screen are just repeaters. This gradually starts to training questions (Weintrit 2008), where the nature of these situations must be emphasized at the very beginning, from the over-reliance point-of-view. Besides strong legislative actions towards illegal ECDIS software and data distributors to be taken, the proper education is essential.

6 CONCLUSION

The paper was written on possible scenarios of electronic charts and associated software theoretical assumptions. After official electronic data protection and encryption elaboration, methods of unauthorized access to ECDIS software and data were described. The results of performed actions were displayed. The detailed procedure of authors' experimental work was not specified. The aim of the paper was not to demonstrate *how* intentional modifications can be performed, but to point out the fact that *they can* be done.

This approach should be also considered in real situations. Easily accessible illegal software is subject to simple and uncontrolled executable code alterations. If it is possible to modify the code, it is also possible to modify the electronic navigational chart content. Such modified software with belonging charts can be easily published on Web and offered to other users. Due to the simplicity of Internet content sharing, whether through torrents or the Web, the user can reach this unverified software, which, in terms of the safety of navigation, is acting as dangerous. If more users were using this software, unaware of their modification, the potential disasters would be only the matter of time. Bearing in mind the availability of decrypted software, and with the aim to provoke maritime accidents, it can be stated that each such misdeed contains terrorism elements.

Nevertheless, the attention should be loudly brought to the usage of illegal software and associated charts, pointing out the consequences that may occur. Besides legislative and educational actions, a strong need appears for better system encryption in order to protect official data. Electronic cartography is currently (still) the area of little interest; however, with time passing, the question is in which extent the focus will be diverted to this domain. And we know that it will.

ACKNOWLEDGEMENTS

Research activities presented in this paper were conducted under the research project *Research into the correlation of maritime-transport elements in marine traffic*, supported by the Ministry of Science, Education and Sports, Republic of Croatia.

REFERENCES

British Admiralty. 2012. *Implementation of ECDIS onboard SOLAS vessels*. Available at: www.thefutureofnavigation, accessed on 1 November 2012.

Budin, L. 1998. *Computing systems security*. Authorised lectures of the Operating Systems course (in Croatian), Faculty of Electrical Engineering and Computing Science, University of Zagreb, Zagreb, Croatia.

Business Software Alliance. 2012. *2011 BSA Global Software Piracy Study*. Available at: www.portal.bsa.org/globalpiracy2011/downloads/study_pdf, accessed on 07.12.2012.

Croatian Hydrographic Institute (HHI). 2008. Catalogue of Charts and Nautical Publications. Split, Croatia.

Duplančić, T. L. 2006. *A New Approach to the Production of Electronic Navigation Charts in Croatia*. Doctoral Thesis (in Croatian). Faculty of Geodesy, University of Zagreb, Zagreb, Croatia.

Duplančić, T. L. & Lapaine M. 2006. A proposal of ENC Cell Distribution of the Croatian Part of the Adriatic. *Cartography and Geoinformation*, 5 (6); 56-67.

Eliam. E. & Chikofsky, E.J. 2007. *Reversing: secrets of reverse engineering*. John Wiley & Sons, Inc. Hoboken, New Jersey, USA.

Kasum, J. 2002. Contribution to Reambulation Optimsation by application of Electronic Information Technologies. Doctoral Thesis (in Croatian). Faculty of Maritime Studies, University of Rijeka, Rijeka, Croatia.

Kasum, J., Bičanić Z. & Perkušić A. 2005. The reliability of the sea charts and nautical publications, influence of law. *Precious Sea*, 52 (3-4); 117-121.

KickAss Torrents (KAT). 2012. Available at: http://kat.ph/usearch/Transas/, accessed on 20.10.2012.

International Convention for the Safety of Life at Sea (SOLAS) 1974, with amendments. 2011. International Maritime Organisation, London, UK.

International Hydrographic Organisation. 1997. Special Publication S-52: Specification for Chart Content and Display Aspects for ECDIS, 5th Edition. International Hydrographic Bureau, Monaco, EU.

International Hydrographic Organisation. 1997. Special Publication S-52: Specification for Chart Content and Display Aspects for ECDIS, 5th Edition. Appendix 3: Glossary of ECDIS – Related Terms, 3rd Edition. International Hydrographic Bureau, Monaco, EU.

International Hydrographic Organisation. 2000. Special Publication S-57: IHO Transfer Standard for Digital Hydrographic Data. Edition 3.1, Appendix B. International Hydrographic Bureau, Monaco, EU.

International Hydrographic Organisation. 2012. Special Publication S-63: IHO Data Protection Scheme, Edition 1.1.1. International Hydrographic Bureau, Monaco, EU.

International Hydrographic Organisation. 2010. IHO Publication S-66e 1.0: Facts about Electronic Charts and Carriage Requirements, Edition 1.0.0. International Hydrographic Bureau, Monaco, EU.

International Hydrographic Organisation. 2012. Recommendations for Consistent ENC Data Encoding. International Hydrographic Bureau, Monaco, EU. Available at: www.iho.shom.fr/COMMITTEES/CHRIS, accessed on 02.11.2012.

International Maritime Organisation. 2006. Resolution MSC.232(82), Adoption of the revised performance Standards for Electronic Chart Display and Information Systems (ECDIS), adopted on 5 December 2006. London, UK.

International Maritime Organisation. 2009. Resolution MSC.282(86): Adoption of Amendments to the International Convention forthe Safety of Life at Sea, 1974, as Amended, adopted on 5 June 2009. London, UK.

Lloyd's Register of Shipping. 2009. IHS Fairplay World's Fleet Statistics 2008. Information Handling Service, Douglas County, Colorado, USA.

International Organization for Standardization. 2003. ISO 19379: Ships and marine technology: ECS databases: Content, quality, updating and testing. Geneve, Switzerland.

Menezes, V.O.P. & Vanstone S. 1996. *Handbook of Applied Cryprography*. CRC Press, Brighton, UK.

PRIMAR ENC Services. 2009. Operational Handbook. Edition 3.0.

Raymond, D. et al. 2004. A survey of reverse Engineering Tools for the 32-bit Microsoft Windows Environment. College of Engineering, Drexel University. Philadelphia, USA. Available at: https://www.cs.drexel.edu/, accessed on 01.12.2012.

Scheiner, B. 1996. *Applied Cryptography*. John Wiley & Sons, Inc. Hoboken, New Jersey, USA.

Scheiner, B. 1994. The Blowfish Encryption Algorithm. Available at: http://www.schneier.com/blowfish.html, accessed on 01.11.2012.

UK Harbour Masters' Association, Maritime and Coastguard Agency. 2012. Marine Information Note MIN 426: *ECDIS – Testing for Apparent Anomalies*. Available at http://www.ukhma.org/mnotices.php, accessed on 13.10.2012.

Vanem, E. et al. 2007. Worldwide and Route-specific Coverage of Electronic Navigational Charts. *International Journal on Marine Navigation and Safety of Sea Transportation*, 1 (2); 163-169.

Ward, R. 2008. ENC Coverage Analysis. International Hydrographic Bureau, Monaco, EU.

Weintrit, A. 2009. *The Electronic Chart Display and Information System (ECDIS): An Operational Handbook*. Taylor & Francis, Abingdon, UK.

Weintrit, A. 2008. Operational Requirements for Electronic Chart Display and Information Systems (ECDIS). Procedural and Organisational Considerations. *Transport Problems*, 3 (2); 67-74.

The Fusion of Coordinates of Ship Position and Chart Features

A. Banachowicz
West Pomeranian University of Technology, Poland

A. Wolski
Maritime University of Szczecin, Poland

ABSTRACT: For the precise navigation of today the conformity of own navigational data and data of chart features from a hydrographic office-approved data base with navigational and hydrographic information is essential. The prerequisite for the compatibility of these data is a uniform reference system used in both navigational-hydrographic data base and in shipboard systems. Then the systems can be utilized for data fusion in order to determine total position coordinates. One particular case is the determination of relative position or an optimal fusion of measurements accounting for constraints. This case is considered in the article.

1 INTRODUCTION

Navigation makes use of many engineering and computing methods to determine position coordinates in an established reference system. Basically, these methods can be divided into three types [1]:
- model, based on a model of navigating object movement – dead reckoning and inertial navigation system (INS),
- parametric, in which a position is determined from a measurement of navigational parameters, that is spatial relations between navigating object coordinates and navigational marks,
- comparative navigation, in which images of measured Earth's physical fields are compared with cartographic images (databases).

Cartographic data are directly used for object position determination in the last mentioned method only. However, these measurements are not combined with other position determination methods. We present herein possibilities of the fusion of data from cartographic database with a running fix (parametric navigation).

The authors got inspired to deal with the issue by the fact that there occurs a statistical incompatibility of ship' position with cartographic data in cases of vessels berthing, docking or proceeding along a fairway.

2 FORMULATION OF THE PROBLEM

All combined navigational data should always be brought to a joint reference system. At present, WGS-84 fulfills this function due to a wide use of the satellite navigational GPS and ECDIS system. For this reason all navigational measurement and cartographic data from a navigational-hydrographic database should be brought to this reference system unless original data have been determined in this system. Failing to satisfy this condition results in a systematic error substantially exceeding random errors of the data.

The following assumptions have been made in the measurement (position) and cartographic data fusion problem to be solved:
- data are determined in the same reference system,
- data are of random character with a specific probability distribution,
- data are not burdened with systematic errors,
- data will undergo fusion by means of the least squares method with or without measurement covariance matrix being considered.

The relative positions of a ship and the pier (chart feature) are shown in Figure 1.

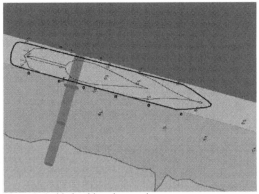

Figure 1. A ship berthing along a pier.

The ship is lying alongside, so the pier line can be regarded as a conventional line of position parallel to the ship's plane of symmetry shifted by a vector from a conventional ship's point, to which all navigational measurements are brought. The vector can be determined by direct measurement or indirectly, calculating its elements on the basis of a known position of conventional point on ship's plane and distance of ship's side to the pier line.

3 DATA FUSION

We will perform a fusion of navigational and cartographic data using the method of least squares. In the method, we will regard the line of a cartographic object (chart feature) as an additional line of position. A Kalman filter can be used if a ship is proceeding. There is also a possibility of measuring the relative position of cartographic objects.

If we do not take data accuracy into account, the method of least squares (LS) can be written in this form [1], [5], [7], [9], [10]:

$$\mathbf{x} = \left(\mathbf{G}^T\mathbf{G}\right)^{-1}\mathbf{G}^T\mathbf{z}, \qquad (1)$$

where

$\mathbf{x}$ – m-dimensional state vector (of ship's coordinates, searched-for position),
$\mathbf{z}$ – n-dimensional vector,
$\mathbf{u}$ – n-dimensional vector of measured navigational parameters,
$\mathbf{G} = \mathbf{f}'(\mathbf{x})$ – Jacobian matrix of the function $\mathbf{f}$ in respect to $\mathbf{x}$.

$$\mathbf{G} = \begin{bmatrix} \dfrac{\partial f_1}{\partial \mathbf{x}_1} & \dfrac{\partial f_1}{\partial \mathbf{x}_2} & \cdots & \dfrac{\partial f_1}{\partial \mathbf{x}_m} \\ \dfrac{\partial f_2}{\partial \mathbf{x}_1} & \dfrac{\partial f_2}{\partial \mathbf{x}_2} & \cdots & \dfrac{\partial f_2}{\partial \mathbf{x}_m} \\ \cdots & \cdots & \cdots & \cdots \\ \dfrac{\partial f_n}{\partial \mathbf{x}_1} & \dfrac{\partial f_n}{\partial \mathbf{x}_2} & \cdots & \dfrac{\partial f_n}{\partial \mathbf{x}_m} \end{bmatrix}, \qquad (2)$$

$\mathbf{f}$ – n-dimensional vector function,
$\mathbf{u}$ – vector of direct measurements,
$\mathbf{z} = \mathbf{u} - \mathbf{f}(\mathbf{x})$ – generalized vector of measurements.

The position $\mathbf{x}$ coordinates vector covariance matrix is expressed by this formula [1], [5], [8], [9]:

$$\mathbf{P_x} = \left(\mathbf{G}^T\mathbf{R}^{-1}\mathbf{G}\right)^{-1} \qquad (3)$$

When we take data accuracy into account, we deal with the method of weighted least squares (*WLS*)

$$\mathbf{x} = \left(\mathbf{G}^T\mathbf{R}^{-1}\mathbf{G}\right)^{-1}\mathbf{G}^T\mathbf{R}^{-1}\mathbf{z}, \qquad (4)$$

where

$$\mathbf{R} = \begin{bmatrix} \sigma_x^2 & \sigma_{xy}^2 & 0 \\ \sigma_{xy}^2 & \sigma_y^2 & 0 \\ 0 & 0 & \sigma_{pier}^2 \end{bmatrix} \text{ - navigational data}$$

covariance matrix.

We make a fusion of positions or lines of position with the pier line following this procedure:
– determine the position coordinates (or lines of position) together with their accuracy assessment (variances and covariances),
– determine the direction and accuracy of berth line (using a relevant chart, or database, and possibly, using the relative error to establish the accuracy of that line [8], [3], [4]),
– shift the berth line parallel towards the ship's position by the vector representing the distance of that line from the assumed reference point connected with ship's position – centre of masses, geometric centre, GPS antenna position or another),
– calculate the ship's position coordinates, regarding the berth line as an additional position line.

The covariance matrix of the running fix in case of a GPS is calculated from a series of positions or from Kalman filter. If there are terrestrial navigational systems, we can use the following relations [2].

An average error of geographic latitude determination for the common middle station

$$\sigma_\varphi = 0,5\sigma_{AD}\,\text{cosec}\,\theta\sqrt{\cos^2 A_{12}\,\text{cosec}^2\frac{\omega_{23}}{2} + \cos^2 A_{23}\,\text{cosec}^2\frac{\omega_{12}}{2}} \qquad (5)$$

where

A_{ij} – average azimuth between the *i*-th and the *j*-th station,

ω_{ij} – base angle between the *i*-th and the *j*-th station,

$\sigma_{\Delta D}$ – measurement error of distance difference.

An average error of geographic longitude determination for the common middle station

$$\sigma_\lambda = 0,5\sigma_{\Delta D} \cos ec\,\theta \sqrt{\sin^2 A_{12} \cos ec^2 \frac{\omega_{23}}{2} + \sin^2 A_{23} \cos ec^2 \frac{\omega_{12}}{2}} \quad (6)$$

The covariance between geographic coordinates for the common middle station

$$\sigma_{\varphi\lambda} = \frac{1}{8}\sigma_{\Delta D}^2 \cos ec^2\,\theta \sin(A_1 + A_2)(\cos ec^2 \frac{\omega_{23}}{2} + \quad (7)$$
$$+ \sin(A_2 + A_3)\cos ec^2 \frac{\omega_{12}}{2})$$

Also, we can change a GPS-obtained position into a system of two position lines by calculating their elements by using a vector of mean coordinates and elements of its covariance matrix. In this case position lines are regression lines running in the same direction (parallel) (tangent near the actual position) [7]:

a) $\quad y = \bar{y} + \dfrac{\sigma_{xy}}{\sigma_x^2}(x - \bar{x}),$ $\quad\quad\quad\quad (8)$

b) $\quad x = \bar{x} + \dfrac{\sigma_{xy}}{\sigma_y^2}(y - \bar{y}) \Rightarrow y = \bar{y} + \dfrac{\sigma_y^2}{\sigma_{xy}}(x - \bar{x})\quad (9)$

c) $\quad (\bar{x}, \bar{y})$ – centre of gravity of the population (mean position).

In the geographical coordinate system these lines are expressed as follows:

$$\varphi = \varphi_{\acute{s}r} + \frac{\sigma_{\varphi\Delta l}}{\sigma_{\Delta l}^2}(\Delta l - \Delta l_{\acute{s}r}) =$$
$$= \varphi_{\acute{s}r} + (\Delta l - \Delta l_{\acute{s}r})\Delta tg\,NR_1, \quad (10)$$

$$\Delta l = \Delta l_{\acute{s}r} + \frac{\sigma_{\varphi\Delta l}}{\sigma_\varphi^2}(\varphi - \varphi_{\acute{s}r}) =$$
$$= \Delta l_{\acute{s}r} + (\varphi - \varphi_{\acute{s}r})\Delta tg\,NR_2, \quad (11)$$

$$\frac{tg\,NR_1}{tg\,NR_2} = \frac{\sigma_\varphi^2}{\sigma_{\varphi\Delta l}}. \quad (12)$$

Let us illustrate the above considerations of the fusion of ship's position and a cartographic line by the following examples.

EXAMPLE 1.

The origin of a local coordinate system $0xy$ is at an established point on the ship (for simplification). In this case the ship is mooring along a pier described by the equation $y = x + 2$ (after a displacement by a vector representing the distance

from pier line to assumed coordinate origin) and accuracy $\sigma_{pier} = 1\,m$. The running fix, determined by GPS on the ship, had this covariance matrix:

$$\mathbf{R} = \begin{bmatrix} 2 & 0 \\ 0 & 2 \end{bmatrix}.$$

Thus we get $\sigma_{xy} = 0, \sigma_x = \sigma_y = \sqrt{2}$. The GPS position can be considered as a point of intersection of a meridian (vertical line) and parallel (horizontal line).

The matrices $\mathbf{G}$ and $\mathbf{R}$ are as follows:

$$\mathbf{G} = \begin{bmatrix} 1 & 0 \\ 0 & 1 \\ -1 & 1 \end{bmatrix}, \quad \mathbf{R} = \begin{bmatrix} 2 & 0 & 0 \\ 0 & 2 & 0 \\ 0 & 0 & 1 \end{bmatrix},$$

while the resultant coordinate vector

$$LS - \mathbf{x} = [-0,667; 0,667]^\mathrm{T},$$

$$WLS - \mathbf{x} = [-0,8; 0,8]^\mathrm{T}.$$

This situation is displayed in Figure 2. We can see that taking into account the accuracy of individual position lines leads to a displacement of *LS* position to the point *WLS* (due to higher accuracy of the cartographic line than that of the GPS-obtained position).

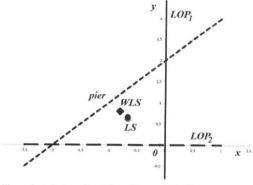

Figure 2. A fusion of a GPS position and pier line.

EXAMPLE 2.

Similarly to Example 1, we adopt $0xy$ at an established point on the ship (for simplification). Now the ship is mooring at a pier described by the equation $y = x + 10$ (after a relevant displacement) and accuracy $\sigma_{pier} = \sqrt{10}\,m$. A running fix was determined on the ship by using a terrestrial hyperbolic system with the following covariance matrix:

$$\mathbf{R} = \begin{bmatrix} 1 & 0,5 \\ 0,5 & 1 \end{bmatrix}.$$

We have then $\sigma_{xy} = 0,5; \sigma_x = \sigma_y = 1$. The matrices **G** and **R** are as follows:

$$\mathbf{G} = \begin{bmatrix} 6 & 1 \\ -5 & 1 \\ -1 & 1 \end{bmatrix}, \quad \mathbf{R} = \begin{bmatrix} 1 & 0,5 & 0 \\ 0,5 & 1 & 0 \\ 0 & 0 & 10 \end{bmatrix},$$

while the resultant coordinate vector

$$LS - \mathbf{x} = \left[-0,161; 3,333 \right]^{\mathrm{T}},$$

$$WLS - \mathbf{x} = \left[-0,115; 7,022 \right]^{\mathrm{T}}.$$

The situation is illustrated by Figure 3. This time the differences between both estimated positions are larger due to another geometric configuration of position lines and other values of their errors.

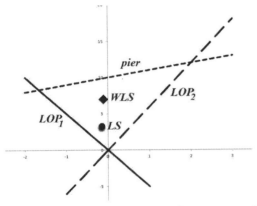

Figure 3. A fusion of positions from a terrestrial radionavigational system with a pier line.

4 CONCLUSIONS

In the above considerations we have shown how additional data, in this case data on cartographic features position and accuracy, can be utilized for estimating ship's position and its covariance matrix. In this way we increase the accuracy and reliability of ship's position being determined. This is of particular importance for ships located in immediate proximity of port and marine facilities and other navigational dangers. Similar situations also occur in aircraft navigation on the airfields. In rail navigation additional requirement is to take into account the spatial position of train or tram tracks, while in truck navigation – position of roads.

Another approach consists in imposing constraints resulting from the Rao-Cramer inequality [1], [6], [8] on the covariance matrix. However, further research into these problems should take into account both deterministic and probabilistic constraints on the coordinate vector, that is the random character of determining coordinates of cartographic objects. The point is to avoid superimposition of two disjoint objects such as a ship and a pier. In the simplest solution the navigator can interfere with the results of calculations. However, such solution is far from satisfactory as it does not take into consideration all possible situations and cannot be automated. It should be borne in mind that cartographic objects are also determined with a definite accuracy.

REFERENCES

[1] Banachowicz A. 1991. Geometry of Linear Model of Parametric Navigation. *Journal of Research AMW* No 109A, Gdynia,. p. 90. (in Polish)
[2] Banachowicz A. The Effect of Matrix Elements of System Geometry on the Accuracy of the Hyperbolic Navigational System's Position Coordinates. *Annual of Navigation* No 2, Gdynia. pp. 13 – 19.
[3] Banachowicz A., Banachowicz G., Uriasz J., Wolski A. 2006. Relative Accuracy of DGPS. *European Navigation Conference*. Manchester, UK, 07-10.
[4] Banachowicz A. 2006. Uogólnione prawo przenoszenia błędów losowych. *Prace Wydziału Nawigacyjnego Akademii Morskiej w Gdyni* Nr 18, Gdynia. ss. 5-16.
[5] Mitchell H.B. 2007. Multi-Sensor Data Fusion. *An Introduction. Springer-Verlag.* Berlin – Heidelberg – New York.
[6] Moore T., Sadler B. 2006. Maximum – Likelihood Estimation and Scoring Under Parametric Constraints. *US Army Research Laboratory.*
[7] Morrison D.F. 1976. Multivariete Statistical Methods. *McGraw – Hill Book Company*, London.
[8] Ristic B., Arulampalm S., Gordon N. 2004. Beyond the Kalman Filter. Particle Filters for Tracking Applications. *Artech House*, Boston – London.
[9] Rogers R.M. 2003. Applied Mathematics in Integrated Navigation Systems. Second Edition. *AIAA*, Virginia, Reston.
[10] Vaniček P., Krakiwsky E.J. 1992. Geodesy: The Concepts. *Elsevier Science Publishers.*

Chapter 3

e-Navigation Concept Development

Hydrographic Data as the Basis for Integrated e-Navigation Data Streams

M. Bergmann

Maritime Industry at Jeppesen, a Boeing Company, Germany

ABSTRACT: The development of e-Navigation on different levels in IMO, IALA, IHO and other arenas is starting to bring together static and dynamic data streams as well as data objects from different origination into a single e-Navigation display. In past years the static data collection found a basis in the IHO GI Register (often known as S-100 Register) and the resulting Common Maritime Data Structure. The paper will summarize the current status of development and the effects on harmonizing static data streams in e-Navigation displays.

In addition to this development more recently dynamic and real time data feeds, like Virtual AtoNs, Tide Gages Data or MSI transmission from VTS stations are bringing dynamic information to the e-Navigation display. The paper will try to explain opportunities and difficulties of merging this different data streams into a single display. It will highlight the need of integrating different data layers seamlessly to increase situational awareness and gain the expected positive results on safety of navigation and environmental protection. It will also highlight that while additional data will enrich the information portfolio of the navigator, the hydrographic data will be the basis of the data streams in foreseeable future.

The paper will focus on the following topics:
– Current development on data in e-Navigation
– Static Data Stream composition and components
– Dynamic Data streams in e-Navigation
– Integrated data for improved situational awareness

1 CURRENT DEVELOPMENT ON DATA IN E-NAVIGATION

Data is a central part of the e-Navigation concept: *"e-navigation is the harmonised collection, integration, exchange, presentation and analysis of maritime information onboard and ashore by electronic means to enhance berth to berth navigation and related services, for safety and security at sea and protection of the marine environment"*

(IMO MSC 85/26 Annex 20)

Based on this IMO (International Maritime Organization) e-Navigation definition the relevant bodies are working on a harmonized data model. Consequently the discussions in the related e-Navigation Committee of the International Association of Marine Aids to Navigation and Lighthouse Authorities (IALA) and the International Hydrographic Organization (IHO) have resulted in a more consolidated view on data architecture. The various working groups have come to the conclusion that a common data structure is necessary to harmonizing the different data streams within the e-Navigation concept. While IALA was originally looking at a new "Universal Marine Data Model", after further review and consultation with experts and the IHO it was agreed that the already established IHO concept of an "Universal Hydrographic Data Model", known as S-100, will be the basis of the "Common Maritime Data Structure". S-100 is based on the International Standardization Organization ISO 19100 series of geographic standards, well established in the GIS (Geographic information system) world. With that S-100 based data is compatible with data created according to the relevant ISO standards, not only within the maritime domain, but also with other GIS areas and is in support of the Spatial Data Infrastructure (SDI) initiatives in various regions of the world. This development provides the necessary harmonized platform for integrated systems. By agreeing on this common GI-Registry (Geographical Information Registry) concept the maritime community prepared the ground for harmonization and interoperability of the different data streams necessary to make e-Navigation a success.

An additional dominant argument unifying all stakeholders to move towards common structures and towards the IHO originated model is the fact that the implementation of S-100 and its related standards is well underway and will materialize shortly. As a consequence ENCs (Electronic Nautical Charts) will follow this data structure, same as associated data streams, like "Inland ENCs" or "Marine Information Overlays".

The IHO has already created and approved S-100 in January 2010 and the first Product specification S-102 (Bathymetric Surface Product Specification) in April 2012. IALA has registered for an own domain in the IHO "GI-Registry" and will conduct a workshop in June 2013 to start the development of IALA product specifications for e-Navigation. As all stakeholders agree that the ENC layer will build the foundation of any kind of advanced navigational systems, it was a natural development to try to align other data stream with this foundation. At the same time this development again verifies that ENCs, or better Hydrographic Vector Chart Data Layers, are the necessary ingredients for any navigational display now and in the foreseeable future.

This development on the data side is accompanied by the necessary review of the regulatory conditions. The concept of e-Navigation is evolving the understanding that future navigation will need constant innovation, and as such will need to change how performance standards are handled. It is widely understood that the current ECDIS (Electronic Chart Display Information System) performance standard is restricting innovation. Its update and certification concept is not geared up to meet the needs of e-Navigation.

The currently being promoted new concept defines a framework in which a growing number of data streams are integrated and harmonized to allow the creation of the necessary information for increased Situational Awareness in an environment of growing complexity.

Figure 1. The regulatory Framework

Figure 1 shows the proposal presented by the author at IMO working groups and currently under discussions. The proposed solution is a regulation which focuses on the definition of "what" conditions are to be met and will not focus on "how" the systems regulated are implementing the defined requirements. The top oval indicates that current ECDIS regulations are spanning the "What" as well as the "How". The blue bottom section describes the desired regulatory framework; whereas the regulation covers the "What", while Industry implementation uses innovative means of implementing the "How".

This concept will also support the IMO e-Navigation focus on "User Needs". The on board and ashore systems to be developed within this framework will have to create a compelling need for their usage by increasing safety and security of navigation (compelling need for coastal administration) and improved efficiency of voyage (compelling need for ship owners and operators).

As outlined, the e-Navigation framework in development through IMO, IHO and IALA is changing the concept in how systems are managed, type approved and more importantly how they handle and render cartographic data. This will change how hydrography and cartography will play together.

2 STATIC DATA STREAM COMPOSITION AND COMPONENTS

The first of two data-pillars for e-Navigation are the static data streams. These are data sets, which are not dynamic in nature but composed beforehand and then used until replaced on a regular or irregular basis.

As ships where sailing the world any type of communication with shore until recently was very limited, if possible at all. As such navigators relayed on those pre-composed, static data to support their navigation. They took those data, usually in form of paper charts or more recently as vector charts, on their voyage for navigational use.

The dominant products to provide a mariner with this necessary static data are and had been navigational charts and nautical publications. The Hydrographic Offices (HOs) around the world are compiling those products using the guidelines of IHO.

Input for those products are often generated by HOs themselves or associated organizations by conducting surveys, which provide bathymetric data layers. Together with information about navigational aids they are building the foundation of navigational data. These basic data sets are enriched by information collecting from organizations like coastal administrations or port masters on navigational relevant objects but also procedural and regulatory information needed for a safe passage.

The cartographers then are combining all of this static data and selecting the necessary data for the intended use of the chart or publication they build. They also compose the data to archive full level of deconfliction and ease of use of their products. The resulting charts and publications are stand alone products with no interoperability as such.

This concept was also adapted in the chart centric paradigm of current ECDIS concept with associated ENCs. HOs are preparing pre-composed ENCs of certain scale bands, which have an intended use and appropriate zoom factor. With the different IHO specifications and their composition of ENCs, HOs are defining the look and feel as well as the appropriate rendering of ENCs in a type approved ECDIS. Any so call "Value Added Data" may form an "Overlay" but cannot integrate with the ENC data.

3 DYNAMIC DATA STREAMS IN E-NAVIGATION

The second data pillar for e-Navigation is dynamic. The data sets are rapidly or constantly changing and as such the data set used is dynamically adjusted as the situation is changing.

Just in recent years the communication between ships and shore has drastically changed and now includes various types of electronic communication, like voice or data transmission. Satellite communications at high seas as well as other communication means like 3G, used in mobile devises like mobile phones or smart phones, or WIFI (wireless local area network), when closer to shore are increasingly enabling ships to receive electronically real time data. The e-Navigation concept builds on these growing capabilities.

Real time transmissions of tidal information are already a reality in some regions. In addition the establishment of AIS (Automated Identification System) AtoNs is under development and is adding another real time data stream to the mix.

These are only two examples of dynamic data streams already available or in development. The general concept of dynamic data is that it allows a view on current reality rather than a generalized composition. In the same category are falling situation centric data like traffic situation information, on board ship sensors like motion sensors for ship movement or propulsion sensors.

e-Navigation as envisioned by IMO and under development at IMO, IALA, IHO and others is including dynamic data sets as part of the concept to improve situational awareness. With this the safety of navigation and environmental protection is intended to be increased, which is the underlying goal of e-Navigation.

Dynamic data streams are essential components of the e-Navigation concept. They are not only "overlays" to support static data sets but integral part of the idea behind e-Navigation.

4 INTEGRATED DATA FOR IMPROVED SITUATIONAL AWARENESS

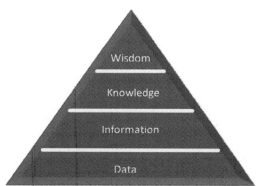

Figure 2. Data-Wisdom Pyramid

Both static as well as dynamic data is, as described, intended to support the mariner on the task of safe navigation. It is important to highlight that, while data is important for situational awareness and as such as an aid to navigation, data by itself is not providing any benefit. Only if the data is received by the navigator as information it is reaching the intent. Information brought to the mariner will enable the expansion of knowledge, which creates the necessary situational awareness to make the right decisions, or in other words create the wisdom which enables the navigator to master the situation on hand (see Figure 2).

In the classic chart paradigm, the hydrographers and cartographers integrated the various raw data they received in a pre composed static data set as described earlier. As we are now providing both static as well as dynamic data sets, the different data streams are not necessarily integrated as they may come from different sources. The growth of data streams are not by default leading to more information, better knowledge and as such wisdom, there is a risk of "data overload". The "data overload" actually results in less information available to the mariner and not in more. When looking at the current view on data as static data sets with dynamic overlays, the mentioned risk of data overload can at least be envisioned, in quite a few cases already been experienced.

To solve this issue the different data streams need to be integrated and prepared for use to ensure they reach the mariner as meaningful information. For example real time water level information will need to be combined with static shore information, static navigational aid information need to be combined with AIS AtoN information and so on. With this concept the different incoming data streams will be integrated in as a "nautical data collection".

The concept of data integration requires technology to group data into meaningful sets and select data sets as needed. Rendering rules and presentation guidelines will allow displaying the selected and grouped data of the "nautical data collection" needed at a certain situation to better inform the navigator while the traditional "overlay" concept would lead to a less and less informative patchwork of data. In the integrated data concept the different available data layer will be rendered based on situational needs, i.e. zooming level as desired by the mariner, and as such will result in a situational centric display.

An aspect not to be ignored is data quality and integrity. The integration of data will need to be done in a way, which ensures that different data streams are not negatively impacting each other and as such reduce situational awareness rather than improve it. In order to ensure this there need to be a clear definition of data rendering priority. The guiding rules for data rendering and integration need to define what is to be used, especially when multiple data streams are providing information on same or associated objects. Secondly the data supply needs to be secure. E-Navigation needs to specify transmission channel security for dynamic data streams and Data Supply Chain Certification (DSCC) for static data supply. This concept, already presented to IHO and worked on in a DSCC correspondence group, can build on years of experience with the DO-200A and DO-201A standard in aviation.

All of this will not reduce the importance of hydrographic data. It is quite the opposite. Hydrographic data will become the base data set and foundation of a situational centric e-Navigation display of an integrated navigational system. But it needs to be integrated and enriched by all the other data streams, even those which will be available in future and which are not yet envisioned.

5 CONCLUSION

The current development of usage of electronic cartography in the maritime world has taken a step towards situational awareness and as such has matured away from simple chart display. This development will intensify and as such will require in future a change in how electronic maritime information are developed, composed, transmitted and stored. The future will focus on a variety of static and dynamic data streams to support data integration and situational centric rendering.

With this development the mariner on board will receive the necessary information, which helps to build up the knowledge needed for the wisdom to ensure a safe and efficient journey at sea.

The Hydrographic datasets are the foundation on which this data and information concept is built on.

While all of that is technically possible and well underway, e-Navigation needs to be supported by a change from a regulatory concept defining the detail implementation aspects to a regulatory framework defining the "What", but allowing innovation to specify the "How" based on state-of-the-art and ever evolving technology.

REFERENCES

IMO, 2009, SOLAS (Consolidated Edition)
IMO, 2008, MSC 85/26 Annex 20
IALA, 2011, Navigation Architecture 'PictureBook' Information Paper
Jeppesen, 2011, ECDIS – What you need to know
IHO, 2010, IHO Universal Hydrographic Data Model, Ed 1.0.0
IHO, 2011, Operational Procedures for the Organization and Management of the S-100 Geospatial Information Registry, Ed 1.0
RTCA, 1998, DO-200A, Standards for Processing Aeronautical Data, Issued 9-28-98
RTCA, DO-201A Standards for Aeronautical Information, Issued 04-19-00
DSCC-CG, 2010, DSCC White Paper
Patraiko D., Wake P., Weintrit A., 2010, e-Navigation and the Human Element. TransNav - International Journal on Marine Navigation and Safety of Sea Transportation, Vol. 4, No. 1, pp. 11-16
Motz F., Dalinger E., Höckel S., Mann C., 2011, Development of Requirements for Communication Management on Board in the Framework of the E-navigation Concept. TransNav - International Journal on Marine Navigation and Safety of Sea Transportation, Vol. 5, No. 1, pp. 15-22, 2011
Bergmann M., 2011, A Harmonized ENC Database as a Foundation of Electronic Navigation. TransNav - International Journal on Marine Navigation and Safety of Sea Transportation, Vol. 5, No. 1, pp. 25-28, 2011

Towards a Universal Hydrographic Data Model

H. Astle & P. Schwarzberg
CARIS, Fredericton, NB, Canada & CARIS, Heeswijk, The Netherlands

ABSTRACT: The International Hydrographic Organization's (IHO [1]) Transfer Standard for Digital Hydrographic Data S-57 [2] standard has been in force for more than a decade, and has successfully been used for official ENCs adopted by Hydrographic Offices around the world and by navigation equipment manufacturers. Additionally S-57 has been used for many additional purposes. However S-57, and especially the administration of the standard, has also experienced limitations. In 2010, IHO released the next generation hydrographic standard called S-100 Universal Hydrographic Data Model [3]. A move that will open up the door to new possibilities to existing S-57 users and potentially broaden the use of IHO standards in the hydrographic community.
This paper will try to explain why an S-57 replacement was needed and give examples on some possibilities with S-100 and its derived product specifications such as S-101.

1 INTRODUCTION

In the 1980s, computer software for geospatial data (better known as GIS software) had been on the market for more than a decade, but hardware, especially "high-resolution" graphic screens were very expensive and most software very specialized. It was still early days in the marine geospatial world, and much would happen before geospatial data would become mainstream.

Around the mid 1980s several national hydrographic offices acquired geospatial software for chart production and internationally there was talk about geospatial standards. Under IHO was a working group responsible for defining new feature classes based on data found on navigational charts. In 1987 was the first work released as "DX-87" (Digital eXchange 87). It didn't get much attention around the world, but the development was (very slowly) rolling.

Electronic navigational charts was coming, but many would argue that it would have taken forever if it wasn't for one special "event": On March 24, 1989, the oil tanker Exxon Valdez ran aground on the Bligh Reef, Prince William Sound, Alaska, USA, resulting in an oil spill estimated to have been around 40,000 tons [4]. There is probably not a person today involved in electronic charts that haven't heard this story.

Whether (oil) accidents acted as a catalyst for the development of electronic charts and chart systems or not is maybe less important! Many people in the industry and government organizations agreed that electronic charts and electronic navigational systems were the future. In 1990 DX-87 had become DX-90 and included in a new IHO standard named S-57 'IHO Transfer Standard for Digital Hydrographic Data'. At this time some argued that electronic charts would replace paper navigational charts within 5 years. More than 20 years later that has still not happened, but it's certainly closer than ever.

2 S-57 TAKING OFF

With the introduction of the IHO S-57 standard in 1990 the world had probably received its first truly international recognised data exchange standard. There were two major purposes with the standard:
1 Exchange of digital paper chart between national hydrographic offices
2 Exchange of electronic charts from national hydrographic offices to mariners' navigational systems on board ships.

The first purpose never took off. The major reason for this is probably that paper charts are graphical products and a geospatial standard focusing on feature encoding of "real world features" like S-57 did, was not suitable for this purpose. Converting graphical data from all the various systems used by hydrographic office around the world is far more complex than most people expected at that time. In fact paper charts today mostly, if not entirely, are still exchanged as raster data and not as vector or feature data.

The second purpose using S-57 for electronic chart exchange had more progress! There was much to be learned and during the first years after the release, the IHO DataBase Working Group that maintained the standard multiple times per year, and the standard was constantly under revision. However in the mid 1990s the standard had matured, and by 1996 edition 3.0 was released. It was decided this edition would remain frozen for 4 years to ensure stability, intended to help hydrographic offices and, maybe especially, navigational equipment manufactures to finalize and release date plus equipment for the market.

Only one year earlier, in 1995, had the International Maritime Organization (IMO [5]) adopted the so-called ECDIS Performance Standard [6], meaning that Electronic Chart Display & Information Systems (ECDIS) now was allowed to be used for navigation, instead of paper charts. Here also were many lessons to be learned, and amendments were frequent in the early years. Work on this standard had taken place internationally for about as long as the data exchange standard and the two have close ties. The ECDIS standard required that charts for ECDIS systems must be Electronic Navigational Charts (ENCs) according to the S-57 standard.

In 2000 edition 3.1 of the S-57 standard was released and this time the standard was frozen "forever". The production of ENCs still lacking behind for many parts of the world, and it was also expensive for the ECDIS manufactures to keep updating their systems. Freezing the standard was intended to help that. Additional it was believed that once an ECDIS was delivered to a vessel there was no guarantee that such system would be updated, e.g. to support new versions of the standard. Neither the ECDIS standard, nor the ENC standard, had means to update ECDIS systems to accept new feature classification or new symbology etc. This worked relatively well for about half a decade, but by mid 2000s there were new requirements from IMO to include new (environmental) features on navigation charts. In January 2007 the so-called 'Supplement 1' to S-57 was released, and it was followed by another supplement two years later. The evidence was clear: A frozen standard would not be able to support future needs.

It was not really a problem with the S-57 standard, which originally included (technical) possibilities to define and encode new feature types. It even included a way to encode class definitions of new features so that navigational systems could "learn" about such new features and present these features to the mariner. However this part of the standard was never utilised.

Additionally S-57 was intended to be used for multiple product types, and had room for various product definitions to be defined. The ENC standard is actually "just" an appendix to the S-57 standard. However that possibility wasn't used; partly because of the frozen state of the standard. At least not under IHO and officially not part of the standard, but more about that below! The S-57 standard and the ENC product specification had become almost inseparable and somewhat synonymous!

3 MULTIPLE PURPOSES

The frozen and somewhat limited S-57 and ENC specification did not stop other S-57 based implementations. One of the earlier ideas for S-57 was for bathymetric survey data exchange, and even though this has been done to some degree, there was much more use of the standard for other purposes. However none of these purposes are developed under IHO, but some of them started using S-57 as early as the mid 1990s.

Some of the other uses of S-57 are:
– Inland ENCs [7]
Electronic Charts for Inland ECDIS, which are used on rivers.
In 2001 the Economic Commission for Europe of the United Nations (UN ECE) adopted the Inland ECDIS Standard as a recommendation for the European inland waterway system; using Inland Electronic Navigational Chart (IENC) data. Outside Europe, other countries also looked at Inland ECDIS and the U.S. Army Corps of Engineers [8] developed the Inland Electronic Navigation Charts. While the European Inland ENC standard extended the original S-57 standard with new features, symbology and rules, the US Inland ENC standard used S-57 more or less as it was. To align the standards the International Inland ENC Harmonization Group (IEHG) formed in 2003 and the standards are now maintained as one standard. Other countries adopted this standard and Inland ENCs covering thousands of river kilometres exists today. Countries using this standard include: Austria, Belgium, Brazil, Bulgaria, China, Croatia, Czech Republic, France, Germany, Hungary, Italy, Netherlands, Peru, Poland, Russia, Serbia, Slovakia, Switzerland, Ukraine, South Korea, USA, Venezuela (plus others).
Note that IENCs are not overlays; they will not be used at the same time as ENCs. They cover different geographic areas and are made for different vessels.

- Ice ENCs [9]
  Using S-57 for encoding of Ice features started in the mid 1990s. These datasets are overlays with additional dynamic (ice) information supplementing ENC data.. Countries around the North Pole, followed by countries around the Baltic Sea are among the players.
- Additional Military Layers (AMLs) [10]
  Situation awareness layers of data supporting military operations, which as the name indicates are overlay of additional data displayed on a warship ECDIS (called WECDIS). Date layers include bathymetric contours, routes, areas & limits for danger and exercise areas, full wreck and major bottom object information, detailed beach and seabed environmental data, etc.

There are more types like Bathymetric Charts and specialised Pilot Charts. Many more possibilities using S-57 for various purposes exist [11] and are often generically referred to as Marine Information Overlays (MIOs). Each typically has their own feature definitions, encoding rules, etc. However the fact that they are all based on S-57 makes it relatively easy to support more of them. For instance a widely used ENC production tool, CARIS S-57 Composer [12], not only allows users to create IHO ENCs but also the other S-57 product types mentioned above. Users can even define their own S-57 product types.

4   WHAT'S NEXT?

There is no question S-57 will be used for many years to come. There is simply too much data and so many systems using S-57 preventing it from dying anytime soon. ECDIS systems will be using S-57 ENCs for many years to come.

However the geospatial world has been evolving. In the 1980s and even in the 1990s electronic geospatial data was not mainstream, but today almost every mobile phone has not only electronic maps, but also navigation. Imagining a world without Google Maps in 2013 is very hard!

The S-57 standard was one of the first geospatial standards, but it focused entirely on marine information, and (at least) two other major initiatives are now more known than S-57 (at least in the non-marine domain):
- GML [13]
  Geography Markup Language (GML) is the flavour of XML defined by the Open Geospatial Consortium (OGC) to encode geospatial information. It is very general and can be used for many purposes. For instance can data types vary a lot between two different data sets both encoded in GML. Meaning there is no guarantee that two systems both supporting GML can exchange data. Google's KML format is often compared to

GML, and can loosely be described as a GML specific flavour.
GML has with the latest major version moved to ISO conformity.
- ISO/TC 211 [14]
  The International Standards Organization's Standards Committee 2011 (ICO/TC 211) is responsible for ISO's standards on geospatial information and many other organizations have been engaged in this work too. OGC and DGIWG (Defence Geospatial Working Group) are among these, and IHO have also been in close contact with ISO in this area for decades. In 2012 a Memorandum of Understanding (MoU) to increase their cooperation was signed by ISO and IHO.
  ISO/TC 211 has released a set of standards know as the ISO 191xx series, each covering different scopes. For instance is 19136 the standard for GML and 19115, which maybe is the best known of these standards, is the standard for metadata.

5   NOT THE NEW S-57

To resolve the issue described above (among other things) IHO has developed the S-100 standard! However it will not immediately replace S-57 and it would be a mistake to call it a new S-57.

The work on S-100 was started by the IHO Transfer Standard and Maintenance Working Group (TSMAD), which previously was known as the DBWG, and this was the group that maintained the S-57 standard. The initial version of S-100 was indeed called S-57 Edition 4.0! However since this new version should resolve some of the issues/limitations with S-57 (both technically and administrative) and it is based on ISO/TC 211 with new terms and models it was decided to give it a new name. Hence S-100 which is following IHO's naming convention for its standards.

With S-100 product specifications are kept completely separate, meaning that feature classes, encoding, etc. are not a part of S-100. S-100 doesn't even dictate what file format to use for the encoding. In line with ISO/TC 211 is GML for instance not mandatory format, but an option depending on what needs to be encoded. It will be the product specifications, which are separately maintained standards, named S-101, S-102, etc. that will contain the encoding and other product implementation rules.

Product specifications are developed under a set of rules defined by S-100, saying for instance that a product specification must consist of (thus define) the following parts:
- product identification
- data content and structure
- coordinate reference system

- data quality
- data capture
- data maintenance
- portrayal
- encoding
- product delivery

S-100 introduces some new data types and structures for modeling and disseminating the data. One of the new abilities in S-100 is the ability to handle more complex attribute situations. A feature can now have multiple values of a given attribute type and a hierarchy of attribute information can be modeled. Also a new concept called Information Types allow common information to be shared or referenced by multiple objects. As part of the development of S-101 a richer encoding of the real world should become available that will enable systems and users to make better use of the information. The hope is that this will be realized by the end user systems in order to provide improved decision support mechanisms.

Figure 1. S-101 example of complex attribute encoding in one feature something that in S-57 requires multiple features.

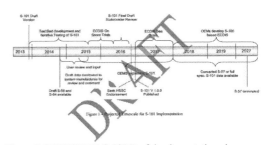

Figure 2. IHO Geospatial Information Register

A major component under S-100 is the IHO Geospatial Information Register [15] where feature catalogues, defining the features to be used in the product specifications, are registered. By registering the feature catalogues (and the features defined in these) in the IHO register it is hoped that feature classes will be shared (and possibly extended), instead of having conflicting feature classes between product specification. It is also hoped that feature classes can be shared between IHO standards and other standards. To help ensure this the IHO registry is open to other non-IHO standard for maintaining

their feature catalogues. Today both IENC and Ice IENC features can be found in the registry too.

6 THE FRUITS OF S-100

The first number in the new line of product specifications defined by the IHO is S-101, which is the "Next Generation ENC Product Specification". S-101 is under development and the first draft version is planned for late 2013. ECDIS systems utilising S-101 ENCs are expected operational in 2018 after shore and sea trials. However even at that time will S-100/S-101 not replace S-57 ENCs, but supplement them. S-57 ENCs are not expected to retire before sometime between 2020 and 2030.

Figure 3. IHO TSMAD S-101 Draft implementation plan (January 2013)

The next IHO product specification in the line of numbers is S-102 [16], which is the Bathymetric Surface Product Specification. This product specification is actually already released (in April 2012). The data/coverage type is a quadrilateral grid coverage together with attributes known as a Bathymetric Attributed Grid (BAG). S-102 is intended for navigational purposes using a digital signature, or for non-navigational purpose (without a digital signature).

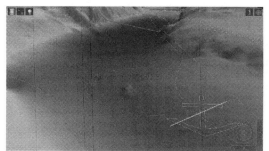
Figure 4. Navigational surface with ENC features.

Using gridded bathymetry in BAG format has for some time been considered; for instance with Port ENCs as described in the 2009 TransNav paper "Enhance Berth to Berth Navigation Requires High Quality ENC's – The Port ENC – a Proposal for a New Port Related ENC Standard [17]" by D. Seefeldt, former Head of the Geographic and Hydrographic Department at Hamburg Port Authority.

S-102 gridded bathymetry is also very suitable for web viewing and download, as early work with S-102 has been shown by the Canadian Hydrographic Service [18].

Besides S-101 and S-102 is IHO also working towards other S-1xx product specifications! Marine Protected Areas is expected to become a new S-100 product specification called named S-103 Geospatial standard for Marine Protected Areas and there are more being mentioned as possible new product specifications e.g.:
- Routes
- Boundaries
- Ice
- Currents and tides
- Etc.

The IMO Sub-Committee on Safety of Navigation (NAV) is looking at e-Navigation as an important topic for the future of nautical navigation. The Sub-Committee on e-Navigation agreed to use S-100 as the baseline for creating a framework for data access and services under the scope of SOLAS (IMO's Safety Of Life At Sea). Plus the International Association of Marine Aids to Navigation and Lighthouse Authorities (IALA) Council has approved registration of IALA [19] at IHO as a Submitting Organization under the IHO GI Registry and as a domain owner (i.e. the IALA domains within the Registry).

7 FLEXIBILITY

As mentioned earlier, S-57 was intended to be a framework for multiple products but IHO only ended up with one product spec. One of the unfortunate results of this and the frozen specifications is that some implementations were designed to be static. When IHO needed to add some new objects to the catalogue it became clear that this would mean software and system upgrades to make it work. With S-100 the intention is to prevent this situation.

S-100 is under a new maintenance regime that should allow new editions to be created as needed. The concept is that S-100 product specifications would be based on a particular edition of S-100, which would mean that new S-100 editions could be released, while product specifications based on earlier editions still would be valid. This will allow S-100 to be updated to support new product specifications requiring new elements not present in the older edition of S-100.

A concept with S-100 is that the more common/expected changes should be treated as just data updates and not require physical changes (software, hardware) to the end systems. To this end S-100 is defining the means to manage Feature Catalogues and Portrayal Catalogues as data that can be distributed and updated. If there is a need to add a new object to the Feature Catalogue an update to the Feature Catalogue would be released along with corresponding updates to the Portrayal Catalogue and when the system encounters the new object in a dataset or update it would recognize the new object and be able to display it.

Only if a product specification needs to include more significant changes such as file encoding or data structures, then a brand new edition of that product specification would be required and only in such cases would it be expected that systems would require corresponding software updates in order to support the new specification.

8 CONCLUSION

The expectation is that S-100 will provide solutions needed by the growing market of Hydrographic products and will allow for the flexibility to grow as new needs are identified. This will not happen without significant effort and involvement by all aspects of the Hydrographic community. It has been evident with the developments of S-100 so far that input and involvement from producing agencies, system manufacturers, governing bodies and end users are necessary for success. Having been involved in S-57 and related technologies, plus in S-100 since its inception, CARIS is excited about the possibilities that S-100 can bring to the world.

REFERENCES

[1] www.iho.int
[2] http://www.iho.int/iho_pubs/IHO_Download.htm#S-57

[3] http://www.iho.int/iho_pubs/standard/S-100/S-100_Info.htm

[4] http://en.wikipedia.org/wiki/Exxon_Valdez

[5] http://www.imo.org

[6] http://www.imo.org/OurWork/Safety/Navigation/Pages/Charts.aspx

[7] http://en.wikipedia.org/wiki/Inland_Electronic_Navigational_Charts

[8] http://www.tec.army.mil/echarts/

[9] http://hgmio.org/information.html

[10] http://www.ukho.gov.uk/Defence/AML/Pages/Home.aspx

[11] http://www.hgmio.org/specifications.html

[12] http://www.caris.com/products/s57-composer/

[13] http://www.opengeospatial.org/standards/gml

[14] http://www.isotc211.org/

[15] http://registry.iho.int/s100_gi_registry/home.php

[16] http://www.iho.int/iho_pubs/IHO_Download.htm#S-102

[17] http://www.transnav.eu/Article_Enhance_Berth_to_Berth_Navigation_Seefeldt,18,280.html

[18] http://www.caris.com/downloads/brochures/The-New-IHO-S-102-Standard.pdf

[19] http://www.iala-aism.org

Software Quality Assurance Issues Related to e-Navigation

S. Lee
Korea Maritime University, Busan, Republic of Korea

L. Alexander
University of New Hampshire, Durham, New Hampshire, USA

ABSTRACT: A key consideration associated with the development and implementation of e-Navigation is software quality assurance. In particular, navigation and communications equipment must reliably indicate that they are functioning correctly. For this to occur, high-quality software that is both stable and complete must be installed and used. Software quality standards have been adapted for many types of safety-critical systems including avionics, automobiles and medical equipment/devices. To date, no software quality standards have been specifically developed for maritime navigation-related systems. This paper describes the development of an IMO guideline on software quality assurance that would apply to any type of e-Navigation related software system. This includes both existing and new development, as well as embedded or software-centered systems. The overall intent is to encourage the installation and use of high-quality e-Navigation software that is fit for purpose.

1 INTRODUCTION

Since 2006, e-Navigation has been developed by International Maritime Organization (IMO) working groups of NAV (Safety of Navigation), COMSAR (Radiocommunications and Search and Rescue) and STW (Standards of Training and Watchkeeping) sub-committees. In addition, the International Hydrographic Organization (IHO) and the International Association of Lighthouse Authorities (IALA) also have supported IMO to achieve the definition of e-Navigation as following:

"The harmonized collection, integration, exchange, presentation and analysis of marine information onboard and ashore by electronic means to enhance berth to berth navigation and related services for safety and security at sea and protection of the marine environment"

In 2008, IMO NAV54 agreed a framework for e-Navigation implementation process including user needs, architecture, gap analysis, cost-benefit analysis, and risk analysis. Currently, user needs, overarching architecture, and gap analysis have been completed. It is expected that the e-Navigation Strategy Implementation Plan will be finalized in 2014 (IMO NAV 57-15).

At NAV58, Republic of Korea recommended that Software Quality Assurance (SQA) was an important factor, and should be considered (IMO

NAV 58-6-4). This was agreed, and the Sub-Committee tasked the e-Navigation Correspondence Group to *consider the issue software quality assurance … and provide comments and recommendations as appropriate.* (IMO NAV 58-14)

This paper describes the development of an IMO guideline for software quality assurance that would apply to any type of e-Navigation related software system. This includes both existing and new development, as well as embedded or software-centered systems. The overall intent is to encourage the installation and use of high-quality e-Navigation software that is fit for purpose.

2 CONSIDERATION OF SOFTWARE QUALITY ASSURANCE ON E-NAVIGATION

There are a number of considerations related to Software Quality Assurance (SQA) that were introduced and discussed IMO NAV 58.

2.1 *Purpose and vision*

IMO NAV 58 agreed on the need for SQA to be part of the development and implementation of e-Navigation. In particular, practical e-Navigation solutions must be based on operational, technical, regulatory and training aspects, including the human element. In addition, COMSAR 14/12 pointed out that both navigation and communications equipment must be able to reliably indicate that they are functioning correctly.

In order to address these concerns, quality assured software should be installed and used in a harmonized manner for both shipborne and shore-based systems. Further, SQA must be considered as a part of an e-Navigation system quality. To do this, practical guidelines on SQA need to be developed and agreed.

Software quality standards have been adopted for many types of safety-critical systems including, avionics, automobiles and medical equipment. In these systems, some defects could have a detrimental impact on human health or safety as well as the environment. Depending on software functionality that is increasingly used, such systems are expected to verify specific qualities, as like reliability, availability, security and safety.

Most likely, software-related issues will become the major challenge for achieving harmonized shipborne and shore-based e-Navigation. The e-Navigation implementation will require connection and integration between existing onboard navigational systems, as well as shore-side equipment. If redundancy was used to provide resilience, the system should be able to transfer automatically to an alternative source, and provide an appropriate indication to the user. In addition, information concerning the source and authenticity of the data is needed as well.

2.2 *ISO standards related to software quality*

ISO has adopted software quality standards for many types of safety-critical systems including, avionics, automobiles and medical equipment/devices. Depending on software functionality requirements, these systems are expected to verify specific qualities including reliability, availability, security and safety. *IEC 61508* is a basic functional safety standard applicable to all kinds of equipment and systems. It defines functional safety as the overall safety-related consideration for any type of electrical equipment being controlled by a safety-related system(s).

The avionic and automobile field has adopted a number of software-related standards to insure product quality necessary for safe operation:

– *ISO 26262,* titled Road vehicles -- Functional Safety, is an adaptation of the *IEC 61508* for Automotive Electric/Electronic Systems. For instance, modern Anti-lock Brake System (ABS) for automobiles relies on software to determine when and how to regulate braking force.
– US Federal Aviation Authority adopted *DO-178B* as a means of compliance for software life-cycle

processes. European statutory regulations specify ED-12B.

Medical equipment and devices also provide typical examples of safety-critical systems. *IEC 62304* is applicable to software used in active implantable medical devices covered by *ISO 14708*. It was revised to *IEC 62304:2006*, according to the change of EU medical device directives' recognition to the need for software quality (IEC 62304:2006).

When using increasingly complex systems, there is the expectation that there will be increased functional capability with no loss in reliability. However, the opposite result often occurs. As such, compliance to ISO standards has become increasingly important.

2.3 *e-Navigation and software quality assurance example*

Associated with the e-Navigation implementation strategy, various systems and services will be the result of developing new software or integrating existent software. One example would be how real-time tidal or water level information will eventually be displayed on ECDIS. Software issues are to be considered for each step:

– water level is measured by tide gauge sensor
– the sensor data is converted into a binary format
– binary data is broadcast from an AIS (Automatic Identification System) Base Station such as an AIS Application-Specific Message (AIS ASM).
– the AIS ASM received from onboard or ashore is provided to other equipment (e.g., via the 'Pilot Plug').
– the tide/water level data is incorporated into a "Dynamic ENC" that is displayed on a shipborne or shore-based electronic chart display (e.g., ECDIS, INS, or a Portable Piloting Unit).

e-Navigation is dependent on hardware, software, and data, all working in harmony. If existing navigation system software is not able to be assured, problems can be occurred. These problems are not just an inconvenience, but can affect navigation safety. This situation often occurs when there is a need to: 1) integrate existing devices, 2) accommodate a new data format, or 3) expand the functions of the current system. Software quality assurance metrics can be used to diagnose the functional capabilities of the existing software, and influence design of new/improved software.

3 GUIDELINE FOR SOFTWARE QUALITY ASSURANCE

There are several aspects that should be considered in the development and implementation of 'suitable' guidelines. This is particularly important in terms of encouraging the installation and use of assured e-Navigation software that is fit for purpose.

3.1 Application and scope

A guideline on software quality assurance should be applicable to any type of e-Navigation related software, including existing legacy, new development, embedded, or software-centered systems. Further, it should be applicable to all types of e-Navigation-related software, from individual components/equipment to large-scale integrated systems.

This guideline is to describe important concepts to consider related to software quality assurance, taking into account e-Navigation development and implementation. It is purposely general and goal-based, and it is not intended to provide specific guidance for individual software systems.

3.2 Definitions

As defined by the Institute of Electrical and Electronics Engineers (IEEE):

– **Software quality** is *"the degree to which a system, component, or process meets specified requirements"*, and *"meets customer or user needs or expectations."*
– **Software quality assurance** is *"a planned and systematic pattern of all actions necessary to provide adequate confidence that an item or product conforms to established technical requirements"*. It can be considered *"a set of activities designed to evaluate the process by which the products are developed or manufactured."*

3.3 ISO software quality standards

There are existing ISO software quality standards which are composed of two parts: product quality and process quality.

3.3.1 ISO/IEC 9126-1

This classifies software product quality in a structured set of characteristics. Each quality characteristic and associated sub-characteristics (e.g., attributes) are to be selected and measured (i.e., a metric) by properties of target software (ISO/IEC 9126-1). Figure 1 and 2 show the characteristics and their sub-characteristics.

– Reliability - A set of attributes that relate to the capability of software to maintain its level of performance under stated conditions for a given period of time.
– Usability - A set of attributes that describe the intended use, and that this use meets the needs of intended users.

– Efficiency - A set of attributes dealing with the relationship between the level of software performance and the amount of resources required.
– Maintainability - A set of attributes that describe what is required to maintain required software performance.
– Portability - A set of attributes dealing with the capability of software to be transferred from one environment to another.

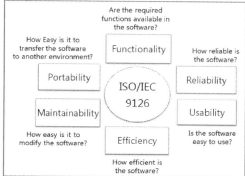

Figure 1. Characteristics of ISO/IEC 9126

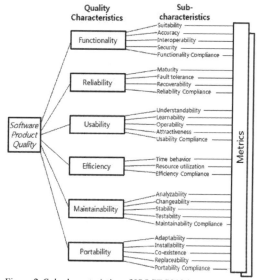

Figure 2. Sub-characteristics of ISO/IEC9126

3.3.2 ISO/IEC 12207

This standard is used to measure the maturity of software development life-cycle based on several criteria: user requirement analysis, design, implementation, test, and maintenance. It includes necessary processes and activities that should be applied during the acquisition and configuration of the system. It also regulates the operation and maintenance of the system, taking into consideration critical aspects such quality assurance.

Figure 3 shows that ISO/IEC 12207:2008 defines software quality assurance processes as a part of software specific process group. There are two software-specific processes: Software Implementation and Software Support. Combined, these processes include project design, integration, testing, and validation. In addition, organizational and technical aspects related to software implementation, software support and software reuse are addressed (ISO/IEC 12207:2008).

3.3.3 e-Navigation solutions related software quality

Collectively both product and process software quality must to be considered in conjunction with the development and implementation of e-navigation. In this regard, software quality assurance has been mentioned as a means to insure that *"quality and integrity verification become a testing requirement for relevant bridge equipment"* (IMO STW 43-6). It was also listed on IMO NAV e-navigation solutions which mentioned as *"shore monitoring of quality/integrity of navigation systems of onboard information and effectiveness of communications."* (IMO NAV 58-14).

Figure 3. ISO/IEC 12207:2008 software specific processes

4 DISCUSSION

A draft Guideline for e-navigation SQA has been developed to help insure e-navigation qualified software. However, there are some additional considerations. Currently, most existing safety or test requirements are usually determined on an individual equipment or system basis. However, for e-navigation this will require additional effort to achieve shipborne and shore-based interoperability.

4.1.1 Consideration of usability, HEAP and CMDS

At NAV58, Japan submitted 'Draft Interim guidelines for usability evaluation of navigational equipment (IMO NAV 58-INF.13), and a 'Proposal for structure of and how to utilize guidelines for usability evaluation' (IMO NAV 58-6-6).

In regard to the Human Element Analysing Process (HEAP) suggested by Australia, this is recognized by IMO Maritime Safety Committee (MSC) as a practical tool that can be used to address the human element for maritime safety and environmental issues (IMO MSC/Circ.878) at 1998. Australia also proposed that HEAP be used as a checklist in the gap analysis process for e-navigation (IMO NAV 58-6-6).

On the other hand, MSC90 approved the development of a Common Maritime Data Structure (CMDS). Further, the IHO S-100 Universal Hydrographic Data Model and associated Registry will be used in conjunction with CMDS as the baseline for creating a framework for data access and services under the scope of e-navigation.

For software systems consisting of user interface, functional software units and data repository, usability evaluation, HEAP and CMDS should also be included in the development process. As such, these also may need to be considered in the guideline for e-navigation SQA.

4.1.2 Level of description depth

The application scope of the guideline for e-navigation SQA can be applied to any type of e-navigation related software, including existing legacy, new development, embedded, or software-centered systems. This applies to all types of e-navigation-related software, from individual components/equipment to large-scale integrated systems. While the guideline on e-navigation SQA is intended to be 'general', some additional, 'specific' guidelines may need to be developed as well.

4.1.3 Testbed strategy

At NAV58, Japan suggested that usability tests be performed to assess the effectiveness of the equipment in performing the intended task. In particular, an 'Example for application of goal-based procedures for test task set-up in case of ECDIS' (NAV 58-INF.12) was given. In this regard, software quality assurance may need to be used to assess the effectiveness of software using new types of e-navigation data. One important example is the future use of ENC data in ECDIS that will be in the new IHO S-101 format, instead of (or in addition to) the current IHO S-57 format. In this regard, testbeds will need to be conducted that include the co-evaluation of both data and software.

5 CONCLUSION

As agreed at IMO NAV 58, an important consideration associated with the development and implementation of e-navigation is software quality assurance. In particular, navigation and communications equipment must reliably indicate that they are functioning correctly. For this to occur, high-quality software that is both stable and complete must be installed and used. Software quality standards have been adapted for many types of safety-critical systems including avionics, automobiles and medical equipment/devices. To date, no software quality standards have been specifically developed for maritime navigation-related systems. The development of an IMO guideline on software quality assurance should apply to all types of e-navigation related software that is used in shipborne and shore-based systems. This includes both existing and new development, as well as embedded or software-centered systems. The overall goal is to encourage the installation and use of high-quality e-navigation software that is fit for purpose.

ACKNOWLEDGEMENTS

The work in this paper was partially funded by Korean Ministry of Land, Transport and Maritime Affairs (MLTM).

REFERENCES

Bergmann, M. (2011) A Harmonized ENC Database as a Foundation of Electronic Navigation. *TransNav-The International Journal on Marine Navigation and Safety of Sea Transportation*, 5(1): pp.25-28

IEC 62304:2006 Medical device software -- Software life cycle processes. Geneva: International Electrotechnical Commission.

IMO
MSC/Circ.878 Interim Guidelines for the Application of Human Element Analysing Process (HEAP) to the IMO Rule-making Process. London: International Maritime Organization.
NAV 57-15 - Report to the Maritime Safety Committee. London: International Maritime Organization.
NAV 58-6-4 - Consideration of software quality assurance issues for e-navigation development (Republic of Korea). London: International Maritime Organization.
NAV 58-6-6 - Proposal for structure of and how to utilize guidelines for usability evaluation. London: International Maritime Organization.
NAV 58-14 Report to the Maritime Safety Committee. London: International Maritime Organization.
NAV 58-INF.10 - The Human Element Analysing Process (HEAP) in e-navigation (Australia). London: International Maritime Organization.
NAV 58-INF.12 - Example for application of goal-based procedures for test task set-up in case of ECDIS. London: International Maritime Organization.
NAV 58-INF.13 Draft Interim Guidelines for Usability Evaluation of Navigational Equipment in final form. London: International Maritime Organization.
STW 43-6 Report of the Correspondence Group on e-navigation to STW 43. London: International Maritime Organization.
ISO/IEC 9126-1:2001 Software engineering — Product quality —Part 1: Quality model. Geneva: International Organization for Standardization.
ISO/IEC 12207:2008 Systems and software engineering — Software life cycle processes. Geneva: International Organization for Standardization.

Visual to Virtual and Analog to Digital – Transformations Leading to Introduction of e-Navigation

D. Filipkowski

Gdynia Maritime University, Gdynia, Poland

ABSTRACT: This article is an attempt to analyze the changes in the industry that is shipping, and especially what has changed navigation process and work on the bridge. Author considers the introduction of e-Navigation as a reference point and tries to find items that led to the start of a system that treats sea transport in a holistic way. The analysis was made in three main areas in which e-navigation is to bring the greatest benefit: the ship, shore and communication. The author presents how the situation in these specific areas looked like in the past, as it looks now, and how a new system will affect the areas in the future. The author presents a possible use of technologies such as the Internet, and Augmented Reality.

1 INTRODUCTION

The sea is constantly changing. Although the vastness of the oceans remains relatively static element, work on merchant vessels changed dramatically in last few years. The process of change is gradual but also inevitable. It is believed that the stones that caused the avalanche called e-Navigation were thrown about 40 years ago. In January 1971 two tankers collided in conditions of poor visibility in the area of Golden Gate in San Francisco. The result of that collision was a spill of about 800 000 barrels of oil in the bay of San Francisco. U.S. Coast Guard was observing the whole event and was completely helpless. They were using an experimental system HARP (Harbour Advisory Radar Project). Subsequent attempts to establish a connection with vessels running into danger have failed and there was a collision. Events described above led to the creation of Vessel Traffic Service (VTS) station in this area. VTS is considered by some as a milestone on the path to create a system that includes shore-based personnel in the process of shipping and treat all aspects of navigation and maritime traffic in the holistic way (Sollosi, 2012).

2 E-NAVIGATION

Works on e-Navigation system is the answer to the need for a strategic vision for the use of new tools, in particular electronic tools. The main objectives of the new system are to improve maritime safety and environmental protection. In 2009, Maritime Safety Comitee (MSC) approved the plan for implementation of e-Navigation. According to this plan the introduction of new technologies should be in correlation with the changes in the legislation and take into account shore-based users as well as those working at sea (MSC, 2009). International Association of Marine Aids to Navigation and Lighthouse Authorities (IALA), created a coherent and at the moment the official definition of the e-Navigation (IALA, 2010):

"The harmonized collection, integration, exchange, presentation and analysis of marine information onboard and ashore by electronic means to enhance berth to berth navigation and related services for safety and security at sea and protection of the marine environment."

According to the above definition e-Navigation system would help ships sail in a safe and efficient manner that is also friendly to the environment. User and his needs would be one of the main drivers of the system. New technologies that enable effective communication and safe navigation would help him send, receive and analyze information and make decisions (Korcz, 2009).

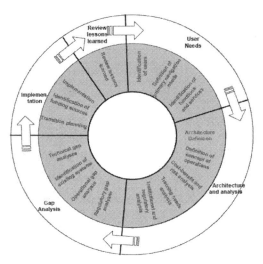

Figure 1. e-Navigation implementation plan diagram (MSC, 2009)

2.1 e-Navigation onboard

It is expected that e-Navigation system is one of the last stages in the integration of navigation systems and equipment on the bridge (Integrated Bridge System, IBS). It is believed that this will allow the future user to obtain additional information, that can be used in decision-making process. The user interface will be standardized. Standardization will also manage risk zones and alarms system. Elements forming the core of such a system are: electronic positioning systems, high reliability electronic navigational charts, ARPA systems and communication systems. It is expected that e-Navigation will reduce the likelihood of human error. Officer would be actively engaged in the process of navigation at the same time system would allow for reduction of the risk of scattering his attention and overwork (Gale, 2007).

In recent years there have been many voices saying that more and more navigators trust to much in electronic devices and no longer look outside the window. Do not forget that visual observation is still primary way of risk assessment and prevention of collisions at sea. The same voices argue that reaching automation with the introduction of e-Navigation only worsen the situation in this regard. No one is able to count how many accidents could be prevented so far by the introduction of radar, ARPA, AIS, etc. No one questions the fact of the usefulness of these devices. e-Navigation is just another step in the integration and harmonization of the information flow. Improvement of voyage planning, precise and frequent obtaining of position, more credible information about the other vessels, this are only examples of benefits coming with new

system . e-Navigation will also provide more complete and actual information to keep the vessel safe. The system in its complexity presents a holistic approach to navigation process. It seems that the disadvantage of the system is the need to supply users in new technology. Another thing is training. It is necessary, because other way the future user won't be able to fully enjoy the benefits of the new system. Training, procedures and equipment cannot be changed overnight, but will be introduced gradually. Hopefully with these changes will increase the level of safety at sea (Patraiko, 2010).

2.2 e-Navigation ashore

It is expected that the management of vessel traffic and related services will be much simply and more efficient than today. Better access to information about the ship and its basic parameters (e.g. course, speed, draft, destination, cargo, etc.) gives shore user wider view on the situation. Better coordination and exchange of information between the ship and the shore will provide comprehensive information on the vessel intentions, its position, speed, etc., but also on the efficiency of equipment, quantity and characteristics of the cargo and information about the crew. The data will be exchanged in the same format, which in turn will be more useful and easier to understand. The user ashore will be able to more efficiently fulfill its advisory and control role (Weintrit, 2007).

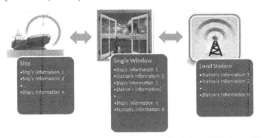

Figure 2. Single Window concept principles (Filipkowski, 2010)

2.2.1 Space management

In the not too distant past navigational aids were placed in order to allow navigators to obtain the position, determine a safe course and avoid unknown dangers. Several types of buoys, lighthouses and COLREG were more than enough to keep safe navigation. Shipping changed since then and challenges standing before the e-Navigation system are much greater. New system will define shipping lanes and areas excluded from the traffic, will take care to preserve the safety and efficiency of navigation. It will be a system that is designed not only for navigator and crew on board, also for the users ashore. System functions allow better planning of sea space in mining areas (offshore rigs and platforms), protected areas (natural reservoirs) or areas where natural resources are being used such as the installation of wind farms. The system will also enable planning when it comes to construction and use of the seabed. United States Coast Guards already faced problems in this area. Emerging on the east coast wind farms necessitate the creation of new traffic separation zones. Experts believe that the planning of the maritime space in many ways resembles airspace planning. Around the airport will never be a tall building. Alone gliders do not even have the right to find near any airport. When it comes to shipping situation is not regulated. Both small fishing vessels and the largest tankers in the world still use the same shipping lanes, and the same approach channels. The only barrier that prevents the tragedy is COLREG, knowledge and experience of officer of the watch. Each year there is a growing number of new built vessels. It makes the surface of the sea is slowly starting to shrink. e-Navigation as the holistic system will have to somehow solve this problem. If the territorial waters of any country for any reason are divided into zones and marine areas are excluded from the shipping, administration must ensure, that all users of these waters are aware of it. Whether this will be done in the traditional way via visual and electronic navigation aids and virtual help, information must be clearly and legibly transferred to. Only such an approach will provide increased safety and environmental protection (Sollosi, 2012).

2.2.2 *Vessel traffic management*

Attempts to control vessel traffic has been ongoing for many years. Thanks to systems such as AIS and LRIT, administrations of the coastal countries can monitor the movements of vessels. It seems to be only a matter of time when the ship's route will be under administration surveillance while sailing from the point A to point B. This is actually possible through various websites presenting vessel's AIS position and the daily reports sent to the owner or charterer. The ability to monitor ships at some point turn into the desire to control traffic for safety and efficiency of navigation. However, there is also a risk that the uncontrolled workload will increase and information flow from multiple sources will grow. Officer of the watch will receive information which are mutually exclusive, which in turn will result in the deterioration of the status quo (Hagen 2012).

2.3 *Communication*

It is not a revealing statement that the Global Maritime Distress and Safety System (GMDSS) is archaic, inflexible and does not leave a large field to make changes and innovation. The creators of the e-Navigation system must therefore ask the question

whether to try to modify existing equipment, or create an entirely new system from scratch. Such a system have to be consistent with the objectives of e-Navigation and be capable of continuous changes resulting from technological advances and growing user requirements. The main function of the GMDSS is to alert in distress and provide communications for safety of navigation. The new system should be automated to a large extent and the process of alerting should be simplified as well. The need to transmit more data not only alphanumeric, but also photos, videos, attachments, etc., make that installed on board GMDSS equipment are used only during the required tests (except VHF). Routine correspondence is maintained thanks to systems based on satellite technology. Usage of the Internet or satellite phone is more often than e.g. Inmarsat because of the reliability and effectiveness of such solutions. Most of this new solutions allow you to send messages with a single click using software well known to users from their own personal computers. (Korcz, 2008)

3 VISUAL AIDS VERSUS VIRTUAL AIDS

Challenges that confronts officer on the bridge every day, in the future, thanks to e-Navigation will be managed by the integration of visual and virtual aids to navigation. Ability of the users to keep up with increasingly sophisticated technology will continue to grow. The needs of the users will be the main factor determining the changes. The system cannot be limited only to the communication and presentation of information, but it must flexibly adapt to existing laws, regulations etc. Analyzes of the information in connection with appropriate regulations should facilitate the navigator to make good decisions thus effectively increase the safety of navigation.

3.1 *Visual aids to navigation*

The introduction of e-Navigation does not mean that the visual navigational aids such as lighthouses, buoys will be withdrawn from use. On the contrary, it is believed that they are a very important and effective source of information, especially in coastal navigation. Nor does it mean that these systems remain unchanged. The integration of new technologies and traditional navigation aids will amount to a new dimension. User will be able to observe this aids not only visually, but also using bridge equipment. Characteristics and quality of the lights would be changed, a well-known racons would be supplemented with AIS transmitters, etc. But no matter how high-tech we use, the possibility of visual references will still be the main feature. First due to the unreliability of technology, and secondly, there will always be users, who for whatever reasons, do not want or are not able to benefit from technical innovations (Sollosi, 2012).

What are the changes in the visual aid involves the introduction of e-Navigation:

- increased amount of information through the use of additional electronic devices, that can provide data on the technical condition of the buoy or collect hydro-meteorological data,
- improvement of optical systems by the use of modern light sources, better performing, more accurate lighting sectors, and lower energy consumption or energy self-sufficient through the use of solar panels,
- associated, synchronized sets of navigation aids, giving information fuller and more readable,
- interactive aids with features such as user-activated audio signals, radar racons , AIS racons,
- use of modern materials, e.g. coatings will help raise the visibility range and durability in harsh environments.

It is important that, in accordance with the objectives of the e-navigation any changes should be in sync with the needs and abilities of users. Whether we are dealing with the interruption of transmission of a sound signal, changing the effective range of the light, or the implementation of new and unproven technologies. When the U.S. Coast Guard settled his first set of buoys with synchronized flashes (San Francisco Bay Channel) it was first tested on the California Maritime Academy. The pilots who every day navigate ships to San Francisco had the opportunity to participate in laboratory studies and simulations. With their suggestions, the functionality gained the more, that they are the group most frequently benefits from usage of this particular project (Sollosi, 2012).

3.2 Virtual aids to navigation

Most people interested in the changes in maritime transport probably heard of virtual aids to navigation. This subject deserves recently more and more attention, and this is due to the possibility of using such aid navigation in the Arctic ice. For obvious reasons, installation, and maintenance of navigation aids in the classic regions of ice causes many problems for offshore services, or is simply impossible because of the lingering ice. Another factor in favor of the use of virtual navigation aids are their relatively low cost. Although this argument should not be decisive. Economic considerations continue to drive changes not only in the maritime transportation but in the industry in general. The balance of benefits and costs can convince a lot of individuals at the level of decision-making, and if it happens for the benefit of safety is a mutual gain. Virtual aids seem to be a great alternative especially when placing classic ones is not possible because of

the hydro-meteorological conditions. Virtual aids also perfectly suited as a temporary or preliminary marking. To quickly determine the nature of the newly created danger or ongoing work on a temporary basis. Use of physical aids not only increase workload but also hamper unnecessarily increased costs. The use of virtual aids, however, need not be limited to the examples described above. Imagine the shipping channel, where the water level, as well as the width of the channel changes with the tides. The use of virtual navigation aids make the width of the channel will also be able to change with tides. Navigator will be available taking into account the depth of the contour, the air draft under bridges and high-voltage cables. His task will reduce the time to maintain the safety of the ship inside the outline created by the virtual "buoy". USCG is already conducting laboratory studies of the possibilities of using this type of assistance. It is at the stage of the simulation and collecting suggestions of users who, as usual, shaping the system at the very beginning of its development are likely to be influenced and make it more functional. Attempt to replace existing beacons with only their virtual representations seems to be a drastic step, but if it is a step harmonized with the needs of users and the availability of reliable devices may be a big step forward when it comes to safety in coastal shipping. One of the main tasks of e-Navigation is to protect the environment. Another example of conditions conducive to the use of virtual assistance, and so different from the Arctic ice, is a tropical coral reef. Where the use of large-scale physical navigation aids would be a crime against the environment. Virtual aids not only would the behavior of the immutability of the ecosystem but also allow for the observation and study of flora and fauna unique coral. The answer to the question of whether virtual aids are actually replacing traditional ones, is not easy. It is certain, that with the introduction of virtual solutions must go hand in hand on the bridge, improve equipment and skills of the personnel operating the machine (Sollosi, 2012).

4 SOFTWARE AND HARDWARE IN E-NAVIGATION

Solutions in the field of computer science are a natural assimilation of existing solutions in the market caused by the need for innovation and the growing demands of users. In this chapter, the author proposes several solutions, which in his view would have wide use in e-Navigation system.

4.1 Computer-basic tool in e-Navigation

"Computer (lat. computare – to calculate) electronic machine designed to process information that can be written in the form of a series of numbers or a continuous signal. Simple computers are so small that they can fit into even the watch and are powered by a battery. Personal computers have become a symbol of the era of information and most of them identified with the word "computer". Calculating machines are the most numerous embedded systems with various control devices - from MP3 players and toys to industrial robots " (Wikipedia.com).

Both personal computers and control systems with various devices, sensors and controllers will form the core of the e-Navigation system. Nowadays everything has a computer chip, though often looks different from our imagination of "computer". Personal computers connected to the ship in a local network, and then communicate with the outside world via the Internet, are not only the points between which the data will be transmitted, but also the base on which the e-Navigation software will be installed. The personal computer due to the fact that we can install any software on it, becomes the ideal solution for e-Navigation. We can use different software, which has the same functions as the actual device. In modern ships there are hundreds of sensors and controllers which gives the amount of information that a man is not able to analyze by his own perception. The information gathered by the sensors is collected, analyzed and clearly presented to the user, which with a one click sends a command executed by the controller. If you think about it a moment, then it is already e-Navigation but no one use that name yet (Filipkowski, 2011).

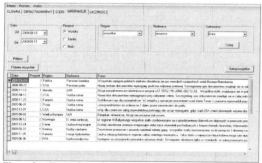

Figure 3. Example of information management system (Filipkowski, 2011)

4.2 Ship network

When producers and customers decide that something must be standardized, the International Organization for Standardization (ISO), is trying to meet their expectations and creates a standard or norm in the field. At present, the ISO forms including guidelines for ship installation of computer networks to improve communication between the devices and systems used on board the ship - ISO/CD. It is believed that this is to some extent the implementation of the e-Navigation ideas and may be the first step to create a basis for a new system. At this stage, the draft prepared by Hiroshi Morono was sent to members of the technical committees and organizations that are in agreement with the ISO in order to comment and pre-approval. This document is the main point and is broadly in line with plan development and implementation of e-Navigation proposed by the IMO. This gives hope that the new standard will also be user-oriented (Filipkowski, 2011).

This standard provides guidance on the installation vessel's computer networks to improve communication between onboard ship systems. Today, there are no guidelines how to connect to the network equipment supplied by different manufacturers. The purpose of this standard is the creation of guidelines for the installation of on-board network. These are the guidelines for issues such as network connectivity, system architecture, data requirements, network administration, operation, start-up, control and monitoring. This standard will allow for uniformity and standardization of devices connected to the network and the network itself. This should result in tangible benefits for all involved parties, including manufacturers, engineering companies, shipbuilders and shipping companies.

5 AUGMENTED REALITY

Augmented Reality (AR) technology is not entirely new. It seems that development of technologies like the Internet, mobile phones, etc. introduced AR to wider range of users. Miniaturization and personalization of computers and other electronic devices (eg, laptops, i-Pads and digital cameras, etc.) have enabled more users to read of AR. Devices that are an integral part of Augmented Reality, in some extent they define the system as well as areas where Augmented Reality has already enjoyed success.

5.1 Augmented Reality Definition

Some researchers define Augmented Reality as the system associated with special displays worn on the user's head. Dr. Ronald Azuma of the University of North Carolina, one of the authorities in this area, to avoid restrictions to specific technologies, defines reality as the system:

– combines real world and virtual reality;
– interactive in real-time;
– with three-dimensional elements (3D).

Figuratively speaking, Augmented Reality is a system that expands and complements the real world with virtual (computer-generated) objects that appear to coexist in the same space and time. Dr. Azuma definition AR is an open system with new technology and is not limited to specific technical solutions. There are several approaches to Augmented Reality and there is no clarity on technical solutions. According to one of the system concept includes the processing and presentation of data, as well as other also data collection and transmission (Azuma, 1997).

5.2 *Application of Augmented Reality*

AR is used in various areas of everyday life such as medical procedures, computer games, exhibitions, museums, air shows, etc. Augmented Reality tools will find their application wherever virtual objects extend the boundaries of user's perception. Below are some examples of fields where AR already found application:

- medicine;
- marketing;
- architecture;
- the military and security forces;
- hydrology and geology;
- teaching;
- aviation;
- entertainment.

AR scope for the use of e-Navigation is very wide.. Augmented Reality tools are used wherever the need is clear presentation of data, or expand the skills of seeing the world by the user:

- safety and security;
- navigation and bridge watchkeeping;
- cargo operations;
- environment protection;
- teaching and training.

6 CONCLUSIONS

e-Navigation can combine all aspects of shipping in one cohesive package. This applies not only navigation, technology and services related to it, but also the activities of maritime administrations and commercial companies related to shipping. e-Navigation, in their assumptions, to enable continuous and stable growth of the safety of navigation and protection of the environment, and at the same time provide a better means of communication and permanently join shore and ship. With its maritime transportation capabilities will become more efficient and competitive, flexible to market volatility and stock market indices, and by all means secure. Changes will not be made overnight, but the process of change leading to the implementation of e-Navigation began some time ago.

REFERENCES

Amato F., Fiorini M., Gallone S., Golino G., 2011, *e-Navigation and Future Trend in Navigation.* TransNav - International Journal on Marine Navigation and Safety of Sea Transportation, Vol. 5, No. 1, pp. 11-14,

Azuma R., 1997, *A Survey of Augmented Reality,* Teleoperators and Virtual Environments, pp. 355–385,

Filipkowski D., Wawruch R., 2010, *Concept of "One Window" Data Exchange System Fulfilling the Recommendation for E-Navigation System,* Transport Systems Telematics, Springer

Filipkowski D., 2011, *Informatyczne elementy systemu e-Nawigacji,* Logistyka Nr 6

Filipkowski D., 2012, *Możliwości wykorzystania rozwiązań rzeczywistości rozszerzonej w systemie e-nawigacji,* Logistyka Nr 3

Gale H., Patraiko D., 2007, *Improving navigational safety. The role of e-navigation.,* Seaways, str.4-8,

Graff J., 2008, *The Role of Operational Ocean Forecasting in E-Navigation.* TransNav - International Journal on Marine Navigation and Safety of Sea Transportation, Vol. 2, No. 3, pp. 259-262,

Hagen E.K., 2012, *Why eNavigation ?,* Seaways, pp.14-16,

IALA e-Navigation Comitee , 2010, *e-Navigation Frequently Asked Questions (Version 1.5),*

Korcz K., 2008, *GMDSS as a Data Communication Network for E-Navigation.* TransNav - International Journal on Marine Navigation and Safety of Sea Transportation, Vol. 2, No. 3, pp. 263-268,

Korcz K., 2009, *Some Radiocommunication Aspects of e-Navigation.* TransNav - International Journal on Marine Navigation and Safety of Sea Transportation, Vol. 3, No. 1, pp. 93-97,

Patraiko D.,2007, *The Development of e-Navigation.* TransNav - International Journal on Marine Navigation and Safety of Sea Transportation, Vol. 1, No. 3, pp. 257-260,

Patraiko D., Wake P., Weintrit A., 2010, *e-Navigation and the Human Element.* TransNav - International Journal on Marine Navigation and Safety of Sea Transportation, Vol. 4, No. 1, pp. 11-16,

Sollosi M., 2012, *Transitioning from traditional aids to navigation* Seaways, str.21-22,

Sub-Committee on Safety of Navigation, Session 85, 2009, *Strategy for the development and implementation of e-Navigation,* IMO, London,

Weintrit A., Wawruch R., Specht C., Gucma L., Pietrzykowski Z., 2007, *Polish Approach to e-Navigation Concept.* TransNav - International Journal on Marine Navigation and Safety of Sea Transportation, Vol. 1, No. 3, pp. 261-269.

e-Navigation Starts with e-VoyagePlanning

G.L. Olsen

Jeppesen Norway AS, Norway

ABSTRACT: Considering the balancing of data/information flow needed in NextGen Navigation (e-Navigation) and decision support already in the VoyagePlanning phase.
The navigational world is getting more and more complex. And the need for intelligent solutions to handle the workflow processes both aboard, in ship-ship and shore-ship collaboration, are in high demand. This includes all tasks; from the information collection in the voyage planning and optimization process, to reporting regimes and berth to berth safe navigation based on up to date and real time situation awareness.
Not only to get more data out to the mariners whom are taking navigational and other operational decisions, but to enhance the ship-ship and ship – shore cooperation, presented as Integrated Intelligent Information. One of the main hazard to (e)Navigation is the availability of good data, but presented and compiled in an "unintelligent" way. The same goes for the workflow for the operators: the process from Planning, Optimizing and Reporting, to berth to berth navigation is only as good as the weakest link of all the marine operators: be it the VTS operator, Pilot, Captain or the Lookout. And with no integrated tools to handle this workflow, the risk for misunderstanding, fatigue and human errors is very much present.
I will in my paper present two central challenges and potentials in the voyage towards eNavigation: 1) More optimized and safer navigation based upon closer ship-ship and ship-shore collaboration and 2) a concept for voyage planning, optimization, collaboration and reporting processes. 3) The impact of e-Navigation on Polar Navigation.

1 INTRODUCTION

As the shipping world is truly moving into the "Digital Ship", international organizations and industry do a joint effort to ensure that the safety and security of both the seafarers and society are maintained and improved.

Saying that "e-Navigation starts with e-VoyagePlanning", one have to look at what information vision IMO has for the Marine Community. The proposed solutions were drafted in IMO NAV 58, annex 4 and 5. Some of the system requirements concerned user-friendliness, automation and graphical integrations, and some functional requirements where information and reporting management, VTS service portfolio, Search and Rescue, and route and voyage plan exchange.

A key consideration is how this information is collected and presented to the marine operators; be it onboard for the mariner, or shore side for a marine VTS/SAR coordinator or manager. Also considerations work to be done to data compilation and compression, and the "postman"; communications.

The work to prepare and execute a voyage plan, is described on high level in IMO Res. A893; "Guidelines for Voyage Planning". Starting with the appraisal, collection of relevant information, updating of charts (be it paper or digital official ENC), publications, and ship particulars are key. Going into planning process, manual or digital tools are available to collect and plan use of navigational aids, weather, and port information etc. Coming into actual execution, the process of navigation begin with maneuvering and ship handling; where one react to changing elements and situations, including COLREG and severe weather, ice and all other conditions that cannot been mitigated in the risk analyzing done in the Passage planning process. Finally for verifying; aids to monitor the process like radar, ECDIS, and most important; lookout need to be used.

The Voyage Plan is sometimes referred to as the "tactic", as it is done usually days or hours ahead of starting the voyage. The process of "e-Navigation" could probably be referred to as "wheel" or a circle; a continuous process of managing nautical information, preparing voyage and passage plans, maneuvering and navigation, optimizing especially in Transocean passages, reporting and voyage analyzing. Looking into the offshore industry; the process is the same, the only difference is that the ports are at sea; complicating the operations further (but mitigated i.e. with use of dynamic position systems).

The process should all collect in the all important voyage plan being available on the bridge. Taking into account more information being available – the risk of "information overflow" is becoming more present. The aviation industry has taken this into account, where Jeppesen has been in the lead to replace the traditional aviation charts with digital means. While in the marine industry; paper charts and publications are still to be maintained in a painstaking process. How do we ensure that the transformation to the digital information age in marine, thus e-Navigation, is happening as effective and user-friendly as possible?

Guidelines for Voyage Planning – process:

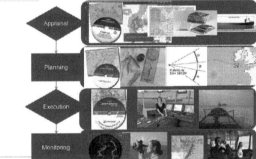

Figure 1.

2 IMO E-NAVIGATION DEVELOPMENT.

For IMO, the visions of e-Navigation are developing into a reality. Even if the current ECDIS is under rigid performance standards, it is considered a key waypoint towards the realization, as part of the "INS" (Integrated Navigational System).

Some of the proposed e-Navigation solutions address potential for integration, graphical display, exchange of "Marine Service Portfolio" information (such as SAR and VTS), route segment exchange between ships (intended route), and collaboration on voyage plans (for authorized parties) for seaside and shore side operators. For the ship operators, the return of investment lies in safer and more efficient operations, and for the global society greener and

safer ship operations with reduced risk of accidents – as well as increased transport efficiency.

The need for a common "language" on information exchange has been solved with the development of the International Hydrographic Organization (IHO) "S100 GI registry" – which in contrast to the current "S57" format is an "open" registry. Meaning that organizations may submit new or improved standards, and is open for the industry to develop, adopt and improve solutions in the registry. This concept is being maintained by cross-organizational work between i.e. IHO, IALA, CIRM and so on –where industry members such as Jeppesen has leading positions and expertise.

Some of the potential for such solutions has already been proven: in 2011 the "EfficienSea" project was closed with a practical demonstrator on VTS and Ship collaboration chaired by the Danish Maritime Authority. In April 2012, as part of the "Marine Electronic Highway" project, a joint sea trial on real time information exchange of MSI (Maritime Safety Information), METOC and Notice to Mariners (NTM) was carried out in the Singapore straight. Systems involved were a Jeppesen eNav prototype based on a Navigational Kernel (Jeppesen SDK) and Kongsberg Norcontrol VTS – "C-Scope system. The sea trial was done on the request from IMO, chaired by Norwegian Coastal Administration and with close support from the Singapore MPA. The process was later replicated under an IMO workshop. It proved that AIS could be one potential carrier for "real time" short range navigational information from i.e. a VTS. In combination with efficient data compilation for NTM (Notice to Mariners) over minimum "internet" access such as WiMax – "S100" can work as a common framework for information exchange (Portrayal and data structure). Based on the project, the "S100" was adopted by IMO as a framework for e-Navigation data exchange.

Figure 2. (From Singapore sea trials on "S100").

Currently the joint EU funded project "MonaLisa2" is being established with the industry as vital contributors. The first waypoint for the project will to be establish of a common route (and potential voyage plan) format, and infrastructure to handle such exchange and collaboration between marine stakeholders. It also aligns with the IMO proposed e-Nav solutions for "exchange of route segment between ships and VTS", and "voyage plan exchange and collaboration for authorized parties". Questions has been raised, especially on "voyage plan exchange and collaboration" – however this is already being done between ship and specialized weather routing services – such as the Jeppesen Vessel Voyage Optimization Service (VVOS). The Monalisa project will look into how more services and expertise can be made available through common platforms for the onboard navigators and shore side operators (marine coordinators/ship operators) to ensure effective collaboration with safer and more efficient marine traffic handling. The objective for the project is to go from "surface based operations" to "voyage based operations". Potential partners are SAAB, Jeppesen, SAM, Kongsberg, Furuno with others.

3 INDUSTRY ALIGNMENT

Jeppesen is one of the companies that already are aligning with the need for improved routing and information exchange; on automatic routing the "SeaRoutes" database is available in selected ECDIS/ECS systems with Jeppesen Chart databases, as well as the legacy "OceanView". The onboard database is based on industrial recommended routes, and is updated weekly together with digital NTM's. For the end user, it reduce time for route planning from hours to seconds, freeing more time available to do quality assurance of the voyage plan. Also additional data such as Tides and Currents, SAR and GMDSS areas, and other information needed for improved voyage planning (Value Added Data) are available in the Jeppesen official ENC (Primar) and derivated chart databases. To further support the need for improved voyage planning and operational support, the "eVoyagePlanner" program is currently being developed. Being another example how technology may help addressing all the aspects that are being involved in the voyage planning process.

As the ECDIS mandate is coming into effect for the IMO mandated fleet, more ship-owners and onboard operators are coming more familiar how technology can help them both to reduce cost and maintenance compared to traditional paper charts management, and improved safety. In parallel, development is being done to help support the transition. Jeppesen's "Nautical Manager" and Chartco's "Passage Planner" are just some of the solutions becoming available to minimize manual work for digital chart ordering and maintenance with highly automatic processes. Last year's partnership between these two companies highlights the

industrial efforts to ensure effective and cost effective transition.

For the navigational officer, effective distribution and loading of officials ENC's and other marine information such as Weather forecasts raise issues due to general communication and performance restrictions. Jeppesen has achieved distribution of Official ENC's, weather forecast and other marine information in a "CM93/3 SENC" format – with a ratio approximately 1:9 compared to the "raw" S-57 and GRID formats. The data however, are in its nature official. But risk in loading into the systems and cost for communication are highly reduced. And probably improvements on the "S100" standard will further mitigate the risk and optimize the transition to the "Digital Ship".

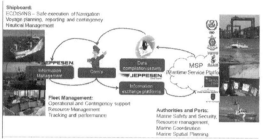

Figure 3. Industrial alignment in the e-Navigation infrastructure.

4 E-NAVIGATION AND POLAR NAVIGATION

In parallel with IMO e-Navigation, the IMO Polar Code is going from a vision to reality as the Polar Regions are being opened for commercial shipping. Questions are raised about the challenges and possible solutions for navigation safety in the areas.

Potential for both improved natural/fossil resource extraction, considerably shorten the Europe – East Asia route, as well as avoiding high security threat areas such as the Indian Ocean are just some of the incentives making shipping and oil and gas operators aware of the potential now opening up. However, marine operations in Polar Regions are facing considerable challenges. Remote areas, communication, icing of vessels, technology development, and access to information and so on, are just some of the issues that need to be addressed. As not only cargo shipping, but also tourist cruises are increasing in numbers.

5 POLAR CODE STATUS

In parallel with the work done on e-Navigation, mandates and regulations are being developed to ensure protection of both human life and environment in the regions under the Polar Code.

Today IMO has issued the "Guidelines for vessels operating in polar waters" (2009) as an important mile stone towards a mandatory code.

This code cover the full range of design, construction, equipment, operational, training, search and rescue and environmental protection matters relevant to ships operating in the inhospitable waters surrounding the two poles.

The obvious challenge is ice coverage in general; a traditional ship with "V-shape" bow will be stuck in the ice attacking it from head on. Today this is solved in ice covered areas such as the Baltic Sea, northern Russia and so on with the support from specialized Ice breaker services. Some ship-operators, such as Knutsen OAS, has contracted specialized LNG carrier vessels to be able to sail the North East Passage route (LNG/C "Ribera del Duero" – classed DNV ICE-1A (up to 0,8 m of ice thickness). However, they still will need help from ice breakers in some of the areas. A more direct approach is i.e. the "Norilskiy Nicke" working in Siberian waters, sailing traditional "bow first" in open waters, but with a special designed rear, going "aft first" up to 1,5 m thick ice.

Another well-known problem sailing in Polar Regions is the icing of the vessels, reducing the stability due to higher center of gravity (CG) For vessels risking this, dynamic stability calculation and mitigation with use of i.e. deck steam are important, as well as solutions for calculating risk of icing.

If an accident happens, both Search and Rescue, evacuation of personnel, as well as environmental impacts are great challenges. Pr. Today the IMO "Guidelines for voyage planning for passenger vessels operating in remote areas", has escort/"tandem" sailing as one of the considerations. Also use of dedicated ice navigators is recommended. Another long term impact is a potential oil spill; as the oil will be trapped inside the ice, oil recovery is practical impossible – and may have greater environmental impact than ever seen before in very vulnerable environments. Norway and Russia have been cooperating to mitigate these potentials, with the "Barents Watch" program. The "Barents Ship Reporting System", coming in effect 1. Of June 2013 are one of the results impacting the marine operations in the Artic as one potential solution to improve Vessel Traffic Services (VTS) in the region.

Taking into account navigational issues – hydrographic surveys and charting is in its nature extremely challenging in these areas, and need special charts. ENC coverage to be used in the "ECDIS" is improving especially due to Norwegian and Russian collaboration, however is still a challenge due to projection issues and quality of data. Organizations like BSH and The Arctic and Antarctic Institute, are distributing special charts and

ice coverage predictions, and navigational data providers such as Jeppesen ensures that the data is available in as efficient means as possible – such as through the "dkart IceNavigator". The Norwegian "Hurtigruten" is one of the pioneers in Arctic/Antarctic tourist cruising. Their vessels doing polar tourist cruises use solutions such as official ENC charts and decision support tool "OceanView" software both from Jeppesen, in combination with a portable survey solution mounted on a tender boat when sailing in unsurvyed areas. Also specialized radars are being developed to ensure better situational awareness sailing in ice covered areas.

For areas surveyed, another issue arise with regards to Aids to Navigation (AtoN) – buoys are well known to drift, and taking into account the heavy forces drifting ice represent and maintenance for short range shore side systems, such solutions are practical impossible. Under the "e-Navigation" development, solutions such as "Virtual Aids to Navigation" and "specialized route exchange and advisory" are only some of the possible solutions that may arise.

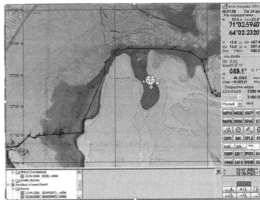

Figure 4. IceNavigator using official ENC and updated Ice information overlay.

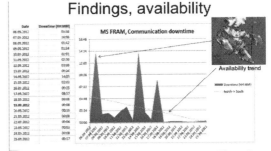

Figure 5. Results from VSAT coverage sailing from Spitsbergen to Greenland and Iceland Sept. 2012.

If e-Navigation is going to be used as a potential solution for polar navigation, communication has to be improved. In an IMO eNavigation workshop in

Haugesund, Norway, results from a sea trial of VSAT coverage done aboard MV "Hurtigruten Fram" in September – coordinated by Nor. Coastal Admin, showed large variations on coverage even within the "A3" area.

Today most SOLAS vessels are under GMDSS "A3" regulations to ensure communication especially for distress situations. For vessels sailing in Polar Regions the GMDSS "A4" regulations will apply. Today this is mostly covered by HF (High frequency – short wave) voice radio carriage, but is subject to effects such as frequency shift throughout the day, atmospheric disturbance and voice/language issues in general. The Iridium system is another available option, but not covering A4 carrier requirements. From the Norwegian hosted workshop, it became clear that both efficient compilations of navigational data, as well as improved communication infrastructure are critical to ensure e-Navigation in the regions.

6 THE WAY FORWARD

As much work are being done in IMO and supporting organizations such as IHO and IALA, with support from the industry, it is clear that navigation in general will change from reaction to changing elements, to better preparation, planning and mitigation of risks. Work is being done to ensure a shared minimum framework in the "S100". It is clear that e-Navigation improve *voyage based operations* through better *voyage planning*. Seafarers will still need to be able to react to changing conditions, but probably technology can help in the processes of risk management and planning both for seagoing and shore side personnel. Information need to be available as early as possible.

Polar navigation represent a new frontier, and will maybe be the ultimate test of e-Navigation solutions for information exchange such as weather and ice forecast, recommended routes, maritime service information and so on.

The industry has already available solutions for efficient data handling, information exchange and management - mitigating risk of system failure or loss of data, and ensuring situation planning and awareness for all marine stakeholders. Be it for the ship side navigators or shore side operators and managers.

Probably the largest challenge lies in communication infrastructure (the "postman") – where joint industrial effort in combination with international collaboration are underway.

Taking all these considerations into account is a huge task. Marine traffic are increasing, and demand for safer and more efficient operations put the whole marine industry under pressure; from the onboard navigators and shore side marine VTS/SAR coordinators, to the Fleet Managers and Authorities.

Simple standards to exchange the knowledge and information between all stakeholders is a necessity and has always been done some way or the other. The change is that e-Navigation will require extended situational awareness. And situational awareness starts with proficient planning and preparation.

That's why "e-Navigation starts with e-VoyagePlanning".

REFERENCES

Bergman, M: *A Harmoniced ENC database as Foundation of Electronic Navigation* in TransNav, the International Journal on Marine Navigation and Safety of Sea Transportation, Vol 5, March 2011.

Bowditch, 1802 (2011) – The American Practical Navigator

IMO, 1999, Annex 24, res. A893 (Guidelines on VoyagePlanning)

IMO, 2007 (Res. A999(25) – Guidelines for passenger ships operating in Remote areas.

IMO, 2009, Res. 1024(26): Guidelines for Ships operating in Polar Waters.

IMO, 2009, SOLAS (Consolidated Edition)

IMO, 2008, MSC 85/26 Annex 20

IMO, 2012, MSC90

IMO, 2012, NAV58 Annex 4-5.

IALA, 2011, Navigation Architecture 'PictureBook' Information Paper

Jeppesen, 2011, ECDIS – What you need to know

IHO, 2010, IHO Universal Hydrographic Data Model, Ed 1.0.0

IHO, 2011, Operational Procedures for the Organization and Management of the S-100 Geospatial Information Registry, Ed 1.0

Kjerstad, Norvald: Fremforing av skip med Navigasjonskontroll – Arktiske operasjoner.

Said, M.H,m A.H Sahruddin: *The Marine Electronic Highway Project in Straits of Malacca and Singapore: Observation on the Present Development*, in TransNav, the International Journal on Marine Navigation and Safety of Sea Transportation, Vol 3, Sept 2009.

Swedish Maritime Administration (SMA): Monalisa information papers

Will Satellite-based AIS Supersede LRIT?

Y. Chen

Shanghai Maritime University, Shanghai, China

ABSTRACT: Recently the satellite-based Automatic Identification System (AIS) system has continuously been developed but also created the debate that whether this system will fully supersede the Long-Range Identification and Tracking (LRIT) system, which is mandatory to be required on certain classes of ships engaged on international voyages to report their position at least every six hours using onboard communication means. This paper, based on the introduction of the satellite-based AIS including the concept and development, the system structure, and the LRIT system, presents the comprehensive comparison between satellite-based AIS and LRIT in terms of the ship's cost, the communication scheme, the monitoring coverage, the information details and the information creditability. The conclusion that the satellite-based AIS should be encouraged to effectively play a complement role to the LRIT system is advanced in the paper.

1 INTRODUCTION

Automatic Identification System (AIS) refers to the technology working on the Very High Frequency (VHF) radio wave that enables the ship's various information such as name, position, type, speed, course, cargo, destination etc. to be automatically exchanged between ship to ship as well as between ship to shore in real time. Since its introduction into the various sectors of the shipping industry, the significant role AIS has played in ensuring the navigational safety, maritime security, marine search and rescue and environmental protection at sea and on shore has been well acknowledged. Nowadays, the use of AIS shore-based station network to capture the ship's information transmitted from the onboard AIS equipment in order to track and monitor those ships close to shore has become a standard practice of the vessel monitoring service for the maritime authorities in many nations. However, this so-called shore station-based AIS ship monitoring system is still limited to satisfy the needs of globally identifying the ships due to the limitation of the VHF horizontal range (approximately 20nm and 100nm for onboard AIS equipment and shore-based AIS station, respectively).

In recent years, the serious situation at sea pertaining to the unlawful acts of piracy, armed robbery and port attack by utilizing ships as a weapon has already drawn many costal states to put the long-range or global ship monitoring system on the top agenda in order to effectively track and identify the ships in the wider horizon and at the earlier time even in the real time. As a result, the ship monitoring system combining the AIS powerful information capacity and the satellite global range detection has been prompted and continued in the progress since the beginning of this century.

2 SATELLITE-BASED AIS

2.1 *Concept and system architecture*

The satellite-based AIS, also referred to as the space-based AIS is of the use of small low orbit satellites carrying the AIS transponder to receive the ship's AIS information from space and relay them to the ground station. Consequently similar to other satellite communication and navigation systems, the satellite-based AIS system consists of five components, i.e. small low orbit satellites in space, shipborne AIS equipment, ground station, user and communication link, see Figure 1.

While the ship's information is automatically exchanged between AIS-equipped ships via VHF communication link, the satellite on which the AIS transponder is installed running on the low earth orbit at the same time is able to receive the VHF signal transmitted for the ship's AIS equipment since the VHF radiowave with the significant signal strength has been proven to be able reach the altitude up to 1000km from the earth surface. The satellite transfers the received VHF signal to the ground station in charge of controlling the whole system. The ground station therefore can distribute the ship's information transferred by the AIS satellite to the authorized user. The communication links between the satellite and the ground as well as between the ground and the user are bi-directional whereas the communication link from the ship to the satellite is uni-directional. Consequently, the satellite-based AIS is capable of globally monitoring the ship's movement in real time if the number of the satellite and the ground station is satisfied. Cain & Meger has shown one of the satellite-based AIS operational results of globally tracking the ships in the paper [1].

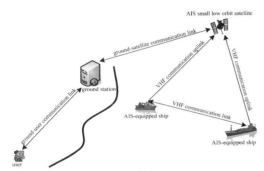

Figure 1. Satellite-based AIS architecture

2.2 Development

The concept of satellite-based has continuously attracted Norway, the United States and Canada etc. to make every effort to carry out the relative research with the great investment since its inception in 2003 [2]. The Radiocommunications and Search and Rescue (COMSAR) subcommittee of International Maritime Organization (IMO) made its debut to receive the proposal regarding the satellite-based AIS in 2005. Subsequently, the IMO Maritime Safety Committee (MSC), Navigation (NAV) subcommittee and COMSAR subcommittee have usually remained the space to discuss the satellite-based AIS topic in the various sessions.

In addition, some commercial companies such as COM DEV and ORBCOMM have gradually and successfully provided the ship's AIS monitoring service based on the satellites. ORBCOMM is scheduled to launch 18 AIS-based satellites and complete the AIS-satellite constellation construction in space over the next few years. Most significantly, International Telecommunication Union (ITU) have allocated two specific VHF Channels (75 and 76) and defined the new Message 27 for the use of the

satellite-based AIS in order to solve the technological difficulty regarding the time slot collision for the shipborne AIS Class A equipment working on the principle of Self-Organized Time Division Multiple Access (SOTDMA) scheme according to the ITU-R M.1371-4 [3]. This new VHF channel allocation for the marine AIS denotes that the shipborne AIS equipment will be upgraded to support the functionality of transmitting the signals to the satellites in the coming years. It is hence estimated that the service provided by the satellite-based AIS will widely accepted by the industry and make the greater contribution in the ship long-range monitoring in future.

3 LRIT SYSTEM

3.1 Mandatory requirement

The Regulation V/19-1 of the International Convention for the Safety of Life at Sea (SOLAS) 1974 coming into effect on January 1 2008 was amended and adopted by IMO in 2006 mandating that ships engaged on international voyages, including all passenger ships, high speed crafts, cargo ships of 300 gross tonnages and above and mobile offshore drilling units are requested to bear the obligation to automatically transmit the Long-Range Identification and Tracking (LRIT) information including their identity, position and date/time of the position to the Flag States at 6-hourly intervals or upon poll requests for an on-demand position report at the interval to a frequency of a maximum of one every 15 minutes. It is therefore quite obvious that the purpose of LRIT system is of providing the Flag State with the global identification and tracking of ships.

3.2 System architecture

The complete LRIT system is comprised of the shipborne LRIT information transmitting equipment, the Communication Service Provider(s), the Application Service Provider(s), the LRIT Data Centre(s), including any related Vessel Monitoring System(s), the LRIT Data Distribution Plan and the International LRIT Data Exchange. The LRIT system architecture is illustrated in the IMO Resolution MSC.263(84) [4].

Each Flag State is required to establish or select a LRIT Data Centre to directly collect the LRIT information transmitted from the ships entitled to fly its flag. The LRIT information is always available to the ship's Flag State while the Data Distribution Plan developed by the IMO in accordance with the Flag State's routing rules and connected with the International LRIT Data Exchange provides the scheme to another Flag State for the valid access of the LRIT information of the ships concerned. The Commercial Service Provider and Application Service Provider play the role of enabling the communication between the satellite and the data centre.

It is also acknowledged that "a robust international scheme for LRIT of ships is an important and integral element of maritime security" and "an active and accurate LRIT system also has potential safety benefits, most notably for maritime search and rescue" [5].

4 COMPARISON BETWEEN SATELLITE-BASED AIS AND LRIT

4.1 *Ship's cost*

The shipborne terminal for the satellite-based AIS is undoubtedly of the AIS equipment onboard. There is therefore no additional cost the ship should currently incur to maintain the normal operation of the satellite-based AIS as long as the AIS equipment has already been installed onboard and operated in normal condition. However, if the new AIS equipment to support the new VHF channel allocation for the use of the satellite-base AIS is required in future, the ship should pay extra expense to upgrade the AIS equipment.

For the LRIT shipborne terminal, the technical means are not specified and it can be any communication terminal on board the vessel that is capable of automatically and on receipt of a specific request from the shore transmitting the ship's LRIT information. In practice, the satellite-based communication mean probably existing Global Maritime Distress and Safety System (GMDSS) equipments such as Inmarsat C terminal with the data-polling service and Ship Safety Alert System (SSAS) is widely applied as the LRIT shipborne terminal on the ships engaged on international voyages. Even though the cost of establishing the LRIT system is mainly borne by the Contracting Government, there is still a cost regarding GMDSS equipment upgrading or standalone equipment installation and testing that ships should incur in ensuring the LRIT equipment on board the ship can respond to the LRIT requirements.

4.2 *Communication scheme*

Subject to the communication scheme between the satellite and the ship, the satellite-based AIS is designed as an uni-directional monitoring system while the LRIT system is maintained as a bi-directional monitoring and communication system. An AIS satellite is able to simultaneously capture all AIS information transmitted from the ships within its footprint but it cannot send any signal instruction to any shipborne AIS equipment within its footprint since there is no communication downlink from the satellite to the ship. From this perspective, the satellite-based AIS is definitely monitoring system.

In contrast, as in most cases the majority of the LRIT shipborne terminals is based on the Inmarsat-satellite mean, each Flag State data center is designated to link to Flag State ship's terminals and vice versa via Inmarsat satellites. In other words, a LRIT ship terminal is able to automatically send information as required to the Flag State data center and the Flag State data center can send signal instruction to a ship because one of most important characteristics for the LRIT system is that the shipborne terminal is able to receive the request instruction transmitted from shore-based data center and make the corresponsive response. The LRIT system is therefore a monitoring and communication system.

4.3 *Monitoring coverage*

As the low-earth-orbit satellites are utilized in the satellite-based AIS, in theory, the satellite-based AIS can globally, including the polar waters, monitor and track the ship's information in real time if the nominal number of the satellites and the ground stations is satisfied. That is why the current satellite-based AIS is able to draw the global ship monitoring picture but still have a little time delay due to too few satellites and ground stations in operation. As for the LRIT system mainly depending upon the Inmarsat Geostationary Orbit satellites for tracking the ocean-going ships, the monitoring geographical coverage is limited to the range between two latitudes of 76° due to the Inmarsat-satellites nominal footprint. Hence, the LRIT system is unable fully to identify and track the ships sailing beyond the range between two latitudes of 76° such as the polar waters.

4.4 *Information details*

Compared to the LRIT system which only three types of ships' information, i.e. their identity, position and date/time of the position are available, the satellite-based AIS has enjoyed quite rich ship's information available since its received information is directly emanated from the any types of onboard AIS equipment. Taking Class A AIS equipment as an example, this equipment is able to exchange four categories of message, static, dynamic, voyage-related and safety-related including more than 20 types of ship's details. Even for the AIS Message 27 which is primarily designed for long range detection of AIS Class A equipped vessels (typically by satellite), it still includes the ship's identifier, position and its accuracy, navigational status, speed and course etc. More ship's monitoring information is available, more choice is also available for more users to manage ships. Moreover, the theoretical capability of the satellite-based AIS receiving the ship's information in real time is also advantageous to the information refresh interval from 15 minutes to 6 hours of the LRIT system.

4.5 *Information credibility*

The LRIT system is established by the international organization and the Contracting Governments from the beginning in order to enhance the navigational safety, security and marine environmental protection. According to the LRIT performance standard, the LRIT information is provided to Contracting Governments and Search and Rescue services entitled to receive the information, upon request, through a system of National, Regional, Cooperative and International LRIT Data Centers. Therefore, the LRIT information can only be used by the governmental organizations for maritime security, safety and marine environmental protection. The confidentiality and sensitivity of the LRIT information are highly stressed by Contracting Governments and not shared with any commercial entities. The LRIT information is of high credibility. However, the satellite-based AIS is launched and developed by some private companies and the ship's information may be widely shared by the commercial users who pay.

5 CONCLUSION

It is obvious that the satellite-based AIS and the LRIT system are developed to provide the service of the ship detection and identification at long range in order to enhance the maritime safety, security, marine environmental protection and the efficient shipping. However, both also have the pros and cons. The satellite-based AIS seems to experience the merits of the ship's global monitoring coverage in theory and the powerful information categories available in contrast to the LRIT system. Nevertheless, the LRIT system is designed as a navigation and communication system and organized by the Contracting Government so that it is able to have a bilateral link between the shore and the ship and enjoy the better information creditability. Additionally, the ship should bear the extra fee to upgrade the current AIS equipment onboard to effectively support the use of the satellite-based AIS. Therefore the satellite-based AIS should not supersede the LRIT system as the LRIT system is more governmental but the satellite-based AIS is focusing on more commercial. And the information provided by the satellite-based AIS is indeed welcome by the industry to promote the efficient shipping, so the satellite-based AIS should be encouraged and developed in order to effectively play a complement role to the LRIT system.

REFERENCES

[1] Cain, J.S. & Meger, E. 2009. Space-Based AIS: Contributing to Global Safety and Security. *ISU 13th Annual Symposium - 'Space for a Safe and Secure World'*. France: Strasbourg.
[2] Wahl, T. & Høye, G. K. 2005. New possible roles of small satellites in maritime surveillance. *Acta Astronautica - ACTA ASTRONAUT* 56(1):273-277.
[3] ITU-R M.1371-4. 2010. *Technical characteristics for an automatic identification system using time-division multiple access in the VHF maritime mobile band.*
[4] IMO Resolution MSC.263(84). 2008. *Revised performance standards and functional requirements for the long-range identification and tracking of ships.*
[5] Popa L.V. 2011. Ships Monitoring System. *International Journal on Marine Navigation and Safety of Sea Transportation*, Vol. 5, No. 4:549-554

Chapter 4

Maritime Simulators

New Concept for Maritime Safety and Security Emergency Management – Simulation Based Training Designed for the Safety & Security Trainer (SST₇)

C. Felsenstein, K. Benedict & G. Tuschling
Hochschule Wismar, University of Applied Sciences – Technology, Business and Design, Department of Maritime Studies & Maritime Simulation Centre Warnemuende (MSCW), Germany

M. Baldauf
World Maritime University Malmö (WMU), Sweden

ABSTRACT: Nowadays more attention must be paid to maritime safety and security challenges. Considering the evidence of accident statistics and that more than 80% of incidents are caused by the human factor so adequate Maritime Emergency Management both on vessels and ashore is essential and must be reviewed in detail. Real time simulators have proved beneficial for ship handling training on well equipped bridges over the last decades. A new simulator has been developed for training and researching specific aspects of Maritime Safety and Security. This simulation system has a design concept developed in 3 D visualization created by Rheinmetall Defence Electronics (RDE) of Bremen in cooperation with Wismar University (HSW) and the Maritime Simulation Centre Warnemuende (MSCW). The MSCW has recently master-minded a new type of simulator called the Safety and Security Trainer (SST7). The innovative concept of the SST covers procedures necessary for integration into the complex environment of full mission ship-handling- and ship-engine simulators and for emergency management training with a complex simulation platform at the MSCW.

1 INTRODUCTION

The SST simulator is dedicated to the "Enhancement of passengers' safety on RoRo-Pax-ferries" and was developed with the support of the Ministry of Education and Research (BMBF). The outcome of this project was so successful that HSW plan a follow-up research project covering Safety and Security over the next three years under BMBF guidance.

An integrated support and decision system, called MADRAS, has been interfaced into the SST to assist officers coping with safety and security decisions during vessel manoeuvres.

The SST simulator is designed with an interface both to the Ship Handling Simulator (SHS) as well as the Ship Engine Simulator (SES). Two types of vessels, a RoPax ferry and a 4500 TEU Container vessel have been modelled in detail for the SST. With the bi-directional system, complex simulation training can be practiced at the MSCW. The entire complex ship is now available for training, the ship handling process on bridge (SHS) combined with engine processes from the engine room (ER) and engine control room (ECR) as well as emergency management and procedures inside the vessel (fire fighting, water inrush and other measures according to ISPS. Functional tests developed for the system

are in progress and running successfully. Two comprehensive training courses with the SST in combination with SHS and SES were carried out in 2011 at the MSCW and successfully tailored for the needs of a large international shipping company for more than 140 masters, nautical- and technical officers. Another course in Emergency Management for Safety and Security is in preparation and will be carried out beginning of 2013 at the MSCW.

2 SIMULATION PLATFORM AT THE MARITIME SIMULATION CENTRE WARNEMUENDE

The Maritime Simulation Centre Warnemuende (MSCW) is one of the most modern simulation centers worldwide encompassing a full mission Ship Handling Simulator (SHS), Ship Engine Simulator (SES) and a Traffic Simulator (VTS) as well as a new type of simulator called the Safety and Security Trainer (SST). This complex simulation platform with four full mission simulators enables the trainee to simulate the entire system ship and offers concrete challenges to officers and crew on board (Figure 1). The simulator arrangement (MSCW) comprises

- a Ship Handling Simulator SHS with 4 Full Mission bridges and 8 Part Task Bridges,
- a Ship Engine *Simulator* SES with 12 Part Task station and
- a Vessel Traffic Services Simulator VTSS with 9 operator consoles
- a Safety and Security Simulator with 10 operator consoles

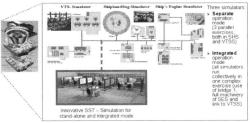

Figure 1. Maritime Simulation Centre Warnemuende (MSCW) – building and structure

2.1 *Integrated use of Safety and Security Simulation for training and research*

The new SST simulator was designed by the manufacturer Rheinmetall Defence Electronics Bremen (RDE) in co-operation with Wismar University, Department of Maritime Studies. The simulator was originally designed in a basic 2D version and is now being developed into a 3D interface. The simulator can be specifically used for "stand-alone" exercising as well as for exercises incorporating both the SHS and SES. Together with the full training material set-up, and including all ships safety plans, it was introduced as the mars[7] concept [1], [2]. The simulation system can be applied in specific simulation based studies and has the potential to help in upgrading existing safety and security procedures in training.

In shipping the situation regards emergency preparedness is generally affected by the following problems:

- crew capability and experience in the event of „disturbed" operation on vessels is limited or even non-existent
- multi-lingual crews cause communication problems in an emergency situation
- reduction of crew members causes lack of personnel available
- complexity of emergency equipment is permanently increasing, but training in emergency handling is not on a par with these developments
- new management systems and regulations of the IMO (ISM/ISPS) mean that new methods in technology for emergency training are necessary. HSW offers simulation based training courses in safety and security at varying levels of complexity, for ratings at a basic level, for

officers and masters at management level – all in accordance with IMO standards.

2.2 *Integration of a new 3D- visualization model into the SST*

One of the most innovative elements at the MSCW is demonstrated on the new three-dimensional implementation of a RoPax-ferry (FS "Mecklenburg-Vorpommern") into the SST-simulator. The 3D-model application has been created according to the relevant ship's safety plans and closely adheres to a series of photo sessions from the vessel used to design within the software system "3D studio – max". For the simulator safety training all available safety equipment on board and safety systems (e.g. CO_2, sprinkler system and water drenching system) have been drafted into the 3D visualization. Figure 2 and Figure 3 illustrate the ship's plan and bridge of M/V with interactive training consoles in 3D visualization.

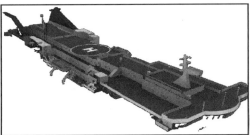

Figure 2. Visualization 3D M/V deck 9

Figure 3. Visualization 3D bridge M/V

In addition to the RoPax ferry another complete model of a container vessel, type CV4500, was drawn up as well as a part task model of the passenger vessel "AIDAdiva". Figure 4 and Figure 5 present sample of visualization of the part task model of the passenger vessel "AIDAdiva" - ship's bridge and engine control room (ECR) with interactive training consoles.

Meanwhile three different types of vessels have been drafted for complex simulation based Safety and Security training. The modelling process has achieved to the highest standard of detailed reality and enables efficient handling of all safety equipment and -systems on board and took six months for each vessel.

Figure 4. Visualization Bridge AIDA

Figure 5. Safety-/security console AIDA

2.3 *Simulation based modules and system for Safety & Security Training*

Generally the SST is designed for procedure training in emergency management. Two modules have been integrated into the SST, a complete fire and fire fighting module as well as a water inrush module.

The fire model (visually adapted) has a module with a number of realistic effects for easy orientation incorporated into the simulation. A modern fire alarm management system with smoke detectors and manual calling points is built into the ship's interior and easily flammable materials are protected by fire resistant A60 walls and doors. This model includes smoke visualization, a fire fighting system with equipment such as fire extinguishers, water hoses and hydrants, breathing apparatus, CO_2 systems and foam. This enables the trainee to simulate a realistic fire fighting situation and interact with support teams as well as the management teams on the bridge and in the engine room. During the simulation a strategic figure's health condition is monitored in regarding oxygen, smoke, temperature and other health influencing parameter. Both modules see Figure 6 / Figure 7.

Figure 6. Fire module in SST (flash over)

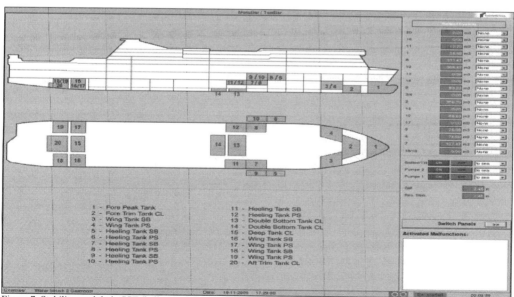
Figure 7. Stability module in SST (ballast tanks)

One further feature of the SST[7] is the module for calculating water inrush and its influence on ship stability. A water ballast system is included and can be called upon during simulation of an emergency in order to stabilize the ship. The trim and stability calculator is adjusted to predict the effect of a water inrush and show the stability parameter. Water-tight doors are built into the modelled vessel. The ballast and stability measuring system can be implemented on the simulator prompting the trainee to take the appropriate counter measures.

A graphic display with selected environmental parameter (temperature, oxygen, gas and other parameter) enables the instructor to control the exercise and evaluate the trainee's performance presenting the results after simulation in the replay mode. For implementation of specific scenarios according to emergency management procedures on board, it was necessary to provide a complex process simulation system with a bi-directional interface for the safety simulator and ship-handling simulator. With these features and combining all simulator resources at the MSCW (SHS/ SES/ VTSS/ SST) it is now possible to visualise the entire complex system ship and to provide training in ship handling and engine simulation processes in 3D quality for most simulation processes in safety and security [3].

2.4 *Decision Support System MADRAS*

The simulation platform includes a new support and decision system called MADRAS. The system was tailored for the SST simulator and superimposes the sensor data from the SST. The control module selection contains the following elements for automatic survey: FIRE, EXPLOSIVES, SECURITY, EVACUATION, GROUNDING and FLOODING. In the event of any sensor alarm the Madras menu opens and displays the affected deck/area with the activated alarm sensor. MADRAS is an interactive system and a helpful tool in critical situations for the Master. Both the SST simulator and the MADRAS system were successfully developed and tested over the last three years within the context of the research project VeSPer (funded by the German Ministry of Education and Research). First pilot courses have been carried out for end-users at the MSCW and are presented in the following chapter.

3 APPLICATION OF TRIAL COURSES ON THE SAFETY & SECURITY TRAINER

In co-operation with the shipping company F.LAEISZ an introductory simulation safety course was held in 2010 and two further trial courses carried out at the MSCW at 2011.

3.1 *Introductory course for Shipping Co. F.LAEISZ on RoPax TRANSEUROPA in 2010*

The aim of the first training course on board was to introduce the Safety and Security Trainer and to implement simulation courses on the SST generally. The scenario of choice for this simulation was a fire emergency on a RoPax ship using the available fire extinguishing equipment available (CO_2, foam, water drenching). The main objective was to offer emergency procedure practice for the officers, crew and service personnel especially measures needed for communication and the evacuation of passengers.

During the 7-day trip several courses were given to the complete crew and finally a "dry training" was carried out on board, mirroring simulation training at the monitor. The result was that the Captain and his crew were able to appreciate a real improvement in the standard of the dry exercise after their experience from the simulation. The company then booked two further demonstration courses at the MSCW during their ships management courses organized by the Warnemuende Technical Academy (WTA) in 2011.

3.2 *Trial courses for F. LAEISZ Shipping Co. at MSCW in 2011*

Company specific emergency scenarios were chosen for the demonstration courses to F.LAEISZ' specifications and were simultaneously run together with the SHS and a second trial included the SES (engine simulator). The courses were tailored to improve emergency management organization on board. The courses at the MSCW with more than 60 participants per course were organized and conducted by MSCW staff in conjunction with a student team and in co-operation with network partners ISV and MARSIG, tailored to requirements of the shipping company (Figure 8). The training was conducted as recommended by the STCW Convention, Manila Amendments and developed and using the required Standard Marine Communication Phrases (SMCP), [4] - [7].

Figure 8. Briefing SHS, bridge 1

The Emergency Management Course was carried out using prepared scenarios. As a sample the schedule of a fire scenario is described starting with a tailored emergency plan for the CV 4500 prepared for trainees to follow exactly the safety regime during the exercise. Standard materials provided to each trainee when performing simulation exercises. Event Schedule Fire Auxiliary E.R. – training carried out acc. to Emergency Plan (Figure 9).

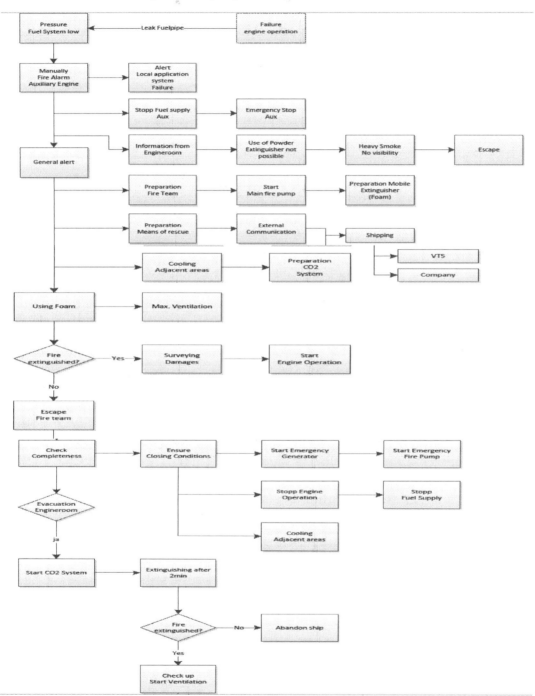

Figure 9. Concept Emergency Plan

3.2.1 *Scenario description:*

On board the container vessel CV4500, Pos. Singapore Strait westbound, loading condition C6 (Hand Out), break out of fire after oil leakage in generator engine room deck 04 PS forward. Ship/shore communication SST carried out by VHF channel 13. Fire alarm is indicated on the alarm panel on the Bridge (SST & MADRAS-system) and ECR. Internal communication held between Bridge and ECR (Master/ 2nd Officer/ Ch.Eng. and 2nd Eng.) with reference to preparation of fire fighting team with protective equipment (breathing apparatus, heat protection suit). Check that fire dampers are closed. Ventilation of affected area cut off to prevent fire spread.

From control point ECR follows advice to start fire fighting. Simultaneously bridge starts preventive evacuation of superstructure while preparation of water and foam supply begins (e.g. fire hoses). VTIS Singapore has to be informed immediately after fire break out via VHF channel 13. VTIS gives order to leave TSS in north direction and for anchorage at „E-Boarding Ground". After the fire fighting team is set up (report ECR to Bridge) fire fighting begins in the engine room. Due to fire spreading intensely (as simulated based physical model indicates) further measures are necessary, e.g. fire fighting with foam. After failure of bilge pump 1 respectively fire pump 1 (malfunctions) start of replacement pumps and repair work (ECR and SST).

Due to intense fire spread the control point (ECR) gives order to the bridge to start evacuation. The fully equipped crew sent to assembly station EGR aft of superstructure. After evacuation is completed in alignment with communication and evacuation procedures according to Bridge Resource Management (BRM) and Crisis Management, crew gathers at assembly station with personal protective suit/life vest. Crew roll call made by 3rd NO and the master gives order to release CO_2. In the event of missing persons a search team is sent through all decks. After a delay of two minutes the engine room is doused with CO_2. With the graphical model and Fire Editor (SST) as well as from the MADRAS working station (ECR) several parameters (temperature, fire spreading, fire fighting) can be checked and adapted. After report back of „fire is extinguished" and an adequate time lapse (20min) the area is checked by a fully equipped fire fighting team (after ventilation of area CO_2), before the signal "all clear" can be given. Further fire watches must be set up.

Table 1. Handout F.LAEISZ 2011

CV 4500 SST combined exercise SST&SES

Chapter	Arrangement	Use stations
Stations	1 Instructor + 6 SST workstations incl. Communication	4 ECR/ ER workstation
Objective	Management + fire fighting measures after fire outbreak D04 PS GR room forward, POS: Spore Strait, W-bound	ECR camera view D04 PS, control of fire fighting, supervisory EUS – system
affected area	DO4 Cell 077, material oil, amount 50kg, extinguish with Water/Foam In case of uncontrolled fire propagation preparation for release CO_2 and fore evacuation measures	ECR workstation camera view. MADRAS: Decision menu Fault tree analysis acc. Bridge management
Documents for Trainees	Safety plan CV4500, ship's particulars, loading and stability manual, ISM/ISPS code, MADRAS manual	
Navigation staff, pos.	Master – SST8 OAC 1. NO – OSC SST 7 2.NO – external communication 3. NO Support SST 6 Support staff distributed to SST stations	EUS- System MADRAS: EUS Fault tree analysis– Malfunctions Leakage D00 PS after 15s. Failure bilge pump1 at 30 cm water level.
Technical staff, pos.	C.E. Command ECR (CO_2 release) 2. TO – Deputy 3. TO – FF Leader SST 5 EL.Eng. – Supp. ECR Fitter fire fight SST2	Failure fire pumps at temp. cell 77 >800°. Failure ME after 10 min. Contr. alarms, ECR/ SST: Crisis Management &evacuation
Graphical display of parameter	Temp. cell 77/ 67 Smoke C.77/ 67 CO_2 C.077 WI - Water Level ER, cell 067/ 077	
Briefing:	Familiarization SST: movement algorithm, orientation, safety arrangement, evacuation, communication, criterion malfunction parameter MADRAS/ SES	
Exercise duration:	2X45 min. = 90 min. Recording of exercise	
De-briefing:	Replay / Evaluation of exercise, Assessment	

Table 2. Master tree – EUS system MADRAS

MADRAS – Master Tree		
Investigation situation on site		
Feedback – extension of fire, what is on fire?		
Yes ⬅ Fire extinguished? ➡		No
Clear location		Fire alarm
Fire watch		Closing procedures
Ventilation		Protection suits
		Communication intern, extern to be secured
		Feedback readiness
		Preparation of measures outside location
		Control of fire parameter MADRAS
		Emergency power supply to be secured
	Proceeding Fire fighting team	
Yes ⬅ Fire extinguished? ➡		No
Evaluation on site		General Alarm
Clear up location		All closing in ECR/ER and superstructure ?
Fire Watch		All crew definitely evacuated?
Ventilation		Release CO$_2$/ Close ventilation
		Monitoring fire extension and CO$_2$ by graphic display SST
Yes ⬅ Fire extinguished? ➡		No
Control fire process: Enter affected area not before 20 min. (protection, CO$_2$ detector)		Secure Communication intern, extern
Clear up location		Start with evacuation procedure
Fire Watch		All crew on Station, well equipped?
Ventilation	Further procedures advice from bridge	

Tab. 1 gives an overview about an emergency combined simulation exercise tested at the SST/ SES, Tab. 2 illustrates an overview about the integrated decision support system MADRAS. Both tables give the trainees support and are used as a guideline to follow up during the exercise procedure.

The scenario is designed for six SST-working stations and as combined exercise with the SES-ECR and ECR- MADRAS working station (five stations). The decision supporting system MADRAS was tested for its effectiveness and practicality during the exercise.

3.2.2 Briefing:

The participants are introduced to the SST as well as to the other work stations (SES). The varying functions and operation algorithms of the strategic SST roles are explained and practiced. The trainees are shown different operational options with their respective parameter (walking, running, turning, crouching etc). The Madras system is sketched out. After the participants' familiarisation with the equipment they are given an outline of the scenario. The intention is to prepare them for specific emergencies that they may encounter on their own ships and to help them achieve awareness of the specific stress problems ensuing. The training is aimed at management level whereby organization, communication, emergency measures and procedures can be realistically rehearsed. Each trainee is given his own specific part on the training platform.

3.2.3 Simulation:

Each trainee is placed at his specific work station (6 SST workstations, 4 SES/ ECR work stations, 1 MADRAS station in ECR). Internal communication via headphones and microphone or on board radio or following bridge commands via public address system. The instructor surveys the parameter windows and adjusts where appropriate. Malfunctions, e.g. non-function of bilge- and fire pumps, have been purposely added into the scenario to be dealt with by the trainees self-sufficiently. The procedure is exercised according to precedents listed in the ship's articles (Muster List) and as given by the central command (Bridge/ ECR). Communication on board and external communication [6], organization/ procedures for using safety equipment and -systems (CO$_2$ system) found as indicated in the safety plan. After the fire has been successfully quenched and the area examined (after proper ventilation) the simulation exercise is complete.

3.2.4 De-Briefing:

During de-briefing the trainee performance is individually evaluated as well as the team co-operation. During the replay of exercise unsatisfactory passages of the exercise, such as orders not correctly interpreted, may be repeated. Extra emphasis is put on adequate communication skills (internal and external). Communication with VTIS, with other ships as well as with the Shipping Co. must conform to STCW standards and Standard Marine Communication Phrases (SMCP) – "Emergency reporting systems". The MADRAS system was tested for its effectiveness and practicality. The data-base stored parameter and processes of the fire fighting procedures were recorded and evaluated in replays together with the participants. The analysis (Tab. 3) shows the only complaint was the fact that there was not enough time given to fully implement such a complex simulation exercise. This will be taken into consideration at further courses in 2013. In summary the SST was well accepted especially at management level.

Table 3. Evaluation SST- Training Course 2011

Course:	LAEISZ - Ship Management Seminar May 2011	
	Number of participants:	56
	Number of courses:	4
	Number of questions:	16
	Questions	Average Evaluation
Number of participants	How do you evaluate....	
1	the organisation and support service of the participants?	1,4
2	the handouts?	2,1
3	the briefing including the performance requirements?	1,7
4	your familiarity with the handling of the SST after the briefing?	1,9
5	the ease of operation of the simulator during the exercises?	2,5
6	the hook up of the SST with the ship handling simulator?	1,7
7	the reality of the simulator as an emergency procedure trainer?	1,8
8	the time limit of the simulation exercises?	3,3
9	the SST as an additional training measure?	1,5
10	the fixed installation of the SST inside the MSCW?	2,1
11	the modeled equipment and the safety systems?	2,1
12	the fire model?	2,1
13	the water inrush model?	2,4
14	the communication system of the SST?	2,5
15	the usage of the SST at management level?	1,1
16	the usage of the SST at basic level?	2,3
	Total Evaluation	**1,9**

Evaluation better than 1,5	
Evaluation in-between 1,5 and 3,0	
Evaluation below 3,0	

4 CONCLUSION

The safety and security trainer has provided new impulses for ship's security while dealing with dangerous accidents in the civil maritime field. New ideas from the analytical examination within the research project can be helpful in the future development of the SST encouraging improved methods for the integration of security measures for safer life and safety awareness on board.

The development and application of simulation-based training supports not only optimization of emergency management training but also improves team performance and collaborative learning [8] as well. Furthermore the sophisticated simulation allows even for the identification of unwanted effects or unforeseen impacts [9] of drafted emergency plans.

Some of the research results presented in this paper were partly achieved during the research project "Enhancement of passengers' safety on RoRo-Pax-ferries" (VeSPer & VeSPer Plus) funded by the German Federal Ministry of Education and Research (BMBF) and supervised by „VDI Technologiezentrum GmbH, Projektträger Sicherheitsforschung".

REFERENCES

[1] C. Felsenstein, K. Benedict (2012): New Concept for Maritime Safety and Security, Simulation based Training designed for the Safety & Security Trainer, 17th INSLC Conference Warnemuende, 3rd to 7th September 2012 (MSCW), Germany

[2] K. Benedict, Ch. Felsenstein, O. Puls, M. Baldauf (2011): Simulation for Navigation Interfacing Ship Handling simulator with Safety & Security Trainer (SST) in A. Weintritt: Navigational Systems and Simulators. pp 101-108, Taylor & Francis, ISBN 978-0-415-69113-0

[3] K. Benedict, Ch. Felsenstein, O. Puls, M. Baldauf (2011) New level of Integrated Simulation Interfacing Ship Handling Simulator with Safety & Security Trainer (SST). TransNav – International Journal on Marine Navigation and Safety of Sea Transportation. Vol 5, No. 1, pp. 105-110

[4] IMO (2010). The Manila Amendments to the annex to the International Convention on Standards of Training, Certification and Watch keeping for Seafarers (STCW), Manila, 2010

[5] Felsenstein, Ch., Benedict, K., Baldauf, M. (2009): Development of a Simulation Environment for Training and Research in Maritime Safety and Security. Journal of Marine Technology and Environment, Editura Nautica 3(2), 2010, pp. 77-89

[6] International Convention on Standards of Training, Certification and Watch keeping for Seafarers, 1978, as amended in 1995 (STCW Convention), and „Seafarer's Training, Certification and Watch keeping Code (STCW Code)", International Maritime Organization (IMO), London, 1996

[7] C. Bornhorst (2011) Safety & Security Trainer SST$_7$ – A new way to prepare crews managing emergency situations. in A. Weintritt: Human Resources and Crew Resource Management. pp 117-121, Taylor & Francis

[8] E. Rigaud, M. Lützhöft, A. Kircher, J.-U. Schröder-Hinrichs, M. Baldauf, J. Jenvald, T. Porathe (2012) Impact: More Than Maritime Risk Assessment. Procedia - Social and Behavioral Sciences, Elsevier, Volume 48: 1848 – 1854

[9] Prasad, R.; Baldauf, M.; Nakazawa, T. (2011), Collaborative Learning for Professional Development of Shipboard Engineers. International Journal of Engineering Science & Technology; 2011 , Vol. 3 Issue 3: 2308-2319

Advanced Ship Handling Using Simulation Augmented Manoeuvring Design and Monitoring – a New Method for Increasing Safety & Efficiency

K. Benedict, M. Kirchhoff, M. Gluch, S. Fischer & M. Schaub
Hochschule Wismar, University of Applied Sciences – Technology, Business and Design, Warnemünde, Germany

M. Baldauf & S. Klaes
World Maritime University, Malmö, Sweden

ABSTRACT: New concepts for on board displays and simulation tools were developed at Maritime Simulation Centre Warnemuende MSCW. A fast time simulation tool box is under development to simulate the ships motion with complex dynamic models and to display the ships track immediately for the intended or actual rudder or engine manoeuvre. The "Simulation Augmented Manoeuvring Design and Monitoring" - SAMMON tool box will allow for a new type of design of a manoeuvring plan as enhancement exceeding the common pure way point planning and an unmatched monitoring of ship handling processes to follow the underlying manoeuvring plan. During the manoeuvring process the planned manoeuvres can be constantly displayed together with the actual ship motion and the predicted future track. This future track is based on actual input data from the ship's sensors and manoeuvring handle positions. This SAMMON tool box is intended be used on board of real ships but it is in parallel an effective tool for training in ship handling simulators: (a) in the briefing for preparing a manoeuvring plan for the whole exercise in some minutes, (b) during the exercise run to see the consequences of the use of manoeuvring equipment even before the ship has changed her motion and (c) in debriefing sessions to discuss potential alternatives of the students decisions by simulating fast variations of their choices during the exercises. Examples will be given for results from test trials on board and in the full mission ship handling simulator of the Maritime Simulation Centre Warnemuende.

1 INTRODUCTION

Within this paper investigations into the feasibility and user acceptance of the new layout of navigation display will be introduced and selected results of simulation studies testing the influence on manoeuvre performance dependent on different kind of prediction functions will be discussed. Examples will be given for results from test trials in the full mission ship handling simulator of the Maritime Simulation Centre Warnemuende.

Normally ship officers have to steer the ships based on their mental model of the ships motion characteristics only. This mental model has been developed during the education, training in ship handling simulator in real time simulation and most important during their sea time practice. Up to now there was nearly no electronic tool to demonstrate manoeuvring characteristics efficiently or moreover to design a manoeuvring plan effectively - even in briefing procedures for ship handling training the potential manoeuvres will be explained and drafted on paper or described by sketches and short explanations. To overcome these shortcomings a fast time simulation tool box was developed to simulate the ships motion with complex dynamic models and to display the ships track immediately for the intended or actual rudder or engine manoeuvre. These "Simulation Augmented Manoeuvring Design and Monitoring" - SAMMON tool box will allow for a new type of design of a manoeuvring plan as enhancement exceeding the common pure way point planning. The principles and advantages were described at MARSIM 2012 (Benedict et al., 2012) specifically for the potential on board application for manoeuvring real ships. This holistic approach goes beyond the prediction tool mentioned e.g. in Källström et al. 1999 and Wilske & Lexell 2011.

This paper presents the potential of the new method to be used on board and for the teaching and learning process at maritime training institutions.

Manoeuvring of ships is a human centred process. Most important elements of this process are the human itself and the technical equipment to support its task (see Figure 1).

However, most of the work is to be done manually because even today nearly no automation support is available for complex manoeuvres. Even worse, the conventional manoeuvring information for the ship officer is still available on paper only: the ship manoeuvring documents are mainly based on the initial ship yard trials or on some other selective manoeuvring trails for specific ship / environmental conditions - with only very little chance to be commonly used in the overall ship handling process situations effectively.

Ship Handling Simulation for simulator training has a proven high effect for the qualification, however, it is based on real time simulation, i.e. 1s calculation time by the computers represents 1s manoeuvring time as in real world. This means despite all other advantages of full mission ship handling simulation that collecting/gathering of manoeuvring experiences remains an utmost time consuming process.

For increasing the effectiveness of training and also the safety and efficiency for manoeuvring real ships the method of Fast Time Simulation will be used in future – Even with standard computers it can be achieved to simulate in 1 second computing time manoeuvres lasting about to 20 min using innovative simulation methods. This allows substantial support in both, the training process and the real manoeuvring process on board ships. A comparison is given in Figure 2 for some essential elements of the real manoeuvring process on ships and in training within the ship handling simulators as well. Additionally, in the right column of Figure 2 some of the Fast Time Simulation (FTS) tools are mentioned and their roles to support each element of the manoeuvring process are indicated: These tools were initiated in research activities at the Maritime Simulation Centre Warnemuende which is a part of the Department of Maritime Studies of Hochschule Wismar, University of Applied Sciences - Technology, Business & Design in Germany. It has been further developed by the start-up company Innovative Ship Simulation and Maritime Systems (ISSIMS GmbH 2012).

2 DESCRIPTION OF THE CONCEPT

2.1 *Fast Time Simulation Modules*

A brief overview is given for the modules of the FTS tools and its potential application:

– SAMMON is the brand name of the innovative system for "Simulation Augmented Manoeuvring – Design, Monitoring & Control", consisting of software modules for Manoeuvring Design & Planning, Monitoring & Control based on Multiple Dynamic Prediction and Trial & Training. It is made for both:

– application in maritime education and training to support lecturing for ship handling to demonstrate and explain more easily manoeuvring technology details and to prepare more specifically manoeuvring training in SHS environment, i.e. for developing manoeuvring plans in briefing sessions, to support manoeuvring during the exercise run and to help in debriefing sessions the analysis of replays and discussions of quick demonstration of alternative manoeuvres and

– application on-board to assist manoeuvring of real ships e.g. to prepare manoeuvring plans for challenging harbour approaches with complex manoeuvres up to the final berthing / unberthing of ships, to assist the steering by multiple prediction during the manoeuvring process and even to give support for analysing the result and for on board training with the Simulation & Trial module.

– SIMOPT is a Simulation Optimiser software module based on FTS for optimising Standard Manoeuvres and modifying ship math model parameters both for simulator ships and FTS Simulation Training Systems and for on board application of the SAMMON System.

– The Advantage and Capabilities of this software is: The Math Model reveals same quality for simulation results as the Ship handling simulators SHS, but it is remarkably faster than real time simulation, the ratio is more than 1/1000, the steering of simulator vessels is done by specific manoeuvre-control settings / commands for standard procedures and individual manoeuvres dedicated for calculation standard ship manoeuvring elements (basic manoeuvres) but moreover for the estimation of optimal manoeuvring sequences of some characteristic manoeuvres as for instance person over board manoeuvres.

– SIMDAT is a software module for analysing simulation results both from simulations in SHS or SIMOPT and from real ship trials: the data for manoeuvring characteristics can be automatically retrieved and comfortable graphic tools are available for displaying, comparing and assessing the results.

The SIMOPT and SIMDAT modules were described in earlier papers (Benedict et al. 2003 and 2006) for tuning of simulator ship model parameters and also the modules for Multiple Dynamic Prediction & Control (Benedict et al. 2009) for the on board use as steering assistance tool.

In this paper the focus will be laid on the potential of the SAMMON software as an integrated system for planning and monitoring of manoeuvres as well as a tool to be used on board and for supporting the teaching and learning process.

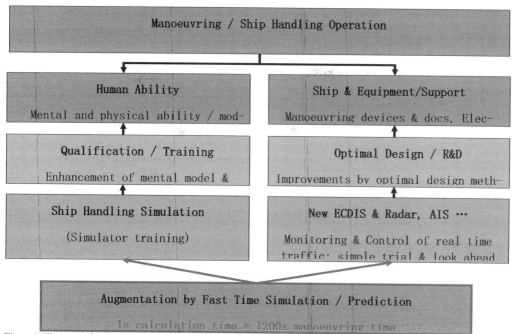

Figure 1. Elements of the manoeuvring process and potential for enhancement by new Simulation / Augmentation methods

Elements of Manoeuvring Process on Ships & in Education/Training and support by Fast Time Simulation Modules / Tools		
Real World/ Ship Operation	**Ship Handling Simulator Training**	**Fast Time Simulation Tools**
Real ship /	Math model of the ship for simulation	SIMOPT & SIMDAT tool for developing & tuning of parameters of math models
Familiarisation runs	Familiarisation Exercises	MANOEUVRING TRIAL & TRAINING tool for Demonstration / Lecturing / Familiarisation
Mission / Planning	Scenario / Briefing	MANOEUVRING DESIGN & PLANNING tool to generate and edit a manoeuvring plan
Manoeuvring Operation	Execution of exercise	MANOEUVRING MONITORING & MULTIPLE DYNAMIC PREDICTION tool to monitor and control the vessels motion
Recording (VDR, ECDIS)	Recording by simulator	SIMDAT tool to display and assess recordings
Evaluation of success	Debriefing	MANOEUVRING TRIAL & TRAINING tool for verification of results by simulation & prediction

Figure 2. Elements of Manoeuvring Process on Ships & in Training and support by Fast Time Simulation Tools for Simulation / Augmentation

2.2 Fast Time Manoeuvring Simulation for Manoeuvring Support in ECDIS environment

The core modules of the fast time simulation tools can be used for the calculation of manoeuvres up to the design of complete manoeuvring plans in ECDIS environment. Some basic functions are shown in the next figures.

Figure 3 explains the operational interface in a sea chart environment which combines the electronic navigational chart ENC window (centre), the status of the current actual ship manoeuvring controls (left) and the interface window for the steering panel of the ship (right).

The ship was positioned in a certain place to demonstrate the ships motion for a very simple manoeuvre kick turn from zero speed. The ships motion can be controlled by the settings in the control panel window where any manoeuvre can be generated to be immediately displayed in the ENC in one second with full length. The length of the track corresponds to the settings in the prediction window (left top corner): the range value represents the duration of the manoeuvre; the interval value controls the number of displayed ship contours on that manoeuvre track. The sample represents a kick turn from zero speed to full ahead with full rudder to Port.

3 FAST TIME SIMULATION FOR DESIGNING MANOEUVRING PLANS

3.1 Principle of fast time simulation of manoeuvres in ECDIS and sample data

The fast time simulation method is used to find out efficient manoeuvres and even more for the design of manoeuvring plans within the briefing for Ship Handling Simulator exercises and practically for route planning process on board (Benedict et al. 2012). The use of this tool will be explained by some sample scenarios:

The sample ship is the RO-PAX Ferry "Mecklenburg-Vorpommern" with Loa=200m, Boa=28.95m, Draft=6.2m, Displacement=22720t and Speed=22kn. She has two pitch propellers and two rudders located behind the propellers and additionally one bow thruster.

The test area is the Rostock Sea Port. The RO-PAX ferry is entering the fairway from north to be steered through the fairway and to be turned at the turning area followed by astern motion to the berth at west pier (as in in the sample Figure 7).

For purposes of demonstration of a complex manoeuvre procedure the ship is initially positioned in the fairway (black contour) and is going to enter the turning area as objective for the first manoeuvring segment. For the planning procedure the ship's motion can be controlled by the settings in the control panel window on the right side. Any manoeuvre can be generated and will be immediately displayed in the ENC in less than one second with full length. In this case the rudders are set 10° to STB to achieve a small turning rate ROT=4.5°/min to port. The length of the simulated track corresponds to the settings in the prediction window (left top corner): the range value represents the duration of the simulated manoeuvre and that means the track length of that manoeuvring segment; the interval value controls the number of displayed ship contours on that predicted manoeuvre track. The selected end position of the manoeuvring segment is indicated by the red ship's contour. Its position can be shifted and adjusted using the slider at the bottom line which is adjusted to 165 seconds after the beginning of the manoeuvre at initial Manoeuvring Point MP 0. If this position is accepted it will be acknowledged as the next manoeuvring point MP 1.

This planning process guarantees the full involvement of the navigating officer: The best version of the manoeuvres can be found by trial and error but it is possible to bring in one's full knowledge and to take advantage of one's skills – it is possible to see and to verify immediately the results of the own ideas and to make sure that the intentions will work. This is import for safety and efficiency, but also for gaining experience for future manoeuvres.

3.2 Sample of designing a full manoeuvring sequence as training concept

The planning procedure for a complete manoeuvring plan follows the principles as described for a single segment in Figure 4 as follows: Figure 5 presents the situation after accepting the manoeuvre previously planned – now the next segment is to be planned from MP 1 to MP 2: the ship is going to enter the turning area and to slow down. Both engines are set to STOP (EOT 0). In Figure 6 the complex turning manoeuvre is to be seen: the ship is using in parallel engines, rudders and the bow thruster to turn as fast as possible. Afterwards the engines have to be reversed and the ship controls are adjusted to go astern to the berth. In Figure 7 the result for the full manoeuvring plan is to be seen with the whole set of Manoeuvring Points (MP) for the complete approach and the berthing manoeuvre.

The different settings of the controls and the track of the planned manoeuvre sequences are stored in a manoeuvre planning file to be displayed in the ENC.

For the execution of the manoeuvre this plan can be activated later to be superimposed in the ECDIS together with the actual position of the ship and, most important, with the prediction of manoeuvring capabilities for effective steering under the actual manoeuvring and environmental conditions.

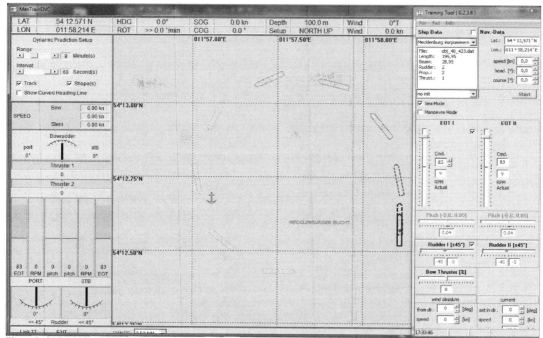

Figure 3. SAMMON Trial & Training Tool Interface with sample for Kick turn with rudder Hard PT from zero speed to EOT Full Ahead port as sample for potential ships manoeuvring capabilities

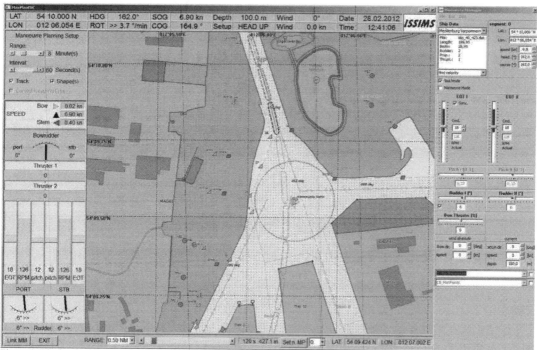

Figure 4. Display for Manoeuvring Design by Fast Time Simulation for immediate presentation of manoeuvring results: Sample for entering the turning area with slight turning to STB from initial conditions in a fairway at initial Manoeuvring Point MP 0

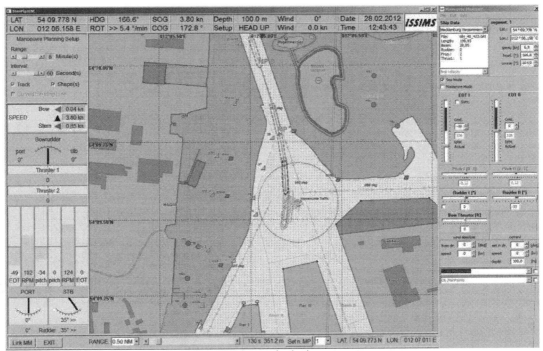

Figure 5. Planning of the next segment from MP 1 to MP 2 – speed reduction

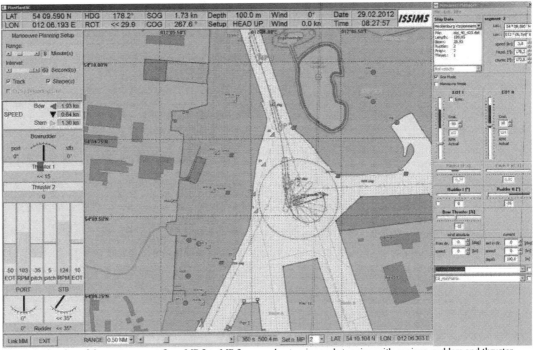

Figure 6. Planning of the next segment from MP 2 to MP 3 – complex turning and stopping with engines, rudders and thruster

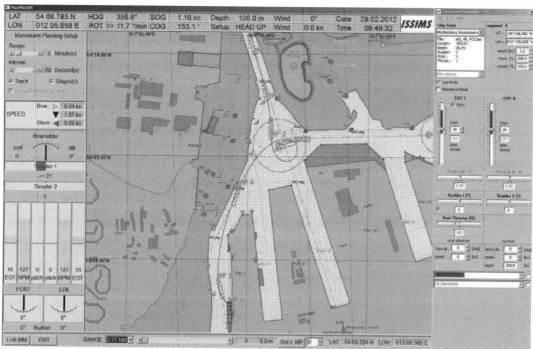

Figure 7. Complete manoeuvring plan for the route segment for passing the turning area and approaching the berth in astern motion

4 MANOEUVRING MONITORING AND MULTIPLE DYNAMIC PREDICTION MODULE - OVERLAID PREDICTION FOR ON-LINE MANOEUVRING DECISION SUPPORT USING MANOEUVRING PLANS

4.1 *Presentation of dynamic predictions in ECDIS environment*

For a compact presentation of information to the captain, pilot and responsible navigating officer respectively a new layout of a conning display was designed and implemented into the equipment installed on an integrated navigation system. For the purpose of testing the technical feasibility and user acceptance the new conning display with the integrated MULTIPLE MANOEUVRING PREDICTION MODULE was implemented in the INS equipment of the large full mission simulator bridge of the ship handling simulator of MSCW. The sample ship is again the RO-PAX Ferry "Mecklenburg-Vorpommern", the test area is the Rostock Sea Port. The RO-PAX ferry is leaving the berth to be steered through the fairway and to leave the port.

The layout of a dedicated prediction display integrated into an ECDIS is shown in Figure 8. It contains conning information together with the prediction and the planned manoeuvring track. The centre window shows the ENC in Head up Mode together with motion parameter for longitudinal speed and transverse speed as well as a circle segment with the rate of turn is shown. The ships position is displayed in the centre of the ENC as ship's contour where also the track prediction can be indicated as curved track or as chain of contours for the selected prediction time. The prediction parameters as range or interval of presentation can be set in the control window at the right side.

The Dynamic Path Prediction with the sophisticated simulation model is shown as chain of ships contours based on full math model (ship contours every 60 sec for 5 min with turning to STB). This dynamic prediction reflects already the effect of the setting of rudder and propeller control parameters shown in the left bottom window: In this sample the two rudders of the ferry used are set to 12° Starboard and the Engine Order Telegraph for the two controllable pitch propellers are set to 50% representing 130 rpm of the propeller. The actual pitch status is 19. This interface allows a presentation of dynamic predictions of steering and stopping characteristics as an immediate response according to the current steering handle or engine order telegraph position. It can be perfectly compared with the planned manoeuvring track as a reference line or curve, shown as blue line in the ENC window along the chain of manoeuvring points MP.

The predicted track for the simplified static path prediction based on of current constant motion parameters (implemented as add-on in some ECDIS

111

solutions) are shown as magenta curve: According to the actual/present small rate of turn to Port the predicted track is presented as a circle segment to the left side.

The use of path prediction with simplified models was already mentioned in previous papers, however, the use of this new multiple predictions based on the full dynamic model including the propulsion / engine process together with the result of preceding manoeuvring design is a great innovation and advantage. It was found that for the application of this dynamic prediction technology new strategies were found to save some minutes in this area which is very important in tight time schedules (Fischer & Benedict 2009).

4.2 SAMMON Manoeuvring Trial & Training Tool

This module combines a full simulation module for the ship manoeuvring process with all the modules above for planning and monitoring in order to test and try out manoeuvring plans and strategies, to be used both:
- as training tool in maritime education
  - in briefing / debriefing sessions for ship handling simulator training,
  - as well as in lectures on ships manoeuvring in classes and
- as training tool on board ships.

In order to control the virtual ship during the simulation process a manoeuvring panel on the screen allows steering the ship in real time along the planned route supported by the Multiple Predictor.

5 INTEGRATION OF SAMMON SYSTEM INTO EDUCATION FOR LECTURING & TRAINING SIMULATION

For training & education the SAMMON System is available as a portable version based on Tablet PCs for Planning of Manoeuvres in Briefing, Instructor stations and use on Simulator bridges Figure 9. The SAMMON system is interfaced to the Rheinmetall Defence Electronics ANS 5000 Ship Handling Simulator (SHS) at the Maritime Simulation Centre Warnemünde by WLAN connection. All ships which are available for the SHS are also ready for use in the SAMMON system for the following Concept of Application for Ship handling simulation:

Briefing:
- Demonstrating ships manoeuvring characteristics by using SIMOPT for familiarisation
- Drafting Manoeuvring Concept as Manoeuvring Plan (using MANOEUVRING DESIGN & PLANNING tool) according to the training objectives
- Optimisation of the concept by several trials of the trainee (using MANOEUVRING TRIAL & TRAINING tool)

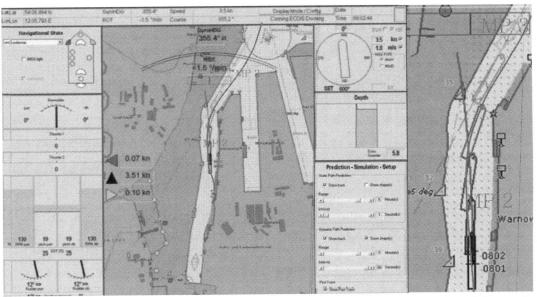

Figure 8. Layout for Manoeuvring Prediction integrated into ECDIS and comparison of static and dynamic predictions together with planned manoeuvring track (blue line)

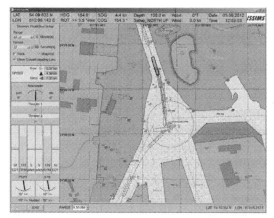

Figure 9. Manoeuvring Prediction integrated into ECDIS and comparison of static and dynamic predictions together with planned manoeuvring track (blue line) and contours at manoeuvring points

Execution of simulator Exercise:
- Training of conventional ship handling procedures and by using the by means of new FTS technology with underlying manoeuvring plan and dynamic prediction (MANOEUVRING MONITORING & MULTIPLE DYNAMIC PREDICTION tool)

Debriefing:
Assessment of the exercise results from full mission SHS by comparison of exercise recordings with trainees own concept or optimised manoeuvring plan by using SIMDAT tool for displaying and assessing the results of the exercise, e.g. comparing the result with the initial concept developed by the student in the briefing session and additionally to discuss alternative manoeuvring solutions by using the MANOEUVRING DESIGN & PLANNING tool).

Figure 10. SAMMON System set up based on Tablet PCs within Ship Handling Simulator environment: as Bridge Version (top), Lecturer System (left) and Instructor Version (right)

ACKNOWLEDGEMENTS

The research results presented in this paper were partly achieved in research projects "Identification of multi variable parameter models for ship motion and control" (MULTIMAR), "ADvanced Planning for OPTimised Conduction of Coordinated MANoeuvres in Emergency Situations" (ADOPTMAN), "Multi Media for Improvement of MET" (MultiSimMan), funded by EU, by the German Federal Ministry of Economics and Technology (BMWi), Education and Research (BMBF) and the Ministry of Education and Research of Mecklenburg-Pomerania, surveyed by Research Centre Juelich PTJ and DLR. Additionally it has to be mentioned that the professional version of the SAMMON software tools has been further developed by the start-up company Innovative Ship Simulation and Maritime Systems GmbH (ISSIMS GmbH; www.issims-gmbh.com).

REFERENCES

Benedict, K., Baldauf, M., Felsenstein, C., Kirchhoff, M. 2003: Computer-based support for the evaluation of ship handling simulator exercise results. *MARSIM - International Conference on Marine Simulation and Ship Manoeuvrability,* Kanazawa, Japan, August 25 – 28 2003

Benedict, K., Baldauf, M., Kirchhoff, M., Koepnick, W., Eyrich, R. 2006: Combining Fast-Time Simulation and Automatic Assessment for Tuning Simulator Ship Models. *MARSIM - International Conference on Marine Simulation and Ship Manoeuvrability,* Terschelling, Netherlands, June 25– 30 2006. Proceedings, M-Paper 19 p. 1-9

Benedict, K.; Kirchhoff, M.; Gluch, M.; Fischer, S.; Baldauf, M. 2009: Manoeuvring Simulation on the Bridge for Predicting Motion of Real Ships and as Training Tool in Ship Handling Simulators. *TransNav - the International Journal on Marine Navigation and Safety of Sea Transportation,* Vol. 3 No. 1 - March 2009.

Benedict, K.; Baldauf, M.; Fischer, S.; Gluch, M.; Kirchhoff, M.; Schaub, M.; Klaes, S. 2012: Fast Time Manoeuvring Simulation as Decision Support for Planning and Monitoring of Ship Handling Processes for Ship Operation On-Board and Training in Simulators. *MARSIM - International Conference on Marine Simulation and Ship Manoeuvrability,* Singapore, 23 -27 April 2012.

Fischer, S., Benedict, K. 2009: "Analyses of manoeuvring procedures on ferry Mecklenburg-Vorpommern in Rostock Sea Port and potential improvements using alternative manoeuvring concepts with Dynamic Predictor" *Internal research report (in German only),* Hochschule Wismar, Dept. of Maritime Studies, Warnemuende

ISSIMS GmbH 2012: Web page for SIMOPT & SIMDAT: http://www.issims-gmbh.com/joomla/index.php/software-products

Källström, C.G., Ottosson, P.; Raggl, K.J.; 1999; Predictors for ship manoeuvring;*12th Ship Control Systems Symposium SCSS,* The Hague, The Netherlands, October 19-21

Wilske, E.; Lexell, O. 2011: Test bed for evaluation of methods for decision support in collision avoidance. *e-Navigation Underway. International Conference on e-Navigation,* Copenhagen, Oslo 31.1. January - 2. February 2011, Proceedings pp. 72-86

Simulator Programs (2-D and 3-D): Influence on Learning Process of BSMT and BSMAR-E Students at Maritime University, Philippines

R.A. Alimen, R.L. Pador & N.B. Ortega
John B. Lacson Foundation Maritime University-Molo, Iloilo City, Philippines

ABSTRACT: This study aims to determine the 2-dimensional and 3-dimensional simulator programs and their influence on the learning process of BS Maritime Transportation and BS Marine Engineering students at maritime university, Philippines. The participants of this study were the 160 BSMT and BSMar-E students of the maritime university (JBLFMU-Molo) for school year 2010-2011. Participants of the study were enrolled at the Deck Simulator Program and Engine Room Simulator (ERS), which intoduced the 2-D and 3-D simulator programs as part of the different tasks for skills development of maritime students (BSMT and BSMar-E) at Maritime University in the Philippines. The researchers instructed the respondents to write down all their comments, suggestions, observations, and remarks on the perceived influence of using the 3D and 2D simula-tor programs. After the gathering the qualitative information, the researchers classified and categorized the write-ups of the respondents into different "categories." The analysis of comparison in relation to learning process brought about by the two (2) simulator programs was processed by the researchers. The "categories" were used towards establishing the concepts/views whether theses simulation programs influence learning process of nautical (BSMT) and marine engineering (BSMar-E) students at Maritime University (JBLFMU-Molo) in the Philippines. The results revealed that the 2-D and 3-D simulator programs are good learning aids which are helpful to marine engineering students. Sustaining the marine engineering students' "competent skill" in performing the different tasks in simulator is needed and likewise be enhanced.

1 INTRODUCTION OF THE STUDY

Video tapes, computer simulations, and multimedia software can encourage the students to think like scientists (Brungart & Zollman, 1996). This kind of instructional technology stimulates students to learn and to like their subject (Harwood & Mc Mahon, 1997; Sumanpan, 2008) even though it seemed difficult to understand. These softwares and technological-instruction activities can facilitate learning process, more likely to those students who are interested with the manipulation and skills. The instructors in higher education institutions should be innovative and creative in dealing with the students in order to convey and translate their ideas to achieve effective learning process.

Studies in the field revealed that simulation activity offers education providers a significant educational tool to meet the needs of today's learners by providing them with interactive and practice-based, instructional technologies. Using simulations in teaching and testing has the following potentials that can enhance the total learning process: more effectively utilize faculty in teaching of basic engineering skills, allow learner to revisit his skill in the simulator a number of times in an environment that is safe, non-teaching and conducive to learning, actively engage students in their learning process where they can display higher-order of learning rather than simply mimicking the teacher role model, contribute to the refinement of the body of knowledge related to the use of simulation in maritime education by providing insights in order to formulate best practices related to design and use of simulation technology (Tumala,Trompeta, Evidente, & Montaño, 2008). Furthermore, the authors underscored the use of virtual environment for instructional use in relation with the learners' characteristics. In this study, the authors stressed that learners benefited from the use of simulator as a learning tool irrespective of the type of cognition. In the same vein, the authors have found out the role of learning program as an indicator of successful learning has now depended on simulation itself. The need to join hands in coming up with program and program design that will best cater to the desired learning outcomes of the learners are well stated in this particular study.

The key issue in successful application of simulator classes is ensuring that simulation serves its purpose. The primary aim of any simulator experience is to create a certain level of skills performance among students. In the study entitled "Attitude, Skills Performance, and Implications of using Simulators Among Marine Engineering Students of JBLFMU-Molo, Iloilo City, Philippines" conducted by Alimen, Ortega, Jaleco, & Pador (2009), it emphasized the following: students do not seem to be sold completely to the use of simulator as indicated by "moderately positive attitude" towards simulator use, sustaining the marine engineering students' 'competent skill' in performing the different tasks in simulator is needed and it should likewise be enhanced, the significant correlation between the attitude and skill performance in simulator is reinforced by several studies which support the relationship between learner attitude and their performance. It is also stated also that technology has been apparent in this regard as it has reached a threshold where virtual or simulated approaches can meet or exceed the learning outcomes of expository (teacher-centered) approaches, the implications suggested that simulator should consist of more than anything else, a set of updated and upgraded computer software to address the observations and comments from the students.

2 STATEMENT OF THE PROBLEM

The present study aimed to determine the use of 3D and 2 D simulator programs and its influence to the learning process of nautical (BSMT) and marine engineering (BSMar-E) students at the Maritime University (JBLFMU) in the Philippines.

To further understand the study, the following questions were advanced:

1 How do marine engineering students perceived the 3D and 2D simulator programs in terms of learning at maritime university?
2 What are the comments, suggestions, and remarks about the 2D simulator program of nautical and marine engineering students?
3 What are the common remarks and suggestions of the nautical and marine engineering students of 3D simulator program?
4 Which are the perceived 2D and 3D simulator influences in the learning process of the nautical and marine engineering students?

3 THEORETICAL FRAMEWORK OF THE STUDY

The present study was anchored on the theory advocated by Alimen, Ortega, Jaleco, & Pador (2010) in their study entitled "Attitudes, Skills Performance, and Implications of Using Simulator Programs Among Marine Engineering Students of JBLFMU-Molo" by employing descriptive-qualitative mode of data collection. Moreover, in terms of the qualitative study, Yamut (2008) employed series of descriptions and information to determine the theme, characteristics, opinions, reflections, and views of the subject of the study. In this study, the researchers allowed the respondents to express their ideas, opinions, and views on 2-D and 3-D simulation programs and their influences on the learning process of marine engineering students at the maritime university in the Philippines.

4 CONCEPTUAL FRAMEWORK

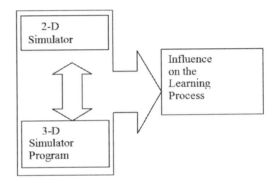

5 THE METHOD

This study used the descriptive research design. The respondents of the study were the nautical (BSMT) and marine engineering (BSMar-E) students at maritime university (JBLFMU) who were using the 3D and 2D simulator programs. The research process involves the description, interpretation, and comparison of the comments, suggestions, and remarks of marine engineering students on simulator programs at maritime university.

6 RESEARCH DESIGN

The present study utilized the qualitative method to achieve the objectives. The answers to the interview questions, comments, and suggestions given by the respondents were used to achieve the objectives of the study and substantiate the data generated from the study.

The qualitative data and information of this study were captured by employing the qualitative method. Qualitative research is a field of inquiry in its own right, it crosscuts disciplines, fields, and subject matters (Denzin, & Lincoln, 2000). This method was used very effectively for social change. This method employed the process of studying real-world situations as they unfold naturally. It is non-manipulative, non-controlling, and provides openness to whatever emerges (Best and Kahn, 1998). The researchers utilized this qualitative method to achieve the objectives of the study.

7 PARTICIPANTS OF THE STUDY

The participants of this study were the 180 nautical (BSMT) and marine engineering (BSMar-E) students of the maritime university (JBLFMU) for school year 2010-2011. Participants of the study were familiar with Deck Simulator and Engine Room Simulator (ERS), which includes the 2-D and 3-D simulator programs as part of the different tasks for skills development of nautical and marine engineering students at Maritime University (JBLFMU) in the Philippines.

8 PROCEDURE

The researchers instructed the respondents to write down all their comments, suggestions, observations, and remarks on the perceived influence of using the 3D and 2D simulator programs. After the gathering the qualitative information, the researchers classified and categorized the write-ups of the respondents into different "categories." The analysis of comparison in relation to learning process brought about by the two (2) simulator programs was processed by the researchers. The "categories" were used towards establishing the concepts/views whether theses simulation programs influence learning process of nautical (BSMT) and marine engineering (BSMar-E) students at Maritime University (JBLFMU) in the Philippines.

9 RESULTS OF THE STUDY

This section of the study focuses on the results and discussion about 2D and 3D simulator programs and their influences on the learning process of marine engineering students at Maritime University (JBLFMU-Molo) in the Philippines.

Table 1. Perceived Influences of 2D and 3D Simulator programs on the Marine Engineering Students Learning Process at JBLFMU-Molo

*2-D and 3-D are educational and can be used for learning in BS Marine Engineering
*More computers should be available for 2-D and 3-D so that learning would be more efficient.
*2-D and 3-D simulator programs are good learning aids which are helpful to marine engineering students
*The 2-D and 3-D simulator programs are great help to improve and add to the learning of the students.
*They are very useful for the students to familiarize the different parts of machine and equipment on-board.
*Extend the number of hours on 2-D and 3-D simulator programs in order to enhance the knowledge and skills of marine engineering students
*2-D and 3-D simulator programs are suitable in the learning process of marine engineering students.

Table 2. Comments/Views towards 3-D Simulator Program

*3-D is more practical than 2-D so therefore it should be given more attention
*I prefer 3-D than 2-D because it is more challenging and it gives critical thinking opportunity to the students.
*The 3-D set-up reflects the reality on-board that gives thorough learning to the marine engineering students.
*3-D is slightly confusing and sometimes difficult to handle
*3-D seems real but the leaking system should be put into higher resolution to achieve more realistic view.
*3-D simulator program is a state-of-the-art learning….it is a great opportunity for the students to experience real engine operation through virtual simulation.
*3-D is higher version of simulator program that needed by marine engineering students in terms of skills-development program of JBLFMU-Molo, Iloilo City

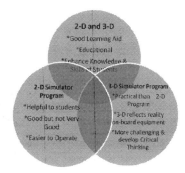

Table 3. Comments/Views on 2-D Simulator Program

*The 2-D simulator program is beneficiary for the students towards shipboard visualization
*it should be imposed to the students as additional learning and must be prioritize among other subjects
* to add more units especially for those students finishing their bachelor degree
*helpful to students
*we suggest to make it more realistic
*make students familiarize with the systems on-board
* it is easier to operate and easy to locate specific systems
* it help the students to identify a valve if it is open or close
*it is good but not so very good
*some system's parts are not working
* I want to see that there is a fluid flowing to the pipes
*The 2-D simulator program shows how to operate the machineries on-board

* It is easy to locate all the valves and machineries when you are operating it.

* the students must be acquainted first on how to use the machineries and they must be aware of the consequences when they failed to follow the correct procedure

*2-D simulator program helps a lot of students of JBLFMU-Molo as a learning material. It stimulates the real situation on-board a ship.

*to improve the 2-D simulator program, it is necessary to expose the students on how to use and maximize all the programs

*it is a great help for us to learn because you could see the actual scenario that are needed on-board a ship, especially the different machineries

*tani ya ang computer sa 2-D duganagan kay gamay lang kag kulang pa (add more computers)

*add more computers in 2-D simulator programs so that the students will have not hard time waiting

*ang masabi ko lang na dapat dagdagan ang computer sa 2-D para hindi mag agawanay ang mga estudyante sa pag gamit nang computer (all I can say is to add more computers on 2-D)

10 CONCLUSIONS

The key issue in successful application of state-of-the-art simulator programs in the instruction to ensure that simulation serves its purpose. The primary aim of any simulator experience is to create a certain level of skills performance among students. In summary, this study has the following conclusions:

The 2-D and 3-D simulator programs are good learning aids which are helpful to marine engineering students. Sustaining the marine engineering students' "competent skill" in performing the different tasks in simulator is needed and likewise be enhanced.

These simulator programs are very useful to the students to familiarize them with the different parts of machine and equipment on-board It is stated also that technology has been apparent in this regard as it has reached a threshold where virtual or simulated approaches can meet or exceed the learning outcomes of expository (teacher-centered) approaches.

The implications found here suggest that simulator should consist of more than anything else a set of updated and upgraded computer softwares and hardware to address the observations and comments of the students.

11 RECOMMENDATIONS

Based on the findings of this study, the researchers arrived at the following recommendations:

In this regard, the following are recommended:

1 The findings of this study revealed that 2-D and 3-D simulator programs effectively enhanced the mastery of desired skills of the marine engineering students at JBLFMU-Molo, Iloilo City. Most of the students preferred the 3-D simulator program, therefore, the administration should look into the advantages of the 3-D simulator program to maximize the applicability of the program. More studies of this kind must be considered to further validate the results of this investigation.

2 The lack of computers of the 3-D simulator program must be addressed through a careful and periodic assessment of the simulation rooms to be conducted.

Navigational trip gives the marine engineering students a chance to meet and talk with people in the field that could provide them with information about their profession (JBLF SPS Manual).

The findings of this study are supported by the objectives formulated by maritime university, specifically John B. Lacson Foundation-Maritime University-Molo, Iloilo City, Philippines regarding the navigational trip or On-the-job Training (OJT) emphasizing that actual sea experience and exposure to the field of marine engineering is productive in the development of students.

REFERENCES

Alimen, R. A. , Jaleco, V.R., Ortega, N.B. , & Pador, R. (2009). Attitude, Skills Performance, and Implications of using Simulators Among Marine Engineering Students of JBLFMU-Molo, Iloilo City, Philippines. JBLFMU Research Research Review (A Refereed Journal) Volume XIX. Number 2. ISSN 1665-8898.

Brungart, J.B. , & Zollman, D. A. (1996). The Influence of Interactive Videodisk Construction using Real Time Analysis on Kinematics Graphing Skills of High School Physics Students. Journal of Research on Science Teaching 32, 855-869.

Denzil, N. & Lincoln, Y. (2000). Qualitative Research Handbook. SAGE Publication Inc., Boonhill Street., United Kingdom.

Harwood, W.S. & Mcmahon, M.M. (1997). Effects of Integrated Video on Student Achievement and Attitudes in High School Chemistry. Journal of Research in Science Teaching. Volume 34, Number 6, pp. 617-631.

Sumanpan, V.O. (2008). Multimedia Instruction and Learning Style: Their Effects on Achievement of High School Chemistry Students. Liceo Journal Higher Education Research. Volume 5, Number 2, ISSN 2094-1064.

Tumala, B.B., Trompeta, G.P., Evidente, L.G., and Montaño, R.C (2008). Impact of Simulator Training on Cognition Among Marine Engineering Students. JBLFMU Research Research Review (A Refereed Journal) Volume XVIII. Number 1. ISSN 1665-8898.

Maritime Simulation Assessment Using Behavioral Framework

P. Vasilakis & N. Nikitakos
University of the Aegean Department of Shipping Trade & Transport, Greece

ABSTRACT: It is a fact that the competence and the skills of seafarers are determining factors for the smooth and reliable operation of merchant ships. S.T.C.W. recommends the maritime simulator as one of the most effective way to train and to increase the competences, the skills and the reaction of participants under difficult circumstances. This research is based on the assumption that maritime simulator represents the reality of each section of the ship in the most effective way. In this virtual environment the participants are called to react and to operate the ship under any condition. The emotional charge of the participants leads them to specific reactions and behaviors. The measures of those behaviors and reactions can improve the skills of the participants and the effectiveness of the training. In the paper a new assessment framework is proposed based on behaviorism. Field studies on seafarers in different countries prove the validity of our approach.

1 INTRODUCTION

The training with maritime simulators is an educational procedure which is usually based on S.T.C.W model courses. As an educational procedure it must follow specific guide lines with specific achievable goals. Moreover the traditional procedure of assessment is based on writing and oral examination in order to define the level of achievement for each participant. None of those methods has taken into account behavioral characteristics. This behavioral assessment which will be proposed in this paper does not come to replace or to criticize any of the traditional assessment methods. On the contrary, it comes to implement a new perception in maritime simulation procedure in order to investigate the behavioral changes of participants during simulation course.

The basic assumption for this behavioral framework is: the changes of human behaviors, during the educational procedure. Our target will be to record those behavioral changes in comparison with the educational goals in order to extract a skills index. A specific prototype educational procedure for simulation course will be established. This prototype procedure will be described in section II. Assessment tools will be established. Those tools have to do with "the categories of educational goals in maritime simulators" and a rating scale tools which will be called "Assessment Levels of simulation". The combination of those tools will be a specific behavioral index for each educational goal. The procedure of measuring will be described in Section II. The reflection amongst the learning environment, thoughts and behaviors of each educational goal can provide specific skills for each individual. In section III the desirable skills will be established for the mariners. From the combination of educational goals the mariners skills will be extracted and will be measured.

The Behavioral Assessment Framework for maritime simulators has been implemented in Wartsilla maritime Academy in Subic–Philippines. The results from the implementations of this model will be presented in section IV. The Behavioral Assessment Framework –B.A.F. is designed to help the instructor of simulation courses to understand the personal side of the change process from the individual point of view in order to determine the strength and the weaknesses of each simulation course.

2 MARITIME EDUCATIONAL PROCEDURE

The training with maritime simulators is an educational procedure which engages different types of learning theories. The learning theories act as an accelerator during the simulation procedure in order to achieve the desirable target of learning. Although the nature of learning must be based in competence based on training as STCW describes in model course 3.12a (I.M.O., 2010) . Experimental learning theory and Problem based on learning is the most common theory which is engaged in competence based training procedure. It is obvious that in learning theories literature can be find a lot of discrimination between those two theories (Kearsley, 2003), (Hmelo-Silver, September 2004), (David A. Kolb, August 31, 1999). As a result of our study in learning theory literature, we conclude that experimental base learning can be applied in basic educational procedure with most effective results during the simulation course. On the other hand, problem based learning, is addressed to the professional mariners which can be engaged in problem solving procedure in a most effective way.

It is a fact that STCW recommend specific model courses (COMMITTEE, 12 February 2010) for each type of simulator. Although the educational procedure is not differentiated in all of those courses. After an extended study of those model courses, a specific educational procedure is established. This procedure has to do with the operation flow of the course. It is very important for the validity and the reliability of behavioral assessment to establish the operation flow and the assessment point during this procedure. The operation flow will be divided into phases. Each phase will be representing a specific procedure which must be executed for the smooth operation of the simulation course. Those phases are: De-Briefing, Introduction, Theory Introduction, Scenario Introduction, Scenario, Briefing. (Figure 1)

In this procedure, an assessment point will be determined. The assessment point will act as a 'sensor' of the educational procedure which will transfer the 'signal' from the output to the inlet in order to improve the procedure. All in all it must be considered that simulation course is a closed educational procedure. The valid feedback at the right time is the key point of the improvement of the course. The capability of continuous improvement

through the behavioral assessment is the strength of this framework.

Figure 1. Simulator course procedure

3 BEHAVIORAL ASSESSMENT FRAMEWORK

The methodology of the behavioral assessment framework starts in the Concern Based Adoption Model (CBAM) and particularly in the Measurement of change in schools with Level of Use (Gene Hall, 2006). Moreover B.A.F. is a modified framework which is adaptive to the needs of maritime education. The basic perception is that during any educational procedure, the behavior of the participant changes. Thus, it is a fact that amongst the team of participants which is monitoring the simulation course some of the individuals will be active members of the procedure and some others will be passive members of the procedure. According to Albert Bandura Theory (Bandura, 1989), behavior, environment and person factors interact to influence learning. They influence and are influenced by each other. Based on this, we can divide the team to the user and non user of the simulation course. The determining factor to separate the individuals will be the behavior.

A table with customized behavior statements has been established according to the mariners behaviors during the simulation courses. Those behaviors are classified into eight levels. Each of those levels represents specific behavior of user or not user (Table 1).Those levels will be called 'Simulation Assessment Level' because they describe the change experienced of the individual during the simulation course.

Table 1. Simulation Assessing Level

SIMULATION ASSESSING LEVELS							
	Non-Use 0	Orientation 1	Preparation 2	Set up 3	Regular use 4A	Refinement 4B	Integration 5	Renewal 6
BEHAVIORS ASSOSIATE WITH SAL	No interest for the simulation course and for the educational object	Begins to gather information and to join the educational procedure	Begins to interact with the simulator with a lot of hesitation	Begins to customize functions of simulator in their own standard in order to use it.	Users know the learning object. They don't make any changes. They have a specific plan of action	They have knowledge of short-term and long term of consequences use.	Users know very well the learning object and try to organize and to coordinate the team	Users are holders of learning object and try to apply new techniques and new way of use.

Furthermore it is a common feeling that behavior statements cannot be identified for each individual. Every individual is a unique personality and has a different perception for the same statement. This exactly was the reason to adapt a rating scale for each behavior statement. This rating scale has to do with how True or not True is the statement for the individual during the simulation course.

It is known that all educational procedure includes educational targets. Usually the individual reflects with the learning environment and products specific behaviors. According to Malone (1990) «…we want to know, given a response, what situation produced it. In all such situations the discovery of the stimuli that call out one or another behavior should allow us to in fluency the occurrence of behaviors; prediction, which comes from such discoveries, allows control…» (Malone, 1990). Those behaviors are closely related with the educational target. That means for each educational target and each Simulation Assessment level a unique behavior will be coming up.

A state of the art review to the model courses of STCW for maritime simulator proves that the educational targets from the one course to another were modified according to the specific needs. Based on the needs of maritime education and particularly on model courses 3, 12a 'Assessment and Examination' (I.M.O., 2010) and to the categories of level of use (Gene Hall, 2006) seven educational goals are established for the maritime simulation course. Those educational targets will be:

Knowledge: This category deals with the understanding of the course and the educational procedure. It determines what the individual knows related with the object of learning.

Acquiring Information: It is a physical behaviorism procedure. When individuals find themselves in a new or unknown environment, then they start interacting with each other. The level of interaction has to do with the cognitive level of the individuals.

Sharing: During the simulation procedure there is an interaction between the individual and the simulator. The exchange of opinion indicates the rating scales of participation and the speculation of individuals related with the learning object. The procedure of sharing reflects directly to the decision maker procedure.

Assessing: The assessing as an educational goal in simulation procedure has to investigate the perception of individuals related with the course or with the function of simulator. From the behaviorism side of view the individuals with the highly critical view and specific suggestions for the improvement of the educational procedure are knowledgeable of the procedure. The target in this category is to separate the individuals with surface knowledge from those individuals with real knowledge of the learning object.

Planning: The planning of function in an effort to address a fact or a specific situation is a usual procedure during the simulation course. The individuals who have adapted to the educational target, can make long term planning without any or few mistakes. On the other hand individuals with low rate of knowledge cannot address the process.

Status Reporting: It is very important for every educational procedure to take into account the perception of the participant. The status reporting indicates the satisfaction of participant during the course. The rate of satisfaction is closely related with to the ability of the individual to learn.

Performing: This category investigates the ability of individuals to manipulate and finally to handle the whole necessary function during the simulation course.

Each of those educational goals will be a unique category which will be measured according to quantities assessment method by the Simulation Assessment Level. As a result of this procedure, a specific measurable behavior will be coming up for each educational goal. The validity and the reliability of the measurement method is based on the functionalism behaviorism. In fact, functionalism allows for an infinite number of physical and mind structures to serve the same functions. Functionalism has its roots in Darwin's Origin of the Species (1859) (Darwin, 1859), and Wittgenstein's Philosophical Investigations (Malcolm, 1954) (Malcolm, 1954). Specifically, according to functionalism behaviorism, the assessment of behavior can be done with direct observation, self-report, computer based assessment, interview and focus group questionnaire.

4 MEASURING THE MARINERS SKILLS

The maritime simulation course is based on competence based education-CBE, as referred in the STCW. It must be clarified that CBE is an educational procedure which focuses on the desired performance characteristics of professional mariners. Specifically, according to the STCW, it can be assumed that competency is the observable, habitual and judicious use of communication, knowledge, technical skills, operational reasoning, emotions, values, and reflection in daily practice for the benefit of the individual and the ships being served. The behavioural competence dictionary (Office, February 2007) indicates that competences and the performing of the individuals can be expressed in terms of behaviors.

The consistency of candidate performance along the simulation course is perhaps the most important issue to test the mariner's competency. Mariners do not perform consistently from task to task, across the educational procedure as it is described in Section I, and it is essential to assess competence reliably. This observation might not be surprising, given the differences in the individuals experiences encountered during training and practice but it challenges during the simulation course to improve and to test the competencies of candidates. Tests of mariners skills have moved into the educational procedure with the development of the objective, structured behavioral examination, consisting of a series of tasks and encounters stations. Many stations and sufficient testing time are essential to achieve adequate intercede reliability for the test.

Based on the assumption that maritime simulators represent the reality of the merchant ships working environment or part of it, we can summarize that the behaviors and the expressions of the individuals during the course will be close or the same with the reality. On the other hand, the structure of educational goals as they are referred in section II with the right combination, can provide the desirable skills for mariners. The pyramid of competence, introduced by Miller is a simple conceptual model, which outlines the issues involved when analyzing validity. This pyramid conceptualizes the essential facets of clinical competence, although the clinical competence has a lot of similarity with the mariner competence. The base of pyramid represents the knowledge components of competence: **knows** (basic facts) followed by **knows how** (applied knowledge). The **shows how**, this factor is a behavioural rather than a cognitive function and involves hands on, not in the head, demonstration. Under graduated student about to start work with the equipment operation must be able to show an ability to operate all the functions and carry out necessary procedures The methodology of skills measurement is based on three assumptions:

1 The Millers pyramid of competences (GE, 1990;65) figure 1, indicates four levels of human competency. Those levels indicate how the individuals perceive their status related with the use of simulation. Thus those levels are customized as follows:
   - Know, the straight factual recall of knowledge.
   - Know how, the applications of knowledge to problem-solving and decision-making. This method might be appropriate in early stages of the simulation curriculum, but, as skill-teaching is more vertically integrated, a careful planning of assessment formats becomes crucial. The ability of individuals to fit their knowledge into scenario and to manipulate the causes of the problems.
   - Show. The skills demonstration related with the functions and the operation of the equipment –simulator.

   - Does. The ability to perform the right functions under working load in collaboration with others.
2 Each of those competencies levels can be mapped with the desirable skills for mariners:
   - Management skills = Know How,
   - Operation skills = Shows
   - Technical skills = Does
3 It is well known that human skills are the combination of knowledge, human reactions and behaviors. The structure of educational goals includes components of all those skills. The combination of educational goals can provide the average score for each of those skills. Thus specific educational goals can be established for each skill. Management skills can be determined from educational goals like Planning, Acquiring information, Performing, Status Reporting. The average score of the combination of those educational goals will be the Management skills index of the simulation course.

The operation procedure during the simulation course requires knowledge of the educational subject, planning of the procedure and finally performing of the function at the right time. From the combination of those educational goals, Operational skills index can be extracted.

The technical skills usually are observable measured skills. Although the behavior of individuals is a determining factor for the performing of the functions. Educational goals like knowledge, sharing, assessing and performing are basic components for Technical skills. The average score of those educational goals can provide the Technical skill index of the participants in simulation course.

The measurement of mariner's skills with the combination of education goals is an innovative procedure. This procedure can provide valuable information to the instructor in order to map out the strength and the weakness of each individual and more generally the course improvement.

Figure 2. Millers Pyramid

5 PROPOSED METHODOLOGY

Reliability is a measure of the reproducibility or consistency of a test and it is affected by many factors such as examiner judgments, cases used, candidate nervousness and test conditions. The implementation of B.A.F. is a simple procedure which is based on the simplicity of individual behavior before the course and at the end of the course. As mentioned in section III, the B.A.F. is based on a self report method of individuals which does not deal with the feelings, emotions or attitudes. B.A.F. deals with what the individuals is doing or is not doing during the simulation course.

Questionnaire forms and statements are completed by the participants. Those statements are matched with the educational goals as they are described in section II. The participants have to complete the questionnaire form and to express how those statements fit or do not with their own behaviors. The rating scale of those statements is amongst zero to seven. The results from this procedure will give us a behavioral score for each of the educational goals. The score of each educational goal will be used in measurement of mariner's skills as described in section IV.

The implementation of B.A.F. took place for the first time in Wartsila Sea Land Academy in Subic Philippines during five simulation courses with different learning objects. That course was Alfa Laval course with simulator, cargo Crane course, and engine simulator course, bridge resource management and ECDIS course. The results from one of those courses will be presented and will be analyzed for the short of time. In order to improve the consistency of the course, two assessment points were established. The first assessment point was during the first day of the course. The second assessment point was on the last day of the course. The comparison between those two assessment points indicates the validity and the improvement of the course.

The data which will be extracted during the first day of the course will be including:
– Demographic information for the participants
– Clear view of the participant's educational level related with the learning object.
– The weaknesses and the strengths of the participants related with the educational goals.

Those results data can be used from the instructors as feedback in order to set up the simulation course according to the participant's level. The improvement of the simulation course will be extracted through the comparison of educational goals before and after the course. It must be considered that the simulation course is addressed to professional mariners with experience in learning objects. That means that the participants have already configured specific behaviors and perceptions related with the learning object. Our goal will be to measure those behaviors and to extract the skills index. After the end of the course it will be recorded as the percentage changes. Those percentage changes will improve the consistency of the B.A.F. and the added value of the course.

Table 2. Pre-Assessment Test

PRE-ASSESSMENT TEST

	Non-Use 0	Orientation 1	Preparation 2	Set up 3	Regular use 4A	Refinement 4B	Integration 5	Renewal 6
KNOWLEDGE								
ACQUIRING					4.03			
INFORMATION					4.13			
SHARING				3.37				
PLANNING				3.06				
STATUS				3.25				
REPORTING								
ASSESSING			2.09					
PERFORMING			2.78					

Table 3. Final-Assessment

FINAL ASSESSMENT TABLE

	Non-Use 0	Orientation 1	Preparation 2	Set up 3	Regular use 4A	Refinement 4B	Integration 5	Renewal 6
KNOWLEDGE						4.72		
ACQUIRING					4.04			
INFORMATION								
SHARING				3.37				
PLANNING				3.06				
STATUS					4.41			
REPORTING								
ASSESSING					4.41			
PERFORMING						4.55		

6 RESEARCH RESULTS

Those results are referred to the cargo crane simulation course which is conducted in Wartsila Maritime sea land Academy in Subic of Philippines. The duration of the course was five days. The number of participants was twelve persons. All the participants were professional seafarers with more than five years sea experience in rank of the 2nd deck officer. In corporation with the instructor, the educational goals were established, with out making any changes in the course curriculum. The implementation of B.A.F. began according to the methodology. The results of pre-assessment during the first day of the course indicate that the participant's moves through preparation, set up, and regular use, were related to the learning object of the course. The rating score of knowledge category and acquiring information category are similar and are classified into Regular use. In categories of Planning, Assessing and Sharing participant statements indicate that usually they look to Set Up the use of cargo crane. Despite the efforts of individuals to set up of cargo crane, the category of performing remains in level of orientation. With a brief look could be considered that the individual initially has a superficial knowledge of use and management of cargo crane with very low performance index (Table 2).

The final measurements which were extracted during the end of the course state majors changes in the behavior of the participants. The increasing of knowledge and performing to the refinement level (Table 3) indicate that during the course the individual engaged in the role of 'cargo handling user' with a few or without any problems. The increased 'status reporting' state persons with a deep knowledge of 'know how', although they are not in the position to plan or to share in corporation with others.

The measurement of educational goals at the end of the course can give a clear view of the educational achievement. To measure the impact of simulation course measurement of mariner's skills will be applied. The combination of educational goal will provide the basic mariners skills index before and after the simulation course (Table 4). Thus it can be concluded that participants initially have low ability, especially in operation of cargo crane. They seem to handle the use of cargo crane with a lot of difficulties and a lot of effort especially in the management sector. It is obvious that at the end of the course they were engaged in all procedures successfully and gain the chance to improve their technical skills and operation skills. However their efforts in management sector failed to increase their competence.

Must be take into account that the implementation of BAF is close related with the culture and the existing experience of participants. It is obvious that in specific implementation the improvement of the course it was close related with the existing experience of the mariner's. Results from the implementations of B.A.F. in basic education of mariners, has a completely different learning out come.

Table 4. Skills index measurement

INITIALY MEASURMENT	SKILLS INDEX	FINAL MEASURMENT
3.06	TECHNICAL SKILLS	4.36
2.99	OPERATION SKILLS	4.37
3.30	MANAGEMENT SKILLS	3.84

7 CONCLUSIONS

This paper presents an innovative assessment tool for maritime simulators course. The focus of this tool is to help the instructor and the change facilitators (Maritime Academies-Training centers) to understand and to realize the changes of participants during the simulation course. B.A.F. is an interactive tool which can easily be fitted into any maritime simulation course and to be presented in real time, during the course feedback to the instructors. The measurement of mariner's skills before and after the course can give a clear view of the effectiveness of the course and to work as an accelerator for the individual improvement.

The BAF has been already installed and is in operation in Merchant Engineer Academy of Chios-Greece and in Wartsila Sea land Academy in Subic Philippines. The presented results are just a small sample of the successful implementation of B.A.F. The results of existing implementations encourage us to continue with the combination of this framework with other forms of computer based assessments in order to improve the validity and the reliability of Behavioral assessment framework.

REFERENCES

Bandura, A., 1989. Social Foundations of Thought and Action: A Social Cognitive Theory. s.l.:Prentice-Hall.

COMMITTEE, M. S., 12 February 2010. TECHNICAL ASSISTANCE SUB-PROGRAMME IN MARITIME. LONDON, IMO.

Darwin, C., 1859. On the origin of species by means of natural se-lection, or the preservation of the favored races in the struggle for life. London: Murray.

David A. Kolb, R. E. B. M., August 31, 1999 . Experiential Learning Theory: Previous Research and New Directions, Cleveland: Department of Organizational Behavior Weatherhead School of Management Case Western Reserve University .

GE, M., 1990;65. The assessment of clinical skills/ competence/performance.. Academic Medical Press , S63(65).

Gene Hall, D. D. G., 2006. Measuring implementation in schols: Level of Use. TEXAS: SEDL.

Hmelo-Silver, C. E., September 2004. Problem-Based Learning: What and How DoStudents Learn?. Educational Psychology Review, Vol. 16, No. 3,, 3(16).

I.M.O., 2010. Assessment Examination and Certification of seafarers. In: Model Courses 3.12a. London : IMO.

Kearsley, C. 1.-2. G., 2003. Explorations in Learning & Instructi on: The Theory Into Practice Database. [Online] [Accessed 2012].

Malcolm, N. (. .., 1954. Wittgenstein's Philosophical Investigation. Philosophical ReviewL, Volume XIII..

Malone, J. C., 1990. Theories of learning: A historical approach . Bel-mont, CA: Wadsworth.. In: s.l.:s.n.

Office, O. R., February 2007 . Behavioural Competency Dictionary. Canada: Organizational Readines Office .

Training on Simulator for Emergency Situations in the Black Sea

F.V. Panaitescu & M. Panaitescu
Constanta Maritime University, Constanta, Romania

ABSTRACT: The Potential Emergency Situations Simulator (PESS) for Constantza Maritime University (CMU) should provide training and practicing of the students or course attendants in choosing the best strategies in a given emergency situation, which is an informational high entropy, multi-tasking, fast changing environment. The simulator is used for the realistic modeling of a crisis situation and it is useful for both marine officers and emergency situation officials. The simulator will be used as an educational instrument enabling the interactive study of the different emergency situations. It has the aim of training students to efficiently react to emergency situations such as a leak from a ship/chemical plant, fire, poisonous gas emissions, or any other situations that could show a potential danger. The trainee must be provided with realistic information and the response of the model on the actions of the trainee must be in accordance with the real conditions and scientific based. It must be possible to accelerate the simulation speed without loss of information or functionalities. The input of the external weather conditions is a must, as well as the trainee-oriented graphic interface. It must be possible to change the chemical and physical properties and characteristics of the different polluting agents. The simulator is also used to evaluate the best strategies to be followed in an ongoing crisis. In order to fulfill this aim, the simulation must have the capability to receive data from various sensors, transducers and servers. The courses are designed to accommodate up to six course participants. Each course includes course material such as course manuals and other documents. The courses include hands-on experience with simulator operations and maintenance. To help the start up of the simulated emergency situations training at Constantza Maritime University, we have made a manual which includes some well-designed exercises with scenarios, initial conditions and relevant documentation. The exercise documentation includes the exercise objectives, exercise guidance, instructor guidance, expected results and all other information to make the exercise successful for an inexperienced instructor.

1 INTRODUCTION

1.1 *The importance of training on simulator*

Involving students in simulation training on PC in the higher education is a constant concern, especially in recent years, being also the best way to save resources-so insufficient. Training students in the field of potential incident and emergency situations could be made with good results using computer simulators.

Students can be trained in daytime / night scenarios, in any weather conditions and terrain, everything is done in a virtual environment as close to the real one, generated by computer and related programs, which include three-dimensional representations of land, objects and locations affected.

Computer assisted instruction allows analysis, programming and training of students, at managerial and operational level for different emergency situations without consuming extremely expensive resources and materials. Familiarizing students with unusual situations, also will permit them to act normally in a real intervention and combat in the future incidents.

Another software advantage is that the application automatically carry a useful tool for creating script carrying tactical exercise (technology based on GPS / GIS), for applications in the field of human resources and materials entrained land. Simulator automatically collect real time data of the position and state emergencies, automatically displays digital map of the terrain and dynamically generates real tactical situation on the ground units, register conducting maneuvers and actions while mechanized units involved in the exercise allows analysis of post-deployment training exercise. This system ensures: managing information about own and colateral resources, personnel and logistics, geographic data and maps, weather situation, radio visibility, preparation of plans, orders and reports, terrain analysis tools, messaging format, logic and computer security, communication possibilities through various media.

1.2 *Objectives*

1 development of pollution scenarios for students (6 workstations) using various types of virtual equipment, in order to limit pollution and recovery / annihilation pollutant;
2 discuss each solution obtained by the students, in order to identify possible errors;
3 the instructor can to assess the effectiveness of each student response to pollution, the assessment of the pollution on the coast, the flora and fauna but also by counting the total cost of equipment used in operations in response to pollution;
4 training on this simulator is recommended for practice management level exchange of documents between institutions / agencies that manage such crises.

2 RESEARCH METHODOLOGY

2.1 *Simulator users*

Many marine companies use this form of "E" training to act quickly and effectively in various pollution situations. In thus saving human and material resources and act towards a sustainable development of marine environment and human resource development in the "E" Era Higher Education. Regular users of this simulator are: Constantza Maritime University students, Faculty of Navigation, Environmental Engineering; Navy officers as recommended by IMO OPRC (Oil Pollution Preparedness, Response and Co-operation); Romanian Naval Authority, Maritime Coordination Centre; ARSVOM-Romanian Agency for Saving Life at Sea; Inspectorate for Emergency Situations Constantza Dobrogea.

2.2 *Methodology applied*

The simulator is organized so that the instructor station can launch to all students pollution scenarios (6 workstations) and they can use various types of virtual equipment, chosen from a library, in order to limit pollution and recovery / annihilate the pollutant.

Simulator has a database for various types of response equipment (booms, dispersants, oil skimmers), the means of intervention (intervention marine division of the types of ships, air and land division), marine and terrestrial species of plants and animals .

Accidents that can be simulated are oil spills pollution at sea and spill of toxic / radioactive in air. The simulator is equipped with a module of crisis management that can be used in cases of forest fires, oil on water pollution, dangerous goods accidents, search and rescue operations or naval air accidents, acts of terrorism. This module serves to exercise managerial level exchange of documents between institutions / agencies that manage such crises.

The mathematical model used for the simulation of the event is the latest generation of pollution and meteorological factors taking into account all data of the sea and air, as well as all physico-chemical parameters of the substances involved (dispersion, emulsification, surface tension).

Simulator is also a powerful forecasting tool in a real accident situation: is coupled with a meteorological station of the Constantza Maritime University (CMU) of GSM LOGOTRONIC providing real time data on air parameters (speed/ wind direction, humidity, temperature and barometric pressure). Also is coupled with a submerged plant that belongs to CMU, mounted the marine central platform Petromar Oil, which provides data for sea state in the location (direction/speed for marine currents, direction/amplitude of waves and water temperature).

Results. Typical class working method is: every student receive same scenario and same response resources (say: two tugs, two booms, 2-3 skimmers, etc.). Then we observe the ability to manage this resources in order to limit the spill effect (e.g. *Training exercise input data, Table 1*).

Table 1. Training exercise on simulator

Entry data of pollutants	Values		
	Values		
	5 hours	4 hours	30 min
Crude oil	10 mt/h	3 mt/h	-
Discharge rate	200 mt/h		
Direction/	290^0 ENE	180^0 S	270^0 SSE
speed of currents	0.29 m/s	0.19 m/s	0.39 m/s
Direction/	20 000	270	-
Wind speed	10 m/s	13 m/s	-
Sea water temp	15^0 C	15^0 C	15^0 C
Waves H = 0.2 m downwind			
Visibility	5	5	5
Sea water density	1015 kg/m3	1015 kg/m3	1015 kg/m3

Each of the obtained solutions during the exercises of training can be discussed with all students (using videoprojector), in order to identify any mistakes. Finally instructor can assess the effectiveness of each student response to pollution (figure 1) (1).

All data obtained are shown graphically in real-time virtual - 3D visualization of oil spill and response resources (figure 2 and figure 3) (2,3).

Conducting a parallel between real and virtual resources for a complete training exercice, considering 138 kg=1baril and ~93 $/baril~311.55 lei/baril (e.g.in month July 2012), we can present below the folowing values for this comparative study (Table 2):

Table 2. Virtual-real values for a pollution incident.

Entry data	Values	
	real	virtual
pollutant discharged	200 t	0 t
costs resources	~100 000 Euro	10000Euro
technical resources (tugs, booms,)	~15 000 Euro	0 Euro

Figure 3. 3D Visualization (3)

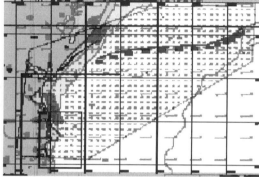

Figure 4. Oil shore impact - 29 hours 30 min from the event (3)

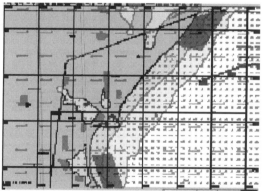

Figure 5. All shore hydrocarbon-H=33 hours from the event (3)

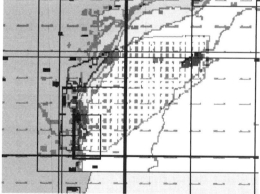

Figure 1. CMU Training simulator for emergency situations (1)

Figure 2. Situation after 9 hours from the event (2)

129

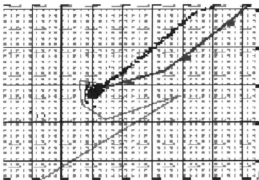
Figure 6. Technical resources no 5 boom ready to water h 6:49

3 CONCLUSIONS

The simulator for emergency situations was designed to evaluate the preparedness to respond effectively to oil spills, in accordance with the requirements of the Oil Pollution Act of 1990 (OPA 90).

The simulator is developed specifically to support the Preparedness for Response Exercise Program (PREP) with the goal of providing an improved training environment for response managers.

The training on emergency simulator provides the exercise participants with interactive information environment based on the mathematical modeling of an oil spill interacting with surroundings and combat facilities (figure 4, 5, 6).

We drafted a system which also includes information-collecting facilities for the assessment of the participants' performance.

The emergency simulator help us to operating modes corresponding to these stages (Forecast, Conduct and Debrief) are used for reproducing the "reality" of the exercise, automation of the instructor's activities and recording of the exercise key events.

We created sets of scenarios (2) to test the responsiveness of the students in real time and effectively.

Software also permits the student skill evaluation:

We establish a cost per hour for each resource, and we receive the total cost for entire operation, for each student.

Effective training simulator consists of lowering real time response, saving human and material resources, low cost price (Table 2) of company staff training costs.

REFERENCES

Ali, A. 2008. Role and Importance of Simulator Instructor. TransNav - International Journal on Marine Navigation and Safety of Sea Transportation, Vol. 2, No. 4, pp. 423-427.

Ali, A. 2009. Maritime Education – Putting in the Right Emphasis, Journal Vol. 3 No. 2 - June.

Arsenie, P., Hanzu-Pazara, R. 2008. Human Errors and Oil Pollution from Tankers , International Journal on Marine Navigation and Safety of Sea Transportation, Vol. 2 No. 4 – December.

Hanzu-Pazara R., Arsenie P., Hanzu-Pazara, L. 2010. Higher Performance in Maritime Education Through Better Trained Lecturers. TransNav - International Journal on Marine Navigation and Safety of Sea Transportation, Vol. 4, No. 1, pp. 87-93.

Kalvaitiene G., Bartuseviciene I., Sencila V. 2011. Improving MET Quality: Relationship Between Motives of Choosing Maritime Professions and Students' Approaches to Learning. TransNav - International Journal on Marine Navigation and Safety of Sea Transportation, Vol. 5, No. 4, pp. 535-540.

Panaitescu, F.V. 2006. Emergency situations simulator - User book, Nautica Publishing, Constantza, Romania.

Panaitescu, F.V. 2006. Emergency situations simulator- Laboratory notebook simulations, Nautica Publishing, Constantza, Romania.

Panaitescu, F.V., Panaitescu M., Ciucur, V., Padineanu, E., Tudor A. 2006. Applications on Constantza Maritime University Emergency Simulator, Nautica Publishing, Constantza, Romania.

Roy, B. 1996. Multicriteria Methodology for Decision Aiding, Kluwer Academic Publishers, Dordrecht.

Stam, A., Sun, M. and Haines, M. 1996. Artificial neural network representations for hierarchical preference structures, Computers and Operations Research, 23(12), 1191-1201.

Stefanowski, J. 1998. On rough set based approaches to induction of decision rules, in: L. Polkowski and A. Skowron (eds.), Rough Sets in Knowledge Discovery, Physica-Verlag, Heidelberg, 500-529.

Simulation of BOTAS Ceyhan Marine Terminals

Ö. Uğurlu, E. Yüksekyıldız & E. Köse
Karadeniz Technical University, Muammer Dereli Campus, Surmene, Trabzon, Turkey

ABSTRACT: Turkey has concentrated its efforts in developing sound projects for economical, safe and environmentally apt transportation of the prominent oil reserves of Iraq Basin markets through the Iraq-Turkey Crude Oil Pipe Line. Turkey as a reliable and stable transit country providing low cost and secure transmission of energy from the Iraq-Turkey Crude Oil Pipe Line Project by benefiting from her geographical advantages.

In this study AWESIM simulation modeling program has been used for the BOTAS Ceyhan Marine Terminal of Turkey. This modeling evaluated for 365 days and each ship has approached the port with intervals of 12-24, 12-36, 24-36 and 24-48 hours. Stormy days in a year have been assumed as 30. Each ship demands tug and pilotage service when approaching and leaving the port. As a result, this research has been done to determine the transportation capacity of the port by evaluating the number of ships approaching the port, the extend of queue of ships waiting for the tug service and space for approaching at the port, the amount of time ships queued by the port. To this end, the minimum and maximum transportation capacity has been spotted.

1 INTRODUCTION

Rapidly developing industry and technology since the 19th century lead to increased oil production which became the managing power of the economic structure and in turn today's large scale commercial oil circulation is emerged (Soylu, 2000). As one of the most significant elements of development, the energy and efficient use of such energy necessitated connecting the countries supplying the energy to demanding centers via various transportation ways, above all, via pipe lines in our world going through a rapid globalization process. Pipe line transportation is fast, economic and safe. Furthermore, the large scale investment is satisfied in a short time. Started at the end of 19th century with small scale and short distance lines, today, the oil and natural gas transportation turned towards for longer distances and at high pressures via pipe with wider diameters in parallel with the increased consumption, demand and technological advancements (Çubuk and Cansız, 2005).

While the pipe line transportation is a high cost investment compared to land and maritime transport, the pipe line transportation has advantages such as being faster, safer and more ecological compared to other transportation modes and not being affected by atmospheric conditions as well as having a shorter return on investment period. Therefore, transporting oil and natural gas to the consumption areas via pipe lines in the most economical way stands out (TUBİTAK, 2003).

Generally, pipe lines are examined in two groups as crude oil pipe lines and natural gas pipe lines. The oil is transported to ports or markets from regions with rich fields via crude oil pipelines (Çubuk and Cansız, 2005). Constituting the basis of the maritime system, ports are the locations where ships and marine vessels berth, and perform operations such as loading, unloading, maintenance and supply. It is difficult to solve the problems of ports analytically. The complexity of port functions has a complex structure dynamically, as in production systems. Utilizing simulation system in analysis of complex structure is inevitable (Demirci et al., 2000).

2 LITERATURE REVIEW

In the study conducted by Alan, B. Pritsker the frequency of vessels arriving at a tanker port in Africa, their duration at the port, days with stormy weather are assessed and the operability of the port and the tugboat activity are evaluated (Pritsker,

1986). Teo (1993) built an animated simulation model of a container port and investigated the movements of containers with automatic guided vehicles (Teo, 1993). Ramani (1996) has developed an interactive computer simulation model in order to support the logistic planning of container operations (Ramani, 1996). In the study conducted by Köse, Başar, Demirci, Güneroğlu and Erkebay, the traffic stream of the Bosphorus is modeled in AWESIM and investigated the effects of the new pipe line to be built on the strait traffic (Köse et al., 2003). In the study conducted by Yeo, Roe and Soak (2007), the maritime traffic congestion potential of Busan port is evaluated using an AWE-SIM simulation model. They concluded that one of the existing mooring berths within the harbor reach needs to be removed and two quays shall be expanded in order to prevent the traffic congestion (Yeo et al., 2007).

3 METHOD

Investigating the system behaviors using simulation technique aims to predict the future behaviors of the existing or future system to be built. In studies conducted using simulations, it is possible to see the results by applying strategies merely on the simulation model without making any changes to the actual system (Ali, 2008; Demirci et al., 2000). On the other hand designing simulation models is difficult and time consuming and allows making predictions regarding the actual system. Simulation studies generally consist of various stages. These well-arranged stages are monitored separately and the relations in each stage are investigated (Demirci et al., 2000). Simulation project adapting the general model to the specific problem situation plays essential role (Neumann, 2011).

This study is prepared in order to determine the handling capacity and usability of BOTAŞ Ceyhan Marine Terminal. Furthermore the following aspects of the BOTAŞ Ceyhan Marine Terminal are investigated: number of incoming vessels, queue values for berthing in the port, vessel waiting times for berthing, usability values of the ports, queue values for tugboat service, the time the vessels wait for getting tugboat service, total tugboat activity. Stay in port durations, frequency of arriving to the port, stormy days and the tugboat service rendered are the most critical criteria of this study. Accordingly AWESIM simulation modeling application is used in the present study.

3.1 *AWESIM*

AWESIM refers to the simulation language for problem solving. It may be used in courses, professional life, industrial engineering, managerial works, operational works and computer sciences.

AWESIM is a simulation language for alternative modeling. High level understanding and compiling of AWESIM lead to an increase in worldwide simulation and modeling utilization (Pritsker, 1996). An AWESIM project consists of one or more scenarios, each of which represents a particular system alternative. A scenario contains component parts. AWESIM incorporates the Visual SLAM modeling methodology. The basic component of a Visual SLAM model is a network, or flow diagram, which graphically portrays the flow of entities (people, parts or information, for example) through the system. A Visual SLAM network is made up of "nodes" at which processing is performed, connected by "activities" which define the routing of entities and the time required to perform operations (O'Reilly and Lilegdon, 1999). Symbols frequently used in AWESIM are shown in Table 1 along with their descriptions.

The create node creates a new entity within the network at intervals defined by TBC (Time Between Creations) and can save the arrival time as an entity attribute. TF; time the first entity enters the system, MA; variable used to maintain mark time, MC; maximum number of entities to create. Activity determines the time of the activities. The duration of an activity is the time delay experienced by the activity. DUR; specifies the duration of the activity using either explicit time or a distribution, CONDITION/PROBABILITY; specifies under what circumstance / probability a particular branch will be traversed by an entity, N; represents the number of parallel identical servers if the activity represents servers, A; is the activity number within the model. A Queue node is location in the network where entities wait for service. When an entity arrives at a Queue node, its disposition depends on the status of the service activity that follows the Queue node. If the server is idle, the entity passes through the Queue node and goes immediately into the service activity. If all servers are busy, the entity waits in a file at the Queue node until a server becomes available. The sequence of occurrences in queue is evaluated in the priority node outside the network. FIFO (First In First Out) is the default priority for files. IQ; initial number in queue, QC; capacity of queue, IFL; file number. The Terminate node is used to delete entities from the network. It may be used to specify the number of entities to be processed on a simulation run. This number of entities is referred to as a termination count or TC value. When multiple terminate nodes are employed, the first termination count reached ends the simulation run. Assign node used as a method to assign values to entity attributes as they pass through the node. Also it can be used to assign values to system variables at each arrival of an entity to the node. VAR defines global or entity variable. The type of Value (expression) must agree with the variable being assigned. A maximum of M emanating activities are initiated. The resource block identifies a resource name or label, RNUM; resource number, RLBL; the initial resource capacity, CAP; number of units of the resource initially available, IFL; file to poll for entities waiting for a resource. Await node used to store entities waiting for UR units of resource to be available or gate to open (use resource or gate label names). Arriving entities are placed in file IFL. QC specifies the queuing capacity of the node. Rule specifies the resource allocation rule. M specifies the maximum branches leaving entities can take. Free node used to release resources previously allocated at an AWAIT node when an entity arrives at the node. Every entity arriving at a FREE node releases UF units of RES resource. A maximum of M emanating activities can be initiated from the node. The alter node is used to change to capacity of resource type RES by CC units. CC can be constant or an expression. If CC is positive, the number of available units is increased. If CC is negative, the capacity is decreased (Pritsker and O'Reilly, 1999, Pritsker et al., 1989).

3.2 Technical Specifications of BOTAŞ Terminal

BOTAŞ Ceyhan Marine Terminal, the termination of Iraq – Turkey Crude Oil Pipeline, located within the district borders of Ceyhan at 36°51,9'N 35°56,7'E is discussed in the present study. BOTAŞ Terminal is owned and operated by BOTAŞ. It is in the BOTAŞ Ceyhan Port Authority management area. The first loading operation of this terminal was performed in 1977. It consists of 4 quays. The loading arms in loading-unloading facilities are hydraulic system operated via the crane tower located on the quay. Quay 1 and quay 2 are suitable for berthing vessels with 100.000-300.000 deadweight tons and quay 3 and quay are suitable for berthing vessels with 30.000-150.000 deadweight tons (Figure 1).

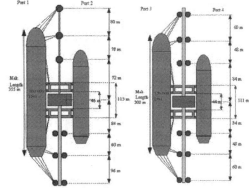

Figure 1. BOTAŞ Ceyhan MarineTerminal quay dimensions

Table 1. Symbols frequently used in AWESIM

Symbol	Node Names	Description
	Create	Creates entities
	Activity	Specifies delay (operation) time and entity routing
	Queue	Holds entities until a server becomes available
	Terminate	Terminates the routing of entities
	Assign	Assigns values to attributes or global system variables
	Resource	Resource definition and initial capacity
	Await	Holds entities until a resource is available or a gate is open
	Free	Makes resources available for reallocation
	Alter	Changes the capacity of a resource

Table 2. Capacities of BOTAŞ Ceyhan Marine Terminal loading arms (BOTAS, 2005)

Port number	Numbers of arm	Diameter of the manifold	Loading rate m³/hours	Maximum Draft (meter)	Maximum LOA (meter)	Minimum LOA (meter)	Maximum Dwt. Tons
1	4	20"-18"-16"	4x5000	23	355	200	300000
2	4	20"-18"-16"	4x5000	23	355	200	300000
3	4	16"-14"-12"	4x2500	18	300	168	150000
4	4	16"-14"-12"	4x2500	17	300	168	150000

4 AWESIM SIMULATION APPLICATIONS

AWESIM simulation network model is used in the present study for Ceyhan Marine Terminal. There are 4 quays in this model. Vessels arrive at the port with 4 different scenarios with intervals of 12-24 hour, 12-36 hour, 24-36 hour and 24-48 hour. It is assumed that same amount of vessels arrive at these four quays. Waiting times of the vessels at the terminal are calculated as 24 hours minimum 64 hours maximum for quays 1 and 2 taking into account the vessel dimensions, ballast capacities, coast loading rate and document processing prior to and following the operation. On the other hand, tankers with smaller dimensions compared to quays 1 and 2 berths at quay 3 and quay 4. Therefore, waiting times of the vessels at the terminal are calculated as 16 hours minimum 64 hours maximum for quays 3 and 4. Tugboat and pilotage service are required for berthing and unberthing. Tugboat and pilotage service cannot be rendered for other vessel before a vessel berthing or unberthing. Tugboat and pilotage service are considered as one hour each for berthing and unberthing operations. This port is unable to offer tugboat and pilotage service on days with stormy weather (Ugurlu, 2006). It is known that there are 30 days with stormy weather in one year for Ceyhan Marine Terminal (BOTAS, 2005).

This simulation program is assessed for each scenario over 8760 hours in total, i.e. 365 days. Accordingly the number of vessels arriving at the port, port queue volume, waiting time for berthing, waiting time for tugboat and pilotage services, operability of tugboat and pilotage services and operability of berths are evaluated.

4.1 Scenario 1

Scenario 1 is simulated as vessel arriving at BOTAŞ Ceyhan Marine Terminal in 12 to 24 hours (Figure 2).

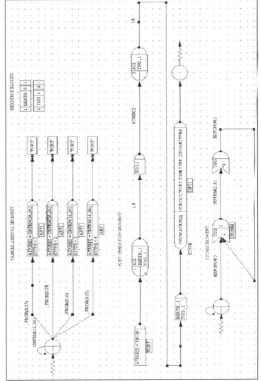

Figure 2. BOTAŞ Ceyhan Marine Terminal Scenario 1 simulation flowchart (12-24 hours)

According to simulation outputs, it is observed that 486 vessels arrived in total being 118 at quay 1, 114 at quay 2, 140 at quay 3 and 114 at quay 4. Any vessel arriving at the berthing is able to commence berthing after an average waiting time of 12 minutes. It is seen that 1 queue is formed for four quays. The quays operate at 2,449 efficiency on the scale of 4; in other words with 61% efficiency. In terms of tugboat service, any vessel arriving at the terminal receives the tugboat service after an average waiting time of 13 minutes and there is a queue of 2 vessels for such tugboat service. The activity of the tugboats is 11% in total (Table 3).

4.2 *Scenario 2*

Scenario 2 is simulated as vessel arriving at BOTAŞ Ceyhan Marine Terminal in 12 to 36 hours (Figure 3).

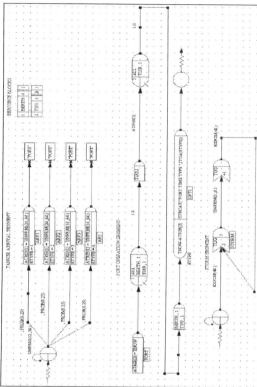

Figure 3. BOTAŞ Ceyhan Marine Terminal Scenario 2 simulation flowchart (12-36 hours)

According to simulation outputs, it is observed that 372 vessels arrived in total being 94 at quay 1, 90 at quay 2, 97 at quay 3 and 91 at quay 4. Any vessel arriving at the berth is able to commence berthing after an average waiting time of 11 minutes. It is seen that 1 queue is formed for four quays. The quays operate at 1,885 efficiency on the scale of 4; in other words with 47% efficiency. In terms of tugboat service, any vessel arriving at the terminal receives the tugboat service after an average waiting time of 10 minutes and there is a queue of 2 vessels for such tugboat service. The activity of the tugboats is 8,5 % in total (Table 4).

4.3 *Scenario 3*

Scenario 3 is simulated as vessel arriving at BOTAŞ Ceyhan Marine Terminal in 24 to 34 hours (Figure 4).

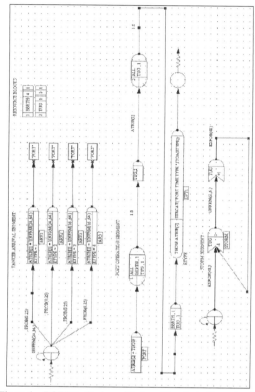

Figure 4. BOTAŞ Ceyhan Marine Terminal Scenario 3 simulation flowchart (24-36 hours)

According to simulation outputs, it is observed that 294 vessels arrived in total being 73 at quay 1, 71 at quay 2, 80 at quay 3 and 70 at quay 4. Any vessel arriving at the berth is able to commence berthing after an average waiting time of 8 minutes. It is seen that 1 queue is formed for four quays. The quays operate at 1,466 efficiency on the scale of 4; in other words with 36,6 % efficiency. In terms of tugboat service, any vessel arriving at the terminal receives the tugboat service after an average waiting time of 13 minutes and there is a queue of 2 vessels for such tugboat service. The activity of the tugboats is 6,7 % in total (Table 5).

4.4 *Scenario 4*

Scenario 4 is simulated as vessel arriving at BOTAŞ Ceyhan Marine Terminal in 24 to 48 hours (Figure 5).

According to simulation outputs, it is observed that 247 vessels arrived in total being 63 at quay 1, 57 at quay 2, 69 at quay 3 and 58 at quay 4. Any vessel arriving at the berth is able to commence berthing after an average waiting time of 11 minutes. It is seen that 1 queue is formed for four quays. The quays operate at 1,251 efficiency on the scale of 4; in other words with 31,2 % efficiency. In terms of tugboat service, any vessel arriving at the terminal receives the tugboat service after an average waiting time of 9 minutes and there is a queue of 1 vessel for such tugboat service. The activity of the tugboats is 5,7 % in total (Table 6).

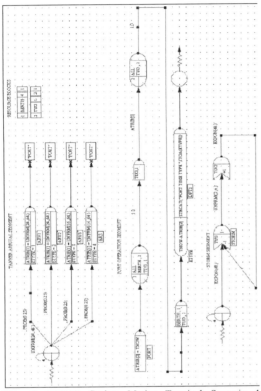

Figure 5. BOTAŞ Ceyhan Marine Terminal Scenario 4 simulation flowchart (24-48 hours)

Table 3. BOTAŞ Ceyhan Marine Terminal Scenario 1 simulation outputs (12-24 hours)

Statistics for Vessels Based on Observation					
Quay Number	Mean Value (hours)	Standard Deviation (hours)	Number of Observations (pcs)	Minimum Value (hours)	Maximum Value (hours)
Quay 1	47,350	11,031	118	28,282	69,093
Quay 2	44,982	11,798	114	26,350	68,383
Quay 3	41,372	12,737	140	20,648	66,776
Quay 4	43,571	13,610	114	18,087	66,974
File Statistics					
File Number	Label/Type	Average Length (pcs)	Standard Deviation (pcs)	Maximum Queue (pcs)	Average Waiting Time (hours)
1	Quay	0,010	0,101	1	0,186
2	Tugboat	0,012	0,112	2	0,217
Resource Statistics					
Resource Number	Resource Label	Average Utilization (pcs)	Standard Deviation (pcs)	Current Utilization (pcs)	Maximum Utilization (pcs)
1	Quay	2,449	0,758	2	4
2	Tugboat	0,111	0,314	0	1
Resource Number	Current Available	Average Available	Minimum Available	Maximum Available	
1	4	1,551	0	4	
2	1	0,809	-1	1	

Table 4. BOTAŞ Ceyhan Marine Terminal Scenario 2 simulation outputs (12-36 hours)

Statistics for Vessels Based on Observation

Quay Number	Mean Value (hours)	Standard Deviation (hours)	Number of Observations (pcs)	Minimum Value (hours)	Maximum Value (hours)
Quay 1	44,787	11,496	94	26,158	66,861
Quay 2	47,568	10,088	90	28,282	65,613
Quay 3	43,689	13,801	97	19,291	65,983
Quay 4	42,121	13,845	91	19,358	65,808

File Statistics

File Number	Label/Type	Average Length (pcs)	Standard Deviation (pcs)	Maximum Queue (pcs)	Average Waiting Time (hours)
1	Quay	0,008	0,089	1	0,187
2	Tugboat	0,007	0,085	2	0,159

Resource Statistics

Resource Number	Resource Label	Average Utilization (pcs)	Standard Deviation (pcs)	Current Utilization (pcs)	Maximum Utilization (pcs)
1	Quay	1,885	0,709	2	4
2	Tugboat	0,085	0,279	0	1

Resource Number	Current Available	Average Available	Minimum Available	Maximum Available
1	4	2,115	0	4
2	1	0,840	-1	1

Table 5. BOTAŞ Ceyhan Marine Terminal Scenario 3 simulation outputs (24-36 hours)

Statistics for Vessels Based on Observation

Quay Number	Mean Value (hours)	Standard Deviation (hours)	Number of Observations (pcs)	Minimum Value (hours)	Maximum Value (hours)
Quay 1	45,738	11,627	73	26,072	65,840
Quay 2	46,352	11,873	71	26,197	67,786
Quay 3	41,893	13,231	80	20,744	65,858
Quay 4	41,331	12,688	70	20,094	66,040

File Statistics

File Number	Label/Type	Average Length (pcs)	Standard Deviation (pcs)	Maximum Queue (pcs)	Average Waiting Time (hours)
1	Quay	0,005	0,068	1	0,136
2	Tugboat	0,007	0,089	2	0,223

Resource Statistics

Resource Number	Resource Label	Average Utilization (pcs)	Standard Deviation (pcs)	Current Utilization (pcs)	Maximum Utilization (pcs)
1	Quay	1,466	0,573	1	3
2	Tugboat	0,067	0,250	0	1

Resource Number	Current Available	Average Available	Minimum Available	Maximum Available
1	4	2,534	1	4
2	1	0,852	-1	1

Table 6. BOTAŞ Ceyhan Marine Terminal Scenario 4 simulation outputs (24-48 hours)

Statistics for Vessels Based on Observation

Quay Number	Mean Value (hours)	Standard Deviation (hours)	Number of Observations (pcs)	Minimum Value (hours)	Maximum Value (hours)
Quay 1	45,490	11,029	63	27,888	68,800
Quay 2	46,161	10,818	57	29,986	67,528
Quay 3	44,079	13,514	69	18,614	65,948
Quay 4	41,707	13,710	58	18,087	66,680

File Statistics

File Number	Label/Type	Average Length (pcs)	Standard Deviation (pcs)	Maximum Queue (pcs)	Average Waiting Time (hours)
1	Quay	0,005	0,071	1	0,180
2	Tugboat	0,004	0,066	1	0,155

Resource Statistics

Resource Number	Resource Label	Average Utilization (pcs)	Standard Deviation (pcs)	Current Utilization (pcs)	Maximum Utilization (pcs)

1	Quay	1,251	0,580	2	3
2	Tugboat	0,057	0,231	0	1

Resource Number	Current Available	Average Available	Minimum Available	Maximum Available
1	4	2,749	1	4
2	1	0,867	-1	1

5 DISCUSSION AND RESULTS

BOTAŞ Ceyhan Marine Terminal is investigated under 2 conditions since the tonnage of the vessels berthing at quays 1-2 differ from the tonnage of the vessels berthing at quays 3-4. There are four different vessel arriving values each being 232, 184, 144, 120 vessels at quays 1 and 3 and 254, 188, 150, 127 vessels at quays 3 and 4 according to four investigated scenarios and based on the arrival frequency of the vessels. It is observed that maximum 232 and minimum 120 vessels will arrive at quays 1 and 2 and maximum 254 and minimum 127 vessels will arrive at quays 3 and 4. When the four scenarios are investigated in terms of queue volume occurring due to setbacks at moorings and tugboat-pilotage service taking into account the values of Scenario 1, Scenario 2, Scenario 3 and Scenario 4, it is seen that a queue of 1 vessel for all four situations is formed. Accordingly, it can be said that there shall be a queue at the port in every case. The queue volume to be formed shall be 1 maximum based on the arrival frequency of the vessels. The fact that vessels arriving at the port encountering waiting times such as 12 minutes, 11 minutes, 8 minutes, 11 minutes based on the 4 scenarios for berthing is in question. It is observed that the operability of the port varies 61% to 31%. The waiting times occurring in the port due to tugboat-pilotage service setbacks are respectively 13 minutes, 10 minutes , 13 minutes and 10 minutes. Any vessel arriving at the port will be required to wait for 13 minutes maximum and 8 minutes minimum. The tugboat activity varies 11% to 5%.

It is seen that a maximum queue of 1 will form for BOTAŞ Ceyhan Marine Terminal regardless of the vessel arrival frequency in four scenarios and the queue volume is not a long value. It is seen that the vessels arriving wait for a short period of time such as 12 minutes maximum for receiving berthing service. Furthermore, it can be said that tugboat-pilotage service shall not be disrupted for a long period of time and accordingly the existing tugboats shall be able to render the terminal service.

It is observed that maximum 486 vessels and minimum 247 vessels shall arrive at BOTAŞ Ceyhan Marine Terminal in total. Scenario 4 values show the minimum number of vessels that shall berth at BOTAŞ Ceyhan Marine Terminal within 1 year. According to Scenario 4 120 vessels berth at quays 1 and 2 in total and 127 vessels berth at quays 3 and 4 in total. Quays 1 and 2 export minimum 12.000.000 ton and maximum 36.000.000 ton of crude oil in 1 year. According to Scenario 4, minimum 3.810.000 ton and maximum 19.050.000 ton of crude oil is exported from quays 3 and 4 export in 1 year. According to Scenario 4 value, it is possible to export minimum 15.810.000 ton and maximum 55.050.000 ton crude petrol from BOTAŞ Ceyhan Marine Terminal in 1 year.

Scenario 1 is the maximum number of vessels arriving at BOTAŞ Ceyhan Marine Terminal in 1 year. Taking into account the fact that according to Scenario 1, 232 vessels arrive at quays 1 and 2 in 1 year in total, it can be said that it is possible to export minimum 23.200.000 ton and maximum 69.600.000 ton of crude oil from the terminal. Taking into account the fact that according to Scenario 1, 254 vessels arrive at quays 3 and 4 in 1 year in total, it can be said that minimum 7.620.000 ton and maximum 37.100.000 ton of crude oil shall be exported. According to Scenario 1 values, taking into account the four quay altogether, it shall be possible to export minimum 30.800.000 ton and maximum 106.700.000 ton of crude oil in 1 year.

6 CONCLUSIONS

Iraq-Turkey crude oil pipeline is in operation for 36 years. It consists of two pipelines. The most significant obstacle of this line is war and political uncertainties. There were setbacks on Iraq-Turkey pipeline from time to time due to political uncertainties. The pipeline has a transportation capacity of 70.000.000 ton provided that it is operated under normal conditions (BOTAS, 2008).

It is seen in the study conducted that it is possible to export minimum 15.810.000 ton and maximum 106.700.000 ton of crude oil via Iraq-Turkey crude oil pipeline. 70.000.000 ton of transportation capacity is among these values. Therefore, it can be said that Iraq-Turkey pipeline can meet this transportation capacity.

It is seen in the four investigated scenarios that there shall not be any intense congestion at BOTAŞ Ceyhan Port in terms of berths and tugboat-pilotage services and BOTAŞ Ceyhan Marine Terminal shall meet this transportation capacity provided that there is no arrest or deceleration in the pipeline. In this regard, the most significant step to take for Iraq-Turkey crude oil pipeline is to eliminate the negative aspects on the pipeline such as political uncertainties.

There shall be a substantial increase in exported crude oil volume in İskenderun Gulf with BOTAŞ Ceyhan Marine Terminal and the ship traffic in İskenderun Gulf shall increase substantially.

This study was funded by the Karadeniz Technical University BAP Project."

REFERENCES

Ali, A. 2008. Role and Importance of Simulator Instructor, *TransNav - International Journal on Marine Navigation and Safety of Sea Transportation*, 2, 4, 423-427.

BOTAS 2005. Pipeline and Marine Terminal Description. Ankara.

BOTAS. 2008. *Iraq-Turkey Crude Oil Pipeline* [Online]. Available: http://www.botas.gov.tr/index.asp [Accessed 05.11.2012]

ÇUBUK, K. & CANSıZ, F. 2005. The Energy Situation between Transport systems in Turkey. *24. Energy Efficiency Week, "Energy Efficiency" Conference* Ankara.

DEMIRCI, E., ARAZ, T. & KÖSE, E. 2000. Trabzon Port Simulation Model *Journal of Industrial Engineering* 11, 2-14.

KÖSE, E., BAŞAR, E., DEMIRCI, E., GÜNEROĞLU, A. & ERKEBAY, Ş. 2003. Simulation of marine traffic in Istanbul Strait. *Simulation Modelling Practice and Theory*, 11, 597-608.

NEUMANN, T., 2011. A Simulation Environment for Modelling and Analysis of the Distribution of Shore Observatory Stations - Preliminary Results. *TransNav -*

International Journal on Marine Navigation and Safety of Sea Transportation, 5, 4, 555-560.

O'REILLY, J. J. & LILEGDON, W. R. 1999. Inroduction to AWESIM. *Proceedings of the 1999 Winter Simulation Conference*. Phoenix.

PRITSKER, A. A. B. 1986. *Introduction to Simulation and Slam II*, Westlafayette, Systems Publishing Corporation.

PRITSKER, A. A. B. 1996. *Introduction to Simulation and Slam II*, Westlafayette, Systems Publishing Corporation.

PRITSKER, A. A. B. & O'REILLY, J. J. 1999. *Simulation with Visual SLAM and AweSim*, New York, John Wiley&Sons.

PRITSKER, A. A. B., SIGAL, C. E. & HAMMESFAHR, R. D. J. 1989. *SLAM II Network Models for Decision Support*, New Jersey, Prentice-Hall.

RAMANI, K. V. 1996. An Interactive Simulation Model for the Logistics Planning of Container Operations in Seaports. *Simulation*, 66, 291-300.

SOYLU, M. S. 2000. *Dogu Akdeniz'in Stratejik Yapısına Bakü Ceyhan Petrol Boru Hattının Etkilerinin Degerlendirilmesi.* MsC, İstanbul Üniversitesi.

TEO, Y. M. 1993. PortSim: Simulation and Animation of Contanier Port. *Proc. of the Summer Simulation Conference*. California.

TUBİTAK 2003. The Scientific and Technological Research Council of Turkey, Panel on Transport and Tourism. *Vision 2023*. Ankara.

UGURLU, O. 2006. *Simulation Of BTC And BOTAS Marine Terminals*. MSc, Karadeniz Technical University

YEO, G., ROE, M. & SOAK, S. 2007. Evaluation of the Marine Traffic Congestion of North Harbor in Busan Port. *Journal of Waterway, Port, Coastal, and Ocean Engineering*, 133, 87-93.

Recommendation for Simulator base Training Programs for Ships with Azimuthing Control Devices

S. Short
South Tyneside College, South Shields, United Kingdom

ABSTRACT: The objective of this document is to compiling into a form that is readily exploitable for use in maritime pilot training and the wider maritime industry.

Specifically, the STCW95 (Standards for Training, Certification and Watch-keeping) code regulates the required competences for all ships operators. According to the STCW95 code, ship masters and chief officers functioning at management level on-board ships more than 500 GT shall possess very specific competences including "be able to respond to navigational emergencies" and "manoeuvre and handle a ship in all conditions".

The document will compile outcomes into a form that may be used to propose a specialised training module dedicated to the operation of ship equipped with azimuthing control devices.

It is expected that the document should result in a deliverable that outlines the necessary information to deliver such a course.

Or, where insufficient information is available, identifies such knowledge-gaps and propose roots to address them.

1 INTRODUCTION

The author was involved in European Project in 2008-2011 "Azipilot" and the paper is based on a work carried during the Project.

In the last two decades, azimuthing control devices (ACDs) have been steadily gaining market share in the shipping industry, particularly in specialist applications; from the thrusters on smaller, but numerous, harbour support vessels; through to the large pod-drives on cruise ships and ocean going liners, ACDs have rapidly established themselves in the maritime industry. The AZIPILOT project brought together the industry sectors responsible for design and testing; simulation and training; the pilots that operate ships fitted with ACDs; and the authorities that regulate them. The project aimed to promote wider understanding across the industry of the contributions made by the four areas of technical expertise, namely: hydrodynamic modelling, marine simulation, maritime training and operational practice; leading to harmonisation of practice and providing recommendations for both policy making and the pilot training process and practice, specifically for ships that use azimuthing manoeuvring devices.

2 ORGANISATION /HISTORY OF SEAFARERS TRAINING, CERTIFICATION AND WATCHKEEPING

Responsibility and requirements for seafarers was originated by International Maritime Organisation (IMO). On the conference in 1978 in London the body of The International Convention on Standards, Training, Certification and Watchkeeping for Seafarers (STCW) was creates which set qualification standards for masters, officers and watch personnel on seagoing merchant ships. This was entered into force in 1984 and significantly amended in 1995.

The **1978 STCW Convention** was the first to establish basic requirements on training, certification and watchkeeping for seafarers on an international level. Previously the standards of training, certification and watchkeeping of officers and ratings were established by individual governments, usually without reference to practices in other countries. As a result standards and procedures varied widely, even though shipping is the most international of all industries. By December 2000, the STCW Convention had 135 Parties, representing 97.53 % of world shipping tonnage.

The Convention prescribes minimum standards relating to training, certification and watchkeeping for seafarers which countries are obliged to meet or exceed.

The Convention did not deal with manning levels: IMO provisions in this area are covered by regulation 14 of Chapter V of the International Convention for the Safety of Life at Sea (SOLAS), 1974, whose requirements are backed up by resolution A.890(21) Principles of safe manning, adopted by the IMO Assembly in 1999, which replaced an earlier resolution A.481(XII) adopted in 1981.

In 1993, the IMO embarked on this comprehensive revision of STCW to establish the highest practicable standards of competence to address the problem of human error as the major cause of maritime casualties. A small number of special consultants developed a document identifying categories of behavioural conditions which, in their view, could be improved to some degree with proper training and enhanced shipboard practices and arrangements. After considering these conditions in terms of the effects of the human element in marine casualties, the consultants prepared a preliminary draft of suggested amendments to the STCW Convention, including a number of proposals directly addressing the human element. They also included a proposal to develop a new STCW Code, which would contain the technical details associated with provisions of the Convention. The amendments were discussed and modified by the STW Subcommittee over the following two years.

The most significant amendments concerned:

1 enhancement of port state control,
2 communication of information to IMO to allow for mutual oversight and consistency in application of standards,
3 quality standards systems (QSS), oversight of training, assessment, and certification procedures,
4 placement of responsibility on parties, including those issuing licenses, and flag states employing
5 foreign nationals, to ensure seafarers meet objective standards of competence, and
6 rest period requirements for watch-keeping personnel.

On July 7, 1995, a conference of parties to the Convention, meeting at IMO headquarters in London, adopted the package of amendments to STCW. The amendments entered force on February 1, 1997.

Full implementation was required by February 1, 2002. Mariners already holding licenses had the option to renew those licenses in accordance with the old rules of the *1978* **Convention** during the period ending on February 1, 2002. Mariners entering training programs after August 1, 1998 are required to meet the competency standards of the new **1995 Amendments.**

Amendments to the 1978 STCW Convention's technical Annex may be adopted by a Conference of STCW Parties or by IMO's Maritime Safety Committee, expanded to include all Contracting Parties, some of whom may not be members of the Organization.

Amendments to the STCW Annex will normally enter into force one and a half years after being communicated to all Parties unless, in the meantime, they are rejected by one-third of the Parties or by Parties whose combined fleets represent 50 per cent of world tonnage.

In aspect of Training the Amendments require that seafarers be provided with "familiarization training" and "basic safety training" which includes basic fire fighting, elementary first aid, personal survival techniques, and personal safety and social responsibility. This training is intended to ensure that seafarers are aware of the hazards of working on a vessel and can respond appropriately in an emergency.

STCW, as amended, will require all training and assessment activities to be "continuously monitored through a quality standards system to ensure achievement of defined objectives, including those concerning the qualifications and experience of instructors and assessors." The *1995 amendments* require those responsible for instruction and assessment of the competence of seafarers to be qualified for the type and level of training or assessment involved. Persons performing these roles are expected to have received guidance in instructional techniques and assessment methods.

A detailed analysis of STCW and other documents has been conducted and "It can be seen that the regulations are reasonably clear that operators of ACD's 'should' have received appropriate training, but experience shows that where this is not mandatory, then only the good operators will ensure that training is carried out and that the majority will not!"

3 MNTB – MERCHANT NAVY TRAINING BOARD

The MNTB has responsibility for setting and approving the education and training frameworks for new entrants into the Merchant Navy. Current frameworks cover:

- Foundation Degree (FD), incorporating the Scottish Professional Diploma (SPD)
- Higher National Certificate (HNC) and Higher National Diploma (HND)

Each framework has been designed to meet international regulations laid down by the International Maritime Organisation (IMO) and enshrined in the Convention on Standards of Training, Certification and Watchkeeping for Seafarers (**STCW**). As well as the international regulations, the frameworks are underpinned by, and referenced to, the industry's National Occupational Standards. In addition, several universities offer MNTB approved Honours degree programmes for new entrants.

The MNTB has produced guidance materials to assist companies and organisations involved in programme delivery. This includes the 'Planned Training at Sea– guidance for companies and seagoing officers'. The MNTB, in conjunction with the MCA has developed the criteria for STCW and International Ship and Port Facility Security (ISPS) mandatory courses. In addition, course criteria have been developed for non-mandatory courses that meet specific industry needs.

The full documents can be obtained from: http://www.mntb.org.uk/education_amp_training_fr amework-16.aspx . Also – MGN 095.

4 ANALYSIS OF STCW WITH REGARD TO SHIP-HANDLING TRAINING

The full code runs to 346 pages and it is quite difficult to read through and reference exactly what and where any requirements may be.

OFFICERS

In (Annex Chapter II) Table A-II/1 (page 27) a 'Specification of minimum standard of competence for officers in charge of a navigational watch on ships of 500 gross tonnage or more' can be found.

It outlines 'Function: Navigation at the operational level' and one of the competencies is 'Manoeuvre the Ship' (Page 35).

MASTERS

In (Annex Chapter II) Table A-II/2 (page 42) a 'Specification of minimum standard of competence for masters and chief mates on ships of 500 gross tonnage or more' can be found.

It outlines 'Function: Navigation at the management level' and one of the competencies is 'Manoeuvre and handle a ship in all conditions' (Page 48).

TUGS

In (Annex Chapter II) Table A-II/3 (page 60) a 'Specification of minimum standard of competence for officers in charge of a navigational watch and for Masters on ships of less than 500 gross tonnage engaged on near-coastal voyages' can be found.

It outlines 'Function: Navigation at the operational level' and one of the competencies is 'Manoeuvre the ship and operate small ship power plants' (Page 65).

PILOTS

Pilots are not covered by STCW95 as they are not deemed to be 'seafarers' although Resolution 10 invites IMO to consider developing provisions covering the training and certification of maritime pilots. From this invitation, IMO Resolution A960 (23) (Recommendations on training and certification and operational procedures for Maritime Pilots other than Deep-Sea Pilots) was developed in 2004.

– Annex 1 - 5.5 ....should be encouraged to provide updating and refresher training.....
   – Annex 1 – 5.5.5 simulation exercises, which may include radar training and emergency shiphandling procedures;
   – Annex 1 – 5.5.6 courses in shiphandling training centres using manned models;
   – Annex 1 – 7.1 .....should demonstrate that he or she has necessary knowledge of the following:
   – Annex 1 – 7.1.12 shiphandling for piloting, anchoring, berthing and unberthing, manoeuvring with or without tugs, and emergency situations;
   – Annex 1 - 7.1.18 manoeuvring behaviour of the types of ships expected to be piloted andthe limitations imposed by particular propulsion and steering systems;
– Annex 2 – 5.4 This exchange of information should include at least:
– Annex 2 – 5.4.4 discussion of any unusual shiphandling characteristics, machinery difficulties, navigational equipment problems or crew limitations that could affect the operation, handling or safe manoeuvring of the ship;

PILOT TRAINING - ETCS (Education, Training and Certification Standards)

ETCS was developed some years ago in Europe as a template for Maritime Pilots and some European countries have or are currently in the process of embracing these standards. There is a current international weakness in ensuring that all pilots are trained to an International Standard. Pilotage is a port state control activity that takes place in territorial waters and is inherently local in nature and therefore difficult, if not impossible, to apply to an International Standard. There are numerous national and international codes and recommendations but little, if anything, is mandatory. STCW95 resolution 10 invites IMO to consider developing provisions covering the training and certification of maritime pilots from which A960 has been developed, we would commend ETCS as an appropriate template to take this forward to be developed, whereby provision for specialist ACD training for pilots could be made.

ISM Code

The company should establish procedures to ensure that new personnel and personnel transferred to new assignments related to safety and protection of the environment are given proper familiarization with their duties. Instructions which are essential to be provided prior to sailing should be identified, documented and given.

We would contend that whilst this may have been appropriate at some point in the past, modern simulator facilities (FMBS and MMS) are far superior to teaching ship-handling and this Model Course needs urgent updating to reflect this. It would be unthinkable that a driving licence could be obtained by simply manoeuvring a wooden model of a car through various scenarios on a table top, but for some inexplicable reason, not only is this acceptable in the shipping industry, but it is still the main method used when examined for a STCW Certificate!

SHIPHANDLING TRAINING

In STCW (Annex Chapter II) Table A-II/2 (page 48) column 2 lists 18 scenarios of 'Knowledge, understanding and proficiency' and column 3 lists 3 'Methods for demonstrating competence';

Examination and assessment of evidence obtained from one or more of the following:

1 Approved in-service experience
2 Approved simulator training, where appropriate
3 Approved manned scale ship model, where appropriate

5 RECOMMENDATION FOR SIMULATOR BASE TRAINING PROGRAMS

There is nothing in the Code that is specific to training on ACD's. However, some parts could be interpreted as such. This is unsatisfactory, as experience has shown that the 'good operators' do take up appropriate training, whilst many do not. It is recommended that a specific specialist training regime should be added to the Code for ACD vessels.

In theory, everyone with a STCW95 certificate should have met the shiphandling requirements listed in the mentioned documents . Experience shows that whilst some shipping companies have met and followed this standard, sadly many have not! Practically, it is very difficult to monitor standards of 'Approved in-service experience', whereas this is easier to achieve with either a FMBS and/or a MMS.

When the Mariner starts his/her sea vocation can't predicted on what ship will be working. Then, during the service at sea, technology progress and a new ships, new propulsion etc. are developed. In the interest of the Safety of the sea and own benefit the Officers should update their knowledge and skills and obtain additional training on the new (for them) ships. The best training is always on the ship but very often it is impossible to have this kind of training.

For reasons above there are some mandatory and additional courses approved by MCA so the Mariner can progress in own sea career.

It is also a question, may be driving be economical reason, do I need that training or not. This question can be answered positively almost all the time: YES. STCW and International Safety Management Code 2002 put the responsibility on the ship-owner to ensure that the mariners on their vessels are competent to carry out the duties they are expected to perform.

Also, type specific training is very useful provided by equipment manufactures. There are so many levels of training available to improve the knowledge in specific area.

From our experience we know that some owners might say: ... well, he is a Captain for so many years so he will manage on the "new" ship – somehow..." How? – without an extra training, without simulation?

Yes, the training to use the ship with Azipod is necessary. Now it is only the issue what kind of training is most effective or most wanted.

Based on existing courses there is still a room for training mariners using the Azipod ship's on Simulator environment. The preparation for the course most of the time started at looking for the company, then what kind of ships (size, purpose, propulsion) are explored by that company and then also some additional requirements. The other factor is the present qualification of the Mariner. Theses preparation let to use/chose appropriate simulation including which most likely matching ship's models and present skills of the user. If the company already using ship's with ACD then is good idea to start preparing the crew using models with AziPods.

There is a need to provide some additional ship-handling courses targeted for Senior Mariners, Pilots and Tug Captains. Although at the moment it is not MCA/STCW requirements for this kind of course, it is beneficial for AziPod users. The outcome is to improve Safety at the Sea and confidence in every day duties on the ship. It should be provided by two different approach courses which will consider:

1 Basic training
2 Specific Requirement (user driven).

Maritime Training – Azimuthing Control Devices

5.1 Basic training

Basic training with ACD should take into consideration steering and alteration of courses using Azimuthing Control Devices in combination with/without rudder. The other concern, very important for safety of the vessel, is to practice crash stop and steering the ship at the low speed. The exercises should be designed to practice all maneuvering aspects of the ship. The next step should be practicing mooring, side stepping, reverse rpm's etc. with extension to maneuvering the ship in ice.

The other considerations taken into account with effects of:
– Propeller momentum
– Diagonal propeller force
– Effort of forces of the AziPod
– Interactions between other AziPod unit(s)
– Interaction effects between AziPod and ship's body (hull)
– Shallow water effect
– Lift and drag forces of the AziPod.

The most shared influencing reasons that affect ships when operating in close quarters are:
– Shallow water effect/Bank effect
– Ship-to ship interactions
– Surface and submerged channel effects
– Steering with Azimuthing Control Devices when towing
– Steering with Azimuthing Control Devices when under tow
– Assisted braking including the indirect mode
– Tugs operating near the stern of AziPod driven ship.

Elements of proposed scenario for basic training can be used in Specific Training to enhance ship-handling procedures for Mariners who have already received ACD training.

5.2 Specific requirement training

It is necessary to stress that experience in conducting training courses shows need for flexibility in arranging course programmes because in many cases programmes should be tailored to meet the particular requirements of the Pilots organizations and ship owners and tug companies. The specific requirement training is very popular and provides training on demand. Usually Senior Mariners, Pilots and Tug Captains come back to marine simulators to improve their abilities in maneuvering ACD ships and to keep up-to-date training with any changes in harbour environment or different ship's models or ACD control handles.

The training can be divided into three groups:
– Harbours – very often Port Authorities would like to find port requirements to minimise damage to ageing infrastructure when using ACD, to assess

port requirements for Towage, to expanded specific requirements for confined waters, busy anchor areas, narrow channels and short track ferry routes.
– Model - Full Mission Bridge Simulators use simplified methods of model coding which is close enough to fulfill the purpose of training.
– Tugs Operations - Escort operations performed over long distances and relatively high speeds where tugs quickly develop high steering and braking forces to a ship when needed.

Avoiding causalities and preventing loss of life at sea are the most important goals and these factors should be included in Marine Training. An important feature that is seriously affected by training is the way of handling a critical situation. A mishap is differentiated into three psychological stages: Action is planned and executed and the system is returned to normal operating status if the action is taken in time, otherwise system fails.

Azipod propulsion is a specialised type of propulsion. Safe operation of ships equipped with Azimuthing propulsion units requires comprehensive acquaintance with this type of propulsion and its specific handling features.

6 RECOMMENDATIONS FOR SAFE OPERATIONAL PRACTICE

Great number of pilots and ship masters expressed the opinion that there is the need for special training courses for ships equipped with Azimuthing Control Devices, in particular in order to enhance knowledge and skill in handling ships in a safe and intuitive manner with Azimuthing propulsion devices in varying critical situations. This is necessary in order to improve safety at sea and especially when close to berth, where predictions of loads are difficult.

There are some references to improve the safe issues for operational practice:
– Thorough training of all operators.
– Users should be fully acquainted with the manoeuvring characteristics of the vessel including emergency and berthing manoeuvres.
– Be trained in the maneuvering of the vessel with reduced capability, e.g. operating with only one ACD.
– A thorough knowledge of the operating control system with particular emphasis on the procedure for changing conning positions.
– Be aware of Shipping Company and Manufacturers recommendations for the safe operation of the propulsion system.
– Before entering restricted waterways be fully aware of all hazards and restrictions which may affect the manoeuvring capability of the vessel, and also be aware of the capability of the vessel

to poses a hazard to other Port users and the Port infrastructure.

7 RECOMMENDATION: SPECIALIST SHIPHANDLING COURSES AND SPECIALIST ACD COURSE

Specialist Shiphandling Courses currently exist that cover Twin Screw operations, emergency shiphandling copping with mechanical failures (engine, rudder) with and without escort tugs, with and without currents and waves, all of which are a logical follow-up on the above Basic Shiphandling Course.

Once we have established at what level a candidate has attained, i.e. an entry level, we can gauge the starting point for a specialist course and develop the course content to deliver a 'generic training course for ACD's' which would satisfy a level of competence and also a 'type specific training course', which is often provided by equipment manufacturers.

It is necessary to stress that experience in conducting training courses shows the need for flexibility in arranging course programmes, because in many cases programmes should be tailored to meet the particular requirements of the Pilots' organizations, ship owners and tug companies, whilst also meeting a recognised standard of competence. Customer specific training is very popular and provides training on demand. Usually Senior Mariners, Pilots and Tug Captains come back to FMBS to improve their abilities in manoeuvring ACD ships, and to keep up-to-date with any changes in harbour environments, different ship types or ACD control systems. A960 also recommends refresher courses every 5 years.

Specialist ACD training delivered by either FMBS or MMS should follow the same fundamental principles of training;
- Revision of theory of shiphandling principles
- Special points/limitations on FMBS and MMS to deliver training
- Generic Training and/or Type Specific Training
- Principles and Theory (pulling, pushing) of ACD's
- Limitations of use and strategies to overcome these limitations
- Many types of ACD's – result can be achieved by different means
- Operational procedures
- Different terminologies currently used and result oriented terminologies
- Series of practical exercises with 'hands on' use
- Relevant Bridge Resource Management training

- Towing for ACD's
- Towing by ACD's
- Safety during towing operations
- Escort Towing
- Emergency scenarios.

Each topic can be broken down further, depending on the type of simulator to be used and addressing the needs of delegates i.e. ship master, pilot or tug operator.

8 CONCLUSION

The paper provides an analysis of existing legal aspects for seafarers' trainings using ACD. A longer term objective in handling ACD's is to do it 'without thinking', where 'it is just natural' for a properly trained and experienced Ship handler i.e. Safe and Intuitive.

Training courses are required to give any new operators the correct theory as to how and why ACD's work.

Although at present there is not a legal requirement for a special training for ASP users, proposed scenarios for basic and advanced training can be used in wide aspect of training to enhance ship-handling procedures. The training centers are already very well equipped for all types of training in use of ACD. Also they are already use basic and advanced programs to fulfill all mariners' requirements. Cooperation between MCA, MNTB, Pilots Associations and training centers can be reflected in creation of appropriate programs and documents which help to enhance safety aspects of maneuvering for ships with AziPod devices.

REFERENCES

IMO STCW Convention code:
ANNEX CHAPTER 1 Regulation I/12 *Use of simulators* (page 31)
Annex 1 Part A Section A- I/12 *Standards governing the use of Simulators* (page 20)
ANNEX CHAPTER 1 Regulation I/14 *Responsibilities of companies* (page 33)
Annex 1 Part A Section A- I/14 *Responsibilities of Companies* (page 24)
ISM Code 2002 (Appendix D)
IMO provisions - regulation 14/V of SOLAS, 1974.
IMO Resolution A.890(21) Principles of safe manning
IMO Resolution A.481(XII) adopted in 1981.
PILOT TRAINING - ETCS (Education, Training and Certification Standards)
IMO Resolution A960 (23) (Recommendations on training and certification and operational procedures for Maritime Pilots other than Deep-Sea Pilots) 2004.
IMO Model Course 7.01 Master and Chied Mate 1999.

Chapter 5

Manoeuvrability

Proposal for Global Standard Maneuvering Orders for Tugboats

A. Ishikura & K. Sugita
Marine Technical College, Ashiya, Hyogo, Japan

Y. Hayashi & K. Murai
Kobe University, Kobe, Hyogo, Japan

ABSTRACT: The use of "Standard Maneuvering Orders" for tugboats, vocabulary and phrases mutually pre-agreed between ships and tugboats, is essential for the ship to provide clear directions for the tug when berthing or unberthing safely. Tugboats will need time to change their posture after receiving the orders from persons responsible for ships' maneuvering. Therefore, when giving directions to change tugboats' posture, persons who handle their ships are required to send out tug orders, taking into account this "delay time," a time lag between the orders from ships and the actions taken by tugboats. "Tug Orders" standardized and used in Japan are composed of the following three factors: tugboat's motion, engine power and direction, but the author's research shows that there are "Non-standard" special maneuvering orders other than those "standardized," which causes such problems as a gap in perception between pilots and tugboat's operators, etc. The purpose of this paper is to research the delay time between orders for and actions by tugboats and consider the appropriate and safe timing of providing instructions to them, and then to propose globally-authorized "Standard Maneuvering Orders for tugboats", discussing a problem involved in the use of the special orders used in Japan, and the way in which tug orders are used in other countries.

1 INTRODUCTION

Maneuvering orders for a tugboat should be given with an advance agreement between the operators of the ship and the tugboat so that the ship can provide clear directions to the tug employed during the operations of entering or leaving the berth. Then it takes some time for the tug to initiate necessary actions and change her positions after receiving any tug maneuvering orders from the ship. Therefore, any ship operators should give the tug orders when the posture change of a tugboat is required, taking into account this delay time, a time lag between when a tug command is given and when any expected action is taken. This is essential for the security of both the tug and the ship supported by her.

The purpose of this study is to research the above-mentioned time lag and to examine a safer timing for a ship to give maneuvering orders to a tugboat. And also the purpose here is to propose the global standardization of maneuvering orders for tugboats, showing the problems caused by using special tug orders in Japan, and weighing how the tug orders are employed in Japan and other foreign countries.

2 THE PRESENT CONDITIONS AND PROBLEMS OF THE TUG MANEUVERING ORDERS IN JAPAN

In harbors of Japan, maneuvering orders for a tugboat are given by using the commands mutually agreed beforehand between a pilot and a tugboat or by using "the Settled Term of Maneuvering Orders for Towage Work" set out by Japan Tug Owners' Association (called "Settled Term" hereafter) in Japanese.

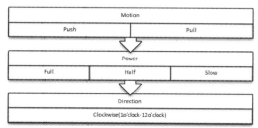

Figure 1. Flow chart of Tug orders in Japan

Fig. 1 shows a flow chart of the tug commands set out in "Settled Term," in which they shall be provided in the following order: the instructions for a tug's motion ("Push" or "Pull), engine power, and direction (given by "Clockwise").

However, there are some special commands for each of tug's motion, engine power, and direction other than those in "Settled Term." Our research reveals that these special orders are more frequently used, and moreover that there is a gap in perception between a pilot giving the orders and a tug operator receiving them. We will begin by considering the problems of the special tug orders in Japan relating to her movement and engine power and by showing the results of our research on the tug orders used in USA and Sri Lanka.

2.1 *Tug orders (Movement) in Japan*

In Japan, as shown in Table 1, there are many special orders requiring a tugboat to prepare for next movements or to shift her positions, and to use her engine at the same time. In addition, there are some different tug orders requiring the same movement for the tugboat.

Table 1. Tug orders (Movement) in Japan

Motion	Tug Orders	
	Phrases (in Japanese)	Their Meaning (in English)
Positioning order only	HIKU-YOUI	Stand by Pulling
	OSU-YOUI	Stand by Pushing
	YOKONINARE	Side by side
Positioning order add in Engine Power	HARIAWASERU	Give a tension
	BURASAGARE	Apply load
	ATAMA-O-TUKERU	Attach the
bow		
	OSITUKE	Keep a pushing
	MOTAREKAKARE	Keep a pushing
	YORIKAKARE	Keep a pushing
	DAKITUKE	Side by side

2.2 *Tug orders (Engine Power) in Japan*

As for engine power, in addition to "Settled Term," there are more than a dozen non-standard tug commands such as "Dead Slow," "Half of Dead Slow," "Quarter of Dead Slow," "Minimum," and "Omega" (driving Clutch), etc., all of which are weaker than so-standardized "Slow" in "Settled Term." In the case of engine power, there are some different tug commands requiring the same power output for a tug, as is the case in orders for her motions. All these special tug commands concerning engine power require the tug to put out the power of less than "Dead Slow" (25%), in which the special orders such as "One Eighth of Dead Slow" or "3% of Maximum Power" are included.

2.3 *Tug orders in USA (Los Angeles, Long beach)*

In USA (Los Angeles, Long beach), the standard tug orders shall be given in the following order: the instructions for motion ("Push" or "Pull") first, followed by engine power and direction (given by "Angle" based on ship).

Compared to the case in Japan, a distinguished difference is that there are only six kinds of engine power orders in USA, as shown in Fig. 2, and that any special tug orders such as used in Japan are hardly employed. Another difference is that, in USA, a tugboat responds to a pilot with a whistle while transceivers are used in Japan.

2.4 *Tug orders in Sri Lanka*

In Sri Lanka, the flow of the standard tug orders is, as follows: the instructions for tug's motion (using "Push" or "Pull"), engine power and direction (given by "Clockwise"). The tug orders relating to the preparation for movements, which are not standardized as "Settled Term" in Japan, are those among the standard tug commands used in Sri Lanka.

There are only six kinds of tug orders for engine power (shown in Fig. 3), as is in the case of USA, and any other special orders as to engine power are hardly employed in Sri Lanka, not as in the case of Japan.

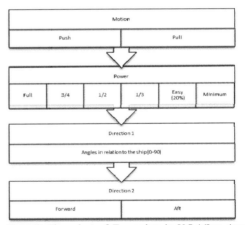

Figure 2. Flow chart of Tug orders in U.S.A(Los Angeles, Long beach)

Table 2. Special tug orders in Japan (Power)

Motion	Tug Orders	
	Phrases (in Japanese)	Their Meaning (in English)
Power based on Engine Telegraph	DEAD SLOW- -NO-HANBUN	Half of Dead slow
	DEAD SLOW- -NO-1/2~1/8	Half of one eighth of Dead slow
	OMEGA	Using OMEGA slipping Clutches
It is not based on Engine Telegraph	KARUKU	Lightly
	MINIMUM	At a minimum level
	GOKUGOKU	Very soft
	BISOKU	Very slow

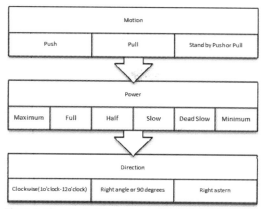

Figure 3. Flow chart of Tug orders in Sri Lanka

2.5 Communication among bridge team

The common language used on the bridge should be English, as required in the International Convention for the Safety of Life at Sea (SOLAS) 1974, and therefore the Standard Marine Navigational Vocabulary (SMNV) was adopted in the English language in 1977 and amended in 1985 by International Maritime Organization (IMO).

In November 2001, the Assembly of the International Maritime Organization adopted the IMO Standard Marine Communication Phrases (SMCPs) as Resolution A.918 (22), intended to revise and replace SMNV. The purpose of SMCPs is to standardize the English terms and phrases commonly used in bridge-to-bridge and bridge-to-shore communication for a safer navigation.

However, in Japan, the communications between a ship and a tug engaged in the operations to enter or leave berth are established in Japanese. When the pilot and the tugboat use their local language in any foreign ports of call, the captain of the ship and her crewmembers cannot understand what the pilot and the tug are talking about, and then have to leave their own ship exclusively under the controls of the pilot. In this case, the captain and bridge personnel in charge of the navigational watch cannot have any doubts about the pilot's maneuvering intentions.

According to Japan Coast Guard (JCG), there were 2,187 marine accidents in total in 2011, except for the accidents beyond human controls (346 vessels in total), and 832 of 2,187 accidents occurred in harbors. And also 1,681 of 2,187 occurred due to so-called "human error," the occurrence rate of which is 77% (JCG 2012).

Many shipping companies in Japan have introduced the Bridge Resource Management training (called "BRM" hereafter) to prevent a chain of "human errors" which is thought of as one of the main factors of a marine accident. In addition, the BRM training has been mandated for seafarers as a minimum requirement for the certification of officers in charge of a navigational watch by the 2010 Manila amendments to the STCW Convention and Code (Table A-II/1).

When pilots explain their navigational intentions to the key personnel on the bridge, their communications should be conducted "in the English language or in a language other than English that is common to all those involved in the operation" (IMO 2004), as pointed out in Annex 2. 6. 2 of the IMO Resolution A.960 (23) Recommendations on Training and Certification and Operational Procedures for Maritime Pilots other than Deep-sea Pilots. However, as in Japan, pilots occasionally use a local language understandable only for tugs and ground staff on the wharf or pier, such as linesmen, when informing their intentions or giving the instructions to them by transceiver. Fig. 4 shows the information flow chart on board the ship entering or leaving berth with the services of tugboats employed. In such berthing operations, the pilot plays a pivotal role as a key person to give orders to the tug or the bridge team on board and to share information with the ground staff on the wharf or pier.

When a pilot uses a local language in the communications with "parties external to the ship, such as vessel traffic services, tugs or linesmen" during berthing operations, "the pilot should, as soon as practicable, explain what was said to enable the bridge personnel to monitor any subsequent actions taken by those external parties" (IMO 2004), as advised in Annex 2. 6. 3 of the Resolution A.960 (23). However, it is actually rather difficult for pilots to explain what they wish to do and what they are talking about with tugs about to the captain and bridge watchkeeping personnel during berthing operations, because the commands and responses between the pilots and the tugs are frequently repeated during such operations.

Generally speaking, standard tug commands, mutually agreed between a pilot and a tug beforehand, should be used in their communications. However, as our research reveals, these commands are not globally standardized but employed only as locally-limited orders agreed between the pilot and the tug. Moreover, some tug commands not agreed between them are occasionally used in Japan. (Ishikura et al. 2011) In these situations, the captain and bridge team personnel cannot understand what the pilot and the tugboat are talking about, and they cannot come to realize what dangerous situations they are in due to the language difficulties, when the pilot may be possibly making a serious human error.

3 POSITION CHANGE ORDERS FOR TUGBOAT AND ITS DELAY TIME

3.1 *Method of research*

We used the data of front-affixed video cameras on bridge and the AIS position data derived from tugboats in the ports of Tokyo and Yokohama from January 2011 to July 2012. These video data include their shooting date, time, and also sound. By using the shooting dates recorded in the video data, we measured the amount of delay time, a time lag between when a position change order for a tug from a pilot is given and when necessary actions for the order are completed.

3.2 *Results of research*

The results of the research on the amount of time for tugs' changing their positions are shown in Fig. 5 The tug orders such as "Push" or "Pull" are generally given after the advance announcement of actions are given to tugboats by using the tug orders such as "Stand by to push" or "Stand by to pull," which are not standardized in Japan. However, the orders such as "Stand by to push" or Stand by to pull" are occasionally be followed by "Push" or "Pull" successively.

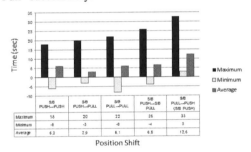

Figure 5. Transition time between tug boat's actions

Each average value of the time lag for tugs' changing positions is, as follows in the descending order: "Stand by to pull" to "Push," "Stand by to push" to "Stand by to pull," "Stand by to push" to "Pull," "Stand by to pull" to "Pull," and "Stand by to push" to "Push." In each case, there is the amount of 5 to over 10 seconds of the delay time on average.

The maximum length of time lag for tugs' changing their positions is the case of "Stand by to pull" to "Push" (more than 30 seconds), and the minimum length is the case "Stand by to pull" to "Pull" (more than negative 5 seconds). The negative vertical axis values in Fig. 5 means that a tugboat has finished her position shift before informing the pilot of the completion of her position shift.

In addition, when a ship is navigating at a certain speed or she is using a long tug line, it takes some more time for a tugboat to change her positions. Taking for instance the case that a tugboat changes her positions from "Stand by to pull" to "Push," the time length for her changing positions is shown in Fig. 6 when a ship is navigating and in Fig. 7 when she is using a tug line of various lengths. It is essential to maneuver a ship, taking into account the amount of time lag, when she is navigating at a speed or a long line is using on board the ship whose freeboard is comparatively higher.

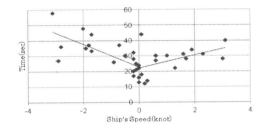

Figure 6. Transition time between tug boat's actions (S/B Pull→Push) by ship's speeds

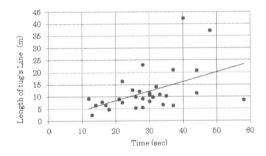

Figure 7. Transition time by every length of tug's line (From S/B Pull to Push)

4 CONCLUSIONS

In Japan, USA, and Sri Lanka, as our research shows above, maneuvering orders for a tugboat are commonly given in the following order: motion, engine power, and direction. As for the orders for tug's movements, "Push" and "Pull" are used as the common terms in ports of all the countries.

However, the tug orders for her engine power and direction varies in each country. Especially in ports of Japan, some special terms of the orders for engine power and direction are used, and, moreover, the pilots and tugboats communicate in Japanese. And it takes a certain amount of time for a tugboat to change her positions after a tug command is given. The length of this time lag tends to become longer and longer when a tug have to change her positions drastically (as in the case of her changes of motions from "Push" to "Pull" or vice versa), when there is a rather long distance between a ship and a tug (in which a comparatively longer tug line is used), or when a ship is moving though very slowly.

5 PROPOSAL

In the communications between a pilot and an operator of tugboat, some standard orders mutually agreed beforehand between them are generally used, but these tug orders, as our research reveals, are not globally standardized yet, and they are nothing but commands locally agreed between the pilot and the tug. In Japan, some maneuvering orders not mutually agreed are occasionally employed.

In the berthing operations, pilots are playing a pivotal role as a key person to provide necessary maneuvering orders to tugboats and ships, and to share information with ground staff on a wharf or pier. And also they are required to handle various operations at the same time, along with maneuvering the ship, or they have to give engine power orders to each of the multiple tugboats and the officers or quartermasters on board the ship at the same time. Moreover, they have to give necessary tug orders, considering a time lag between when an order is given and when any necessary actions are taken and completed, so it is absolutely essential for the bridge team to double-check the pilots' maneuvering intentions in order to ensure the navigational safety. Here we recommend and propose that globally-standardized orders for maneuvering a tug are settled and included in SMCPs for the bridge team to serve its function for maritime safety.

REFERENCES

International Maritime Organization (IMO) 2002, IMO Standard Marine Communication Phrases.
---. 2004, IMO Resolution A.960 (23) Recommendations on Training and Certification and Operational Procedures for Maritime Pilots other than Deep-sea Pilots.
http://www.imo.org/blast/blastDataHelper.asp?data_id=27137 &filename=A960(23).pdf
Ishikura,A., Sugihara,S., Hayashi,Y. & Murai,K. 2011, A Research on Maneuvering Orders for Towage Service in Japan. *The Journal of Japan Institute of Navigation*, Vol. 125, 91-97
Japan Coast Guard (JCG) 2012, Kainan no genjyo to taisaku ni tuite: the 2011 Edition.
http://www.kaiho.mlit.go.jp/info/kouhou/h24/k20120322/

Feasibility Analysis of Orthogonal Anchoring by Merchant Ships

J. Artyszuk
Maritime University of Szczecin, Poland

ABSTRACT: The method of orthogonal anchoring performed merchant ships, in particular very large crude oil carriers, for which this method proved to be superior, has been popularised in McDowall (2000) published by The Nautical Institute. This booklet scientifically explains some aspects of this manoeuvre and benefits thereof. However, some simulation and analytical studies reveal difficulties connected with: a) significant reduction of forward speed (as expected by this method) after initiating a turn against the wind with the kick ahead, b) effectiveness of this turn in terms of duration and final heading, c) keeping the constant heading against wind/wave/current while a ship is freely drifting. The latter sideways drift on a steady course is crucial for keeping the orthogonal direction of anchor cable during the paying out process. The aim of this paper is to contribute to a general discussion on the performance and range of application of the orthogonal anchoring method, by defining conditions and assumptions under which the method is successful, and when it cannot be used.

1 BACKGROUND

It is commonly known that anchor equipment for large vessels according to the IACS requirements (International Association of Classification Societies) is relatively weak, see e.g. OCIMF (1982), Clarke (2009). This regards either the situation of restraining a large ship having an initial kinetic energy at the time of the anchor being let go, due to ship's non-zero speed over ground, or the situation of resisting the environmental forces.

In the former case one deals with dynamic, potentially high loads in the anchor equipment. They generally depend on the elasticity of the devices, including a catenary shape of an anchor cable, the cable/anchor material properties (made of special steel or iron), and the kinetic energy to be absorbed. In the latter case static loads are developed, which are always equal to environmental forces generated by wind, waves, and current. The latter is very crucial in shallow water areas, where ship's hull hydrodynamic forces significantly increase, especially if the depth (h) to draft ratio (T), denoted by h/T, is generally less than 1.2-1.5. The possible static and dynamic loads in the anchor cable and anchor are also taken into account in the design process of anchoring equipment, e.g. the anchor windlass, but these efforts are rather limited (Morton et al. 1987).

It is also worthwhile to state at this point the often forgotten factor such that anchors used for merchant ships have holding power generally below the breaking strength of a new anchor and cable. So in case of excessive loads, whatever their nature is, static or dynamic, the anchor should drag. The holding power, dependent i.a. on seabed type, is usually expressed in term of multiple of anchor's weight. The highest risk is imposed by rocky seabed, or other equivalent bottom with some obstructions, where this multiplier for a standard stockless anchor assumes values much above ten. The latter may already create the anchor breaking loads. The general uncertainty of actual seabed conditions, even partially revealed and marked on a nautical chart, makes the master apply a safety margin at each stage of anchoring operation.

The problem of dynamic/shock loads calls for a caution and/or special procedures during anchoring operations of large ships. Therefore, it leads either to a very careful anchoring with the conventional method of in-line approach with wind/tide, or to the so-called U-turn approach (or orthogonal anchoring), popularised by McDowall (2000) and Priest (2005). The latter topic has been undertaken by some local authors in each country, e.g.

Jurdzinski (2005), as to promote the orthogonal procedure, though rarely providing their own operational/technical feedback. Furthermore, Clark (2009) finds the orthogonal method to be very personal of Cpt. McDowall and limits himself to only quote statements and opinions of the original inventor/initiator.

This author admits that some time ago had a private phone conversation with Cpt. McDowall, who strongly emphasised that simulators, or in other words various modelling and simulation methods, were not capable to prove the benefits of orthogonal anchoring. According to Cpt. McDowall, only some empirical observations and practical experiences of ship masters involved in such anchoring procedure are the only proper source of information. He personally does not believe in the simulation results, which within the state of the art in ship hydrodynamics, generally cannot confirm the advantages of the orthogonal anchoring. Moreover, he was afraid of undertaking by this author the problem of scientifically justifying, or not, the merits of orthogonal anchoring. Of course, this is not a scientific way of solving the problem. The detailed experience gained by Cpt. McDowell is also not written down and discussed deeper in the literature. Moreover, such rare full scale trials under various environmental conditions are not typical tests that can be fully repeated and commented on.

Simulator or hydrodynamic analyses are, however, indispensable in improving new ideas or concepts, as is the case with orthogonal anchoring. They provide more operational details and are very useful in analytical and simulator training of nautical officers.

The aim of this paper is to start a wider discussion since, serving as the simulator instructors and lecturers, we are often forced to face the problem of this anchoring procedure. Because the procedure is described in very popular books, we owe some reliable explanations to various participants of shiphandling training courses. The present paper investigates selected aspects of the orthogonal anchoring

2 PROCEDURE OF ORTHOGONAL ANCHORING. GENERAL REMARKS

The procedure of orthogonal anchoring is well described, though in a slightly different, systematic way and with some surprising modifications, in many references, e.g. McDowall (2000), Priest (2005), Clark (2009). Cpt. McDowall was also a consultant to Priest (2005), which means the latter source should serve as the ultimate reference for the orthogonal anchoring. Clark (2009), in the aspect of orthogonal anchoring itself, is rather a pure quotation of McDowall (2000).

However, to be well understood, some necessary general key points with the author's comments on this manoeuvre shall be reminded of within the present section.

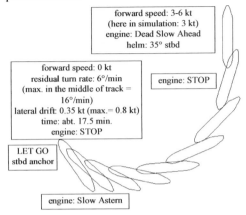

Figure 1. Sketch of McDowall (2000) turning procedure (based on the author's simulation of 270m length loaded tanker in no wind and tide conditions).

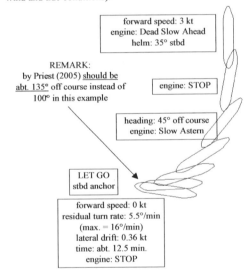

Figure 2. Sketch of Priest (2005) turning procedure (based on the author's simulation of 270m length loaded tanker in no wind and tide conditions).

Before studying the present and following sections, the reader is somehow encouraged to refer to the original publications for more details on the orthogonal anchoring, in particular to two figures illustrating the procedure of orthogonal anchoring, namely - McDowall (2000, Fig. 4, pp. 12) and Priest (2005, pp. 19), and to the corresponding remarks. Both, seemingly equivalent, schemes are rather different. They are reproduced in above Figures 1-2, but are exemplified on a simulation of real tanker

(270m in length, 135kDWT) in non-wind and non-tide weather conditions. The author's developed simulation software SMART (Artyszuk 2005) and his library of ship manoeuvring four-quadrant mathematical models have been here used. For comparative purposes, equal approach speed of 3 knots is assumed in both cases, and the ship's contour is plotted every 2 minutes. Analysing Figures 1-2, one can easily notice a large difference in final ship's positions before letting go anchor, which calls for attention while planning the approach to the anchor position. This aspect, however, will not be further discussed due to the scope of the present paper. This very concerning effect is expected to be undertaken in author's future research

In the following paragraphs of this section, this author's attempts to discuss particular elements of the orthogonal anchoring procedure with regard to McDowall (2000) and Priest (2005).

Since a ship has a tendency to turn to starboard while making astern, due to the propeller's lateral effect (paddle-wheel effect), the U-turn also to starboard and the use of starboard (inner) anchor are the only option. This can be easily presumed, though McDowall (2000) also considers turning to portside. However, turning to starboard is indirectly adopted in Priest (2005). This means that the starboard anchor and cable are subject to a higher wear-down.

According to the procedure, a ship approaches an anchorage downwind and/or downtide at a speed of about 6 knots, but the latter can be reduced to 3 knots (McDowall 2000) in crowded anchorages, though the a bit later book of Priest (2005) just states 2-3 knots as the standard without further discussion. There is no influence of wind and tide conditions upon the speed of approach in the cited references. However, keeping initially 6 or 3 knots will essentially impact the final stopping phase with engine making astern in that a ship would travel various stopping distances comparable to ship's length or a fraction thereof.

When the starboard hawse-pipe is reaching the beam direction of an intended anchor position, a full rudder with engine put dead slow ahead is recommended for a short while just to initiate a turn. As the turn is developing, augmented by the wind action (if this exists, see next chapters), the engine has to be stopped. The speed is naturally dropping to about 50-60% of the initial magnitude at the time the heading change equals 90° (McDowall 2000). In the case of initial 6 knots thus we get around 3 knots, which shall further diminish, up to about 2 knots as a ship continues to turn against the wind.

However, one should be aware that a turning performed/initiated the above way, with only a short 'kick ahead', is referred to as the coasting turn in the theory of ship manoeuvring hydrodynamics. In order to keep a sharp and quick turn in such conditions,

i.e. with engine stopped, so a ship may swing much more than 90° towards the wind/tide, such a ship shall be highly directionally (dynamically) unstable. The latter manoeuvring property is frequently seen on full form ships, e.g. tankers, and means that a ship is practically maintaining the initially excited yawing motion after the rudder is put amidships (with the engine still running). The fundamental cause of this behaviour is a high hull hydrodynamic yaw moment due to non-zero drift angle, which supports the initial turning. In case of engine STOP, quite independent of the helm angle, keeping on the forced turning on such ships will be much more a matter of fact. In the latter circumstance we namely lose the straightening effect of the rudder (at zero angle) streamlined by the propeller's race.

Nevertheless, McDowall (2000) advocates the execution of the mentioned turn up to about 135°, then advises to put the engine slow astern, since the forward speed is about 2 knots as stated before. A ship shall naturally keep the constant heading of 20° off the wind/tide, drift sideways, and start paying out the anchor cable in a normal direction to the ship's bow. The final equilibrium angular position of the ship against the wind/tide, namely 20° off, is also repeated in Clark (2009), but Priest (2005) alters this value to 45° off.

Moreover, Priest (2005) recommends to start slow astern immediately after a change of heading equal to 45° from the approach direction has been reached, during the stage of initial turning. This may be held as another essential modification to the original McDowall (2000).

It is true that McDowall (2000) suggests for a non-standard procedure of orthogonal anchoring to combine initiating a turn with subsequent and sudden making astern, even full astern. In those situations the whole ship's forward movement should be stopped after 90° (the wind/tide abeam) or 135° (the wind/tide 45° off a bow), dependent on ship's type, and the ship would drift only sideways. At this moment we shall let go the anchor. McDowall (2000) also states the necessity of the strong wind condition for this non-standard method to be effective. He tacitly assumes, too, that a ship will maintain a constant heading against the wind/tide, 90°, 45°, or even the former 20° off, which, however, cannot be taken for granted.

According to McDowall (2000) and Clark (2009) the final pure sideways drift decreases due to hull lateral force, thus allowing to secure the chain by the baron brake. This can be only valid for performing a turn in calm weather, since the wind induces almost a steady drift through the water, while the tide/stream does the same, but over the ground. In addition, for the same direction of wind and tide, a ship is experiencing various directions of the corresponding wind-driven, water-related drift and the ground-related drift driven by the tide. Though the tide-drift coincides with the tide direction, the wind-drift does not often match the wind direction. Both drifts are then to be geometrically combined.

The frequently forgotten or missed fact is that the anchor and cable will always <u>follow</u>, of course in the opposite direction, the instantaneous resulting drift over the ground of the hawse-pipe location point, see Artyszuk (2003a). If a ship's heading is kept steady, then the local drifts of particular points along the ship's centre plane are identical.

The orthogonal anchoring manoeuvre, from the viewpoint of the performed procedure and dynamic problems involved, may be divided into two general stages. Stage 1, being essentially a U-turn, comprises most of the transient phase, in which ship's directly controlled motions, i.e. forward and yaw velocities, are not yet steady and hence the anchor is not released (though being ready for letting go). Stage 2 consists of developing the pure sideways drift under a constant heading and the final steady-state of the drift. Within stage 2 the anchor cable is being paid out at the right angle to a ship's side, usually with the speed of drift, or a bit higher in order to keep no tension in the cable. So, in terms of passed time and effort, stage 1 is exactly a transient phase (but not the whole of it), while stage 2 quickly recovers from an initial unsteady lateral drift and keeps its steady value for most of the time. Within stage 2 some minor engine/rudder movements are allowed to compensate for any disturbances. These actions are neglected in the present investigation since other more fundamental issues arise.

The usually applied anchor cable length is defined by the scope ratio ensuring a practically zero angle between the anchor shank (cable) and the seabed. The angle at the seabed is thus called a scope angle. The absolute anchor cable length l is calculated by adopting the common scope ratio (h/l) equal to 5.0. This is a rough output from the catenary theory.

Depending on water depth, the distance x travelled by a ship under the lateral drift with no tension in the cable is $x = l - h$. Assuming further that the cable is being paid out, controlled by means of a brake, at the speed of drift, see e.g. Clarke (2009), we can also calculate the time of stage 2. This is gathered in Table 1.

Table 1. Time of paying out a cable.

Water depth h[m]	Cable length l[m]	Lateral drift speed			
		0.5	1.0	2.0	[kt]
		0.25	0.50	1.00	[m/s]
15	75	4m00s	2m00s	1m00s	
20	100	5m20s	2m40s	1m20s	
30	150	8m00s	4m00s	2m00s	
40	200	10m40s	5m20s	2m40s	
50	250	13m20s	6m40s	3m20s	

Table 1 figures are almost independent of ship's size.

3 THE CASE OF NO WIND/WAVE/CURRENT

Conditions of no wind, wave, and current pose some difficulties for performing the orthogonal anchoring. A ship here acts on its own, using the engine and/or the rudder. A ship's lateral drift is being developed mostly owing to angular speed during turning. Responsible for that is an inertia force in ship's rotational motion, which can be interpreted roughly as a centrifugal force or more precisely as Coriolis force. This inertia force and resulting drift are always directed outside the ship's curvilinear track. The linear lateral speed is proportional to ship's forward speed, though the drift angle itself is independent of the forward speed. The lateral drift can be considered as a kind of subordinate effect to the yaw motion induced by the rudder or the propeller making astern. This is because of coupling in manoeuvring motions. The role of rudder/propeller lateral force in an increase of the lateral drift is relatively low.

A ship's drift is practically independent of the ship's size given the similar geometric designs. A certain impact of ship's type, owing to specific variations in hull design and appendages, on the lateral drift can be however noticed.

It is also normal that the yaw motion during ship's turning is developing much faster than the corresponding linear lateral motion. A difference between the angular inertia and the lateral translational inertia can be deemed responsible for that. In the other situation of yaw checking, the similar phenomenon also takes place, in which at the first moment of zero angular velocity there is a residual drift angle. This drift often causes a significant hydrodynamic hull yaw moment, which makes a ship restore the previous direction of turning if the yaw motion has not been before adequately checked by the helmsman. An experienced helmsman should take allowance for a non-zero drift angle.

If a ship stabilises on a constant heading, i.e. when the turning motion is significantly decreased by the counter-rudder, the lateral drift speed gradually drops down due to hull hydrodynamic lateral resistance. However, in the case of a prolonged counter-rudder action, allowing a ship to turn a bit the opposite way, the drift may be suddenly cancelled. The magnitude of the initial drift, and its transient phase of how quickly it reduces, influence the performance of stage 2 of the orthogonal anchoring. The time and distance of sideways movement for efficient paying out the cable is whatsoever limited here.

In the case of calm weather, the problem of orthogonal anchoring and solutions to it belong to the classical theory of ship manoeuvring hydrodynamics modelling and testing. The issues involved in the orthogonal anchoring of particular interest to ship hydrodynamics include, among others: a) coasting turn (i.e. turning with the subsequent engine stop), b) ship's forward speed reduction with the engine making astern, c) propeller lateral action (also referred to as paddle-wheel effect) while rotating astern, which brings the ship's stern usually to portside, and hence turning the ship to starboard.

Some aspects regarding the coasting turn phase have been already dealt with in the previous section. Since a relatively high lateral drift is desired at the end of turning - in Priest (2005) this turning is essentially a sequence of a coasting turn and an additional turn induced/supported by a propeller running astern - a quite fundamental question arises whether the corresponding drift is high and durable enough for paying out the cable. The high drift speed is achievable for high turning rates, which are not always the case. According to this requirement, a ship does not need to turn more than 135° of the approach direction, but 90°, or even less, seems to be sufficient. Dependent on the extent of effective turning, the approach direction to an anchorage may be thus adjusted.

4 THE CASE OF WIND AND WAVE ONLY

With regard to the last paragraph of the previous section, the wind and wave can assist (e.g. for a loaded tanker) or prevent (e.g. for a ballasted tanker or loaded container carrier) making a good turn according to the sequence of actions in the orthogonal anchoring.

The wind and wave will certainly generate an additional lateral drift due to the corresponding aerodynamic force and the second-order hydrodynamic wave force, sometimes called the wave drift force. This shall be treated as advantageous. Nonetheless, there are also yaw moments from both excitations, which create essential difficulties in keeping a constant heading by means of short rudder/engine kicks. This especially refers to final wind/wave directions as standard for the orthogonal anchoring, namely 20° or 45° off the wind, dependent on the source reference.

The next paragraphs within this section will consider only the effect of wind and wave, under an equilibrium of forces and moment, upon a free steady drift.

The wind-waves in a steady state phase, i.e. when fully developed, can be considered to be very similar in action to the wind itself, both in magnitude and

dependence on the ship's incidence angle. So sometimes, in rough analyses, in order to achieve a proper ship's manoeuvring motion response under combined wind and wave, it is enough to double the wind forces.

If a ship under no power conditions is subject to the joint wind and wave excitation, or even to pure wind, it usually initially assumes an unstable motion. If an appropriate time is allowed, this transient phase, more or less long-lasting, as dependent on initial conditions, simply decays. A ship is then entering a steady-phase of motion in the form of a free drift on the constant heading. The equilibrium heading is a relative direction to wind/waves for which the wind and wave yaw moments compensate with the hydrodynamic hull yaw moment due to ship's motion through the water. The hull yaw moment depends on a ship's drift angle in ship body axes.

The below results, in Figure 3 and Table 2, are taken from the author's previous work (Artyszuk, 2003b). They provide some guidance on steady drift parameters of a tanker. These numerical computations also include such factors as: ship size, loading condition - loaded or ballast, wind force/sea state in Beaufort scale. A ship's size is represented by the ship's length L. The method adopted in Table 2 for presenting data - triple values - is as follows. The first value in the sequence is the wind/wave incidence angle γ (see Figure 3), the second value denotes the drift angle β, and the final value means the drift velocity in m/s.

It is very interesting that the wind and wave in the equilibrium condition is not exactly abeam (abt. up to ±15°), as well as the resulting drift is not perpendicular to a ship's hull (much higher deviations up to 30° towards a ship's bow are observed). In none of the cases the wind direction reaches the aforementioned 20° or 45° off the wind/wave as characteristic one for the orthogonal anchoring. With regard to the specified angles for the orthogonal anchoring, one can suspect a strong requirement of frequent rudder and engine usage, but without a guarantee of being successful. This calls for a modification of the orthogonal anchoring in that a ship does not initially approach an anchorage downwind (and makes nearly a 150° turn - i.e. the actual 'U-turn'), but sails with a wind just abeam and only reduces the forward speed (so there is no turn at all). The latter method is also mentioned by McDowall (2000), but is somewhat hidden in this publication (included in checklists at the end of this book) and called a 'tentative' method. Cpt. McDowall suggests the range 20-90° for the direction off the wind in the tentative method. One can expect that the upper value 90° is more valid for anchoring in wind and/or wave only, while the lower value 20° is specific and better achievable in pure tide or in the wind/wave combined with a strong tide.

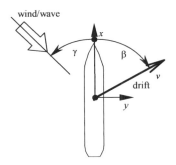

Figure 3. Frame of reference for drift under wind/wave.

Table 2. Parameters of steady drift in wind and waves.

Weather condition [°B]	ballast	
	L=100 m	L=300 m
5	088°/067°/1.01 m/s	097°/058°/1.76 m/s
5 (pure wind)	078°/069°/0.83 m/s	
7	086°/068°/1.49 m/s	094°/062°/2.49 m/s
7 (pure wind)	078°/069°/1.35 m/s	
9	084°/068°/2.07 m/s	092°/064°/3.05 m/s
9 (pure wind)	078°/069°/1.96 m/s	

Weather condition [°B]	loaded	
	L=100 m	L=300 m
5	085°/078°/0.79 m/s	079°/082°/0.93 m/s
5 (pure wind)	075°/113°/0.31 m/s	
7	083°/082°/0.96 m/s	082°/081°/1.46 m/s
7 (pure wind)	075°/113°/0.51 m/s	
9	083°/082°/1.12 m/s	083°/081°/1.83 m/s
9 (pure wind)	075°/113°/0.74 m/s	

The parameters of free drift in pure wind, when the sea state is not yet developed, are generally independent of the wind strength (except for the drift speed) and a ship's size.

The validation and application range of the results of Table 2 for investigating the orthogonal anchoring largely depends on whether a ship quickly starts a steady velocity drift. Some simulation studies to examine a transient phase of drift are scheduled in the nearest future. But if the required cable length to be paid out is sufficient, e.g. when a ship anchors in relatively large depths, see Table 1, then the steady drift movement will essentially dominate the phase of paying out the cable.

5 THE CASE OF TIDE ONLY

Since a weak tide is sometimes relatively stronger in terms of exciting forces and/or motions than a medium wind, the drift of a whole ship will much more follow the direction and speed of tide. Thus the tide will significantly affect the performance of the whole orthogonal anchoring. The wind/wave may sometimes be even neglected in such case. On the other hand, the pure current can also exist, e.g. in tidal areas, when good weather is prevailing.

From the viewpoint of ship manoeuvring dynamics , two equivalent concepts for modelling and simulation of a uniform, stationary current (i.e. revealing the constant velocity vector over the water space) exist. These are:
1 deriving equations of motions vs. ground, where all hydrodynamic forces, originated from a ship's hull, propeller, rudder, and added masses are however to be computed for the velocity vector through the water,
2 maintaining the standard equations of motion, i.e. for the rest water, where ship's motions in relation to water and ground are virtually the same, and finally at each discrete time step the requested ground-related velocity vector is calculated as the geometrical sum of the velocity vector through the water (from motion equations) and current velocity vector.

The latter idea, referred to as the kinematic approach, seems to be more beneficial and easier in studying the orthogonal anchoring.

However, one should be aware that trajectories against the water and ground are different. The ground-related trajectory of a ship, with regard to the midship or arbitrary point, has a direct nautical interpretation. Though both trajectories start from the same reference point in typical considerations, they further diverge or sheer. The ground-fixed trajectory is computed as the time integral of the ground-related velocity vector. This, according to the well known rule of integration, can be next decomposed into the sum of independent integrals for the water-related vector and the current vector.

For the anchor cable to be perpendicular to the hawse-pipe in the conditions of a tide, a ship has to stay on heading 90° versus the tide, if it is not making the way, or to make a significant headway, in the term of advance/forward speed, for angles less than 90°. The required forward velocity v_x and the resulting lateral drift velocity v_y, see Figure 4 and Table 3, are proportional to the water stream velocity v_C. The proportionality constants are dependent here on the tide direction γ_C and are equal to $\cos(\gamma_C)$ and $\sin(\gamma_C)$, accordingly.

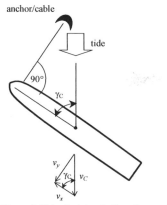

anchor/cable

tide

Figure 4. Velocity triangle for orthogonal anchoring in tide.

Table 3. Required forward velocity and final drift in tide.

Tide relative direction γ_C [°]	Forward velocity v_x/v_C [-]	Lateral drift v_y/v_C [-]
10°	0.98	0.17
20°	0.94	0.34
30°	0.87	0.50
40°	0.77	0.64
50°	0.64	0.77
60°	0.50	0.87
70°	0.34	0.94
80°	0.17	0.98
90°	0.00	1.00

This exact yet simple analysis is of course made for steady-state drift over ground in current, in which the current itself does excite hydrodynamic forces upon a ship's hull. In other words a ship moves together with the water mass of the tide. This is true if a ship was not initially restrained by either a mooring line or anchor cable, in which situation a sufficient time is necessary to develop a steady drift. This stipulation does not apply to the investigated problem of orthogonal anchoring nor to other classical problems of a ship manoeuvring in wind and tide.

There is one unquestionable benefit of pure tide condition that makes easier the orthogonal anchoring. Namely, the tide/current does not apply any yaw moment upon a ship during free manoeuvring, so we have no problem with keeping a constant heading. Only the propeller/engine setting needs to be adjusted to ensure the required forward component of a ship's velocity according to Table 3.

Hence the recommended procedure of orthogonal anchoring for areas with pure or dominating current conditions is essentially different from the primary procedure of McDowall (2000). It is much more similar to his tentative or secondary method as already discussed in the previous section 4. In other words, the specific U-turn phase is redundant. So, the objective here is just to relevantly reduce a ship's speed almost at arbitrary angle off the tide (of course, except for the in-line direction), maintain a proper headway and allow the ship to proceed with current. Any initial yaw movements can be checked by rudder application.

6 FINAL REMARKS

Another interesting question, not tackled within the present study, is whether the orthogonal method is also beneficial for smaller ships as compared with the conventional method. A risk or reliability assessment and cost-benefit analysis are required for both methods. This is worthwhile to do for decision-support systems or dedicated simulation-based shiphandling training of deck personnel. The latter is often focused on providing quantitative/technical details in given conditions of ship, area, and weather rather than giving a qualitative/general advice.

The idea of controlling the angular momentum instead of the linear fore and aft momentum, being the basis for the orthogonal anchoring, is very advantageous for the safety of anchoring operation. Additionally, keeping the anchor cable at right angle to a ship's side while paying out is a good visual indicator of a ship's forward speed. This is another advantage of the orthogonal anchoring. However, we pay for it with both the allocation of a lot of space at anchorage for the safe curvilinear track of a ship and the necessity of controlling a ship's swing after the anchor cable is tight. Also, the accuracy of reaching the target anchor position is not high.

The written original procedure of orthogonal anchoring has been evolving for the last years. Although expected to be very concrete/precise in terms of time moment for initiating particular action and corresponding engine/rudder command, this shall be rather treated as a very general/rough guidance, where there is only one undoubted goal. The goal is to initiate a sideways drift combined with ship's practically steady heading, the latter supported, if necessary and possible, by rudder/engine orders.

Given particular circumstances in form of water depth, approach waterways, hydrometeorological conditions and manoeuvring characteristics of a ship, the required sequence of detailed actions may and will significantly vary. Such spectrum of ship manoeuvring behaviour, especially during the turning and stopping phase in different weather conditions, can be investigated using simulation (e.g. full-mission) models and/or analytical models of ship manoeuvring. The analytical models, though comprising by nature a small piece of reality, allow sometimes more effective (less time-consuming) generalisation of the results. A comprehensive program of research on orthogonal anchoring by both simulation and analytical methods, as based on the author's developed SMART software (Artyszuk

2005), is scheduled soon and will supplement the initial considerations of the present paper by providing a parameterisation of the orthogonal anchoring.

It is also widely known that masters of many relatively large ships have skills in attempting to anchor their ships not in-line with wind/wave/tide, but with a certain amount of turning. However, their ability and actual performance of keeping a cable perpendicular to the hull, as well as the sideways drift velocity and especially the speed of paying out the cable, are not certain due to lack of recorded data. Hence it not possible to appraise the effectiveness of this 'turn-supported' anchoring, since using in these circumstances the name 'orthogonal' might be confusing. This practical method can be sometimes treated as a mix of both the conventional and the firm orthogonal method (according to McDowall). To exactly appraise its successful use, more detailed feedback data of practical seamanship experience is highly appreciated.

REFERENCES

Artyszuk, J. 2003a. Anchor Forces in Ship Manoeuvring Mathematical Model. *Annual of Navigation*, no. 6: 5-18.

Artyszuk, J. 2003b. Drift of a Disabled Ship Due to Irregular Waves. *Scientific Bulletin* (Marine Traffic Engineering Conference MTE' 2003), *Maritime University of Szczecin*, no. 70: 11-23.

Artyszuk, J. 2005. Revision and Quantification of Anchor Assisted Ship Manoeuvring Practice. In S. Gucma et al. (eds), *Seas&Oceans (2nd International Congress of Seas and Oceans ICSO'05), vol. 1 (Monographs)*, Szczecin-Swinoujscie, 20-24 September 2005: 13-26. Szczecin: Maritime University of Szczecin.

Clark, I.C. 2009. *Mooring and anchoring ships. Vol. 1 - Principles and practice*. London: The Nautical Institute.

Jurdzinski, M. 2005. *Anchoring of Large Vessels*. Gdynia: Foundation for Development of Gdynia Maritime University (in Polish).

McDowall, C.A. 2000. *Anchoring Large Vessels - a New Approach*. London: The Nautical Institute.

Morton, A.J. & Baines, B.H. & Ridgway K. 1987. Stopping and Anchoring Large Ships - a Feasibility Study. *IMarE* Trans. 99: 2-17.

OCIMF, 1982. *Anchoring Systems and Procedures for Large Tankers*. London: Witherby & Co..

Priest, J. 2005. *Anchoring Safely. A Videotel Production*. London: Videotel Productions/Steamship Mutual Underwriting Association (Bermuda) Ltd.

Chapter 6

Global Navigation Satellite Systems (GNSS)

Single-frequency Horizontal GPS Positioning Error Response to a Moderate Ionospheric Storm Over Northern Adriatic

R. Filjar, D. Brčić & S. Kos
Faculty of Maritime Studies, University of Rijeka, Croatia

ABSTRACT: Assessment of the real ionospheric impact on GNSS Positioning, Navigation and Timing (PNT) services is essential for identification of vulnerabilities and risks to GNSS performance and operation, as well as to performance and operation of the GNSS-reliant technological and socio-economic systems. Here we present the assessment of the impact of a moderate ionospheric storm on the horizontal component of the single-frequency GPS positioning error in Northern Adriatic. While clearly identifying the effects of a moderate ionospheric storm on northing, easting and horizontal components of GPS positioning error, the results of our analysis show the inexpensive monitoring of the lower ionosphere may provide a feasible means for fast identification of approaching ionospheric-originated GNSS positioning performance disturbances.

1 INTRODUCTION

Modern technological and socio-economic systems increasingly rely upon the operation and performance of satellite navigation systems (Thomas *et al*, 2011, Filjar & Huljenić, 2012). The GNSS vulnerabilities and risks assessment has become a major subject of numerous field studies in various positioning environments (Filjar & Huljenić, 2012). The GNSS performance response to space weather and ionospheric disturbances is one of the cornerstones of such an assessment (Hapgood & Thomson, 2010, American Meteorological Society, 2011, Filjar, 2008).

Here we report the results of an assessment of single-frequency horizontal GPS positioning error response to a moderate ionospheric storm on 7 March, 2012, based on the analysis of the single-frequency GPS positioning estimates' time series taken in the Northern Adriatic region (Hughes & Hase, 2010). Different sources of relevant data were used in the assessment, including GPS positioning samples and lower ionosphere screenings taken at GNSS Space Weather Laboratory, Faculty of Maritime Studies, University of Rijeka, Croatia. As the result, the time series of horizontal GPS positioning errors observed on both sides of the Northern Adriatic were compared both mutually and with the time series of the observed parameters of local ionospheric disturbances during the moderate ionospheric storm.

The paper is organised as follows. After the presentation of problem description and previous research (Chapter 2), an outline of the applied methodology is given (Chapter 3). The presentation of the results of the experimental data analysis (Chapter 4) are followed by discussion of the results of the analysis (Chapter 5). The paper concludes (Chapter 6) with the assessment's substance and proposals for future research.

2 PROBLEM DESCRIPTION AND PREVIOUS RESEARCH

The quality of service and robustness of the GNSS-based technological and socio-economic systems rely heavily on sustained and robust GNSS positioning, navigation and timing (PNT) services (Filjar and Huljenić, 2012, Thomas *et al*, 2011). Those are affected by various natural and artificial sources of errors and disruptions, with the space weather and the ionospheric effects as the major single source of GNSS PNT performance degradations and operation disruptions (Parkinson & Spilker, Jr, 1996).

The GNSS PNT performance degradation results from the satellite signals' propagation effects in the ionosphere, which increase the satellite signals' travelling time thus introducing the ionospheric GNSS PNT error (Davis, 1990, Helliwell, 2006, Parkinson & Spilker, Jr, 1996). The Maxwell Equations (Davis, 1990, Helliwell, 2006):

$$\nabla \times E = \frac{-\partial B}{\partial t}$$
$$\nabla \times B = \mu_o J \qquad (1)$$

give the following definition of frequency-dependent refractive index n (Davis, 1990, Helliwell, 2006):

$$n^2 = 1 - \frac{X}{1 - jZ - \left[\dfrac{Y_T^2}{2(1 - X - jZ)}\right] \pm \left[\dfrac{Y_T^4}{4(1 - X - jZ)}\right] + Y_L^2} \qquad (2)$$

with:

$$X = \frac{\omega_N^2}{\omega^2} \qquad Y = \frac{\omega_B}{\omega}$$
$$Y_L = \frac{\omega_L}{\omega} \qquad Y_T = \frac{\omega_T}{\omega} \qquad Z = \frac{v}{\omega} \qquad (3)$$

where ω_N = angular plasma frequency; ω_B = electron gyrofrequency; ω_L = longitudinal component of ω_B of direction of propagation; ω_T = transversal component of ω_B of direction of propagation; v = electron collision frequency; ω = angular wave frequency.

By definition, the GNSS ionospheric delay can be expressed as (Parkinson & Spilker, Jr, 1996):

$$\Delta t = \frac{1}{c} \int (1 - n) dh \qquad (4)$$

Using the approximation:

$$n \approx 1 - \frac{X}{2} \qquad (5)$$

the GNSS ionospheric delay can be expressed as (Parkinson & Spilker, Jr, 1996):

$$\Delta t = \frac{40.3}{c \cdot f^2} \int N(h) dh \qquad (6)$$

$N(h)$ represents the so-called vertical ionospheric profile, an ionospheric parameter highly affected by space weather and geomagnetic conditions. Using the definition of the Total Electron Content (Parkinson & Spilker, Jr, 1996):

$$TEC = \int N(h) dh \qquad (7)$$

the expressions (6) and (7) yields:

$$\Delta t = \frac{40.3}{c \cdot f^2} TEC \qquad (8)$$

The ionospheric-related GNSS PNT performance deteriorations can be mitigated using differential satellite navigation techniques (DGNSS) or a GNSS ionospheric correction model (for instance, Klobuchar model for GPS, or NeQuick model for Galileo) (Davis, 1990, Parkinson & Spilker, Jr, 1996). However, due to usually global nature of the GNSS ionospheric correction models, single-frequency GNSS receivers often cannot compensate the ionospheric impact in full, or, even worse, increase the amount of GNSS ionospheric error by utilising the corrections inappropriate for local ionospheric status (Davis, 1990, American Meteorological Society, 2011, Filjar, 2008, Filjar, Kos & Kos, 2009).

3 METHODOLOGY

Various sources of observations were used in the assessment of the moderate ionospheric storm on 7 March, 2012 in the Northern Adriatic region, as follows:

– IGS RINEX-organised data taken at Medicina, Italy reference station (IGS, 2012),
– US NOAA space weather indices (NOAA, 2012),
– single-frequency GPS position estimates and SID monitor-based (Stanford Solar Center, 2010, Filjar, Kos & Kos, 2010) lower ionosphere data collected at GNSS Space Weather Laboratory, Faculty of Maritime Studies, University of Rijeka, Croatia.

The GPS RINEX observed pseudo-ranges collected at a 12 channel-GPS reference station in Medicina, Italy were used for reconstruction of the single-frequency horizontal GPS positioning performance (dynamics of GPS-based positioning estimates). The GPS RINEX data allowed for reconstruction of the positioning estimates time series with the 30 s-sampling period. The standard (Klobuchar) GPS ionospheric delay correction model was applied with the transmitted Klobuchar model coefficients. The true position of the Medicina reference station was noted in the GPS RINEX file, thus allowing for definition of the horizontal GPS positioning error components with the reference to true position (Parkinson & Spilker, Jr, 1996), as follows:

$$\varphi_E[m] = \frac{(\varphi_M - \varphi_T) \cdot \pi \cdot R}{180} \qquad (9)$$

$$\lambda_E[m] = \frac{(\lambda_M - \lambda_T) \cdot \pi \cdot R \cdot \cos(\frac{\varphi_M \cdot \pi}{180})}{180} \qquad (10)$$

where φ_E = northing error; φ_M = measured latitude; φ_T = true latitude; λ_E = easting error; λ_M = measured longitude; λ_T = true longitude; R = 6378137 m; π =

3.1415927; latitudes and longitudes expressed in [°] units.

An R-based (R Development Core Team, 2010) software was developed for RINEX data processing and positioning estimates time series analysis.

Reconstruction of the space weather environment for satellite positioning was conducted using the NOAA SPIDR data, as shown in Figure 1.

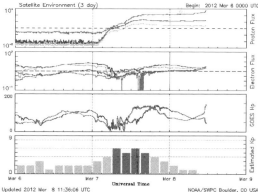

Figure 1. Space weather environment, 6 - 9 March, 2012 (SPIDR-NOAA, 2012).

The parallel time series of GPS positioning estimates were collected with the 2 s-sampling period using a single-frequency maritime GPS receiver at the GNSS Space Weather Laboratory, Faculty of Maritime Studies, University of Rijeka, Croatia, with the Klobuchar correction model applied. Again, a dedicated R-based software was developed for processing the GPS data in NMEA-0183 format and positioning estimates time series analysis. An advantage of a commercial Garmin GPS receiver's ability to provide the northing, easting and horizontal positioning error observables through a proprietary NMEA protocol extension has been exploited.

The local low-ionospheric status was reconstructed from time series of the Sudden Ionospheric Disturbance (SID) monitors' observations of two signals:
– 23.4 kHz signal broadcast from Rhauderfehn, Germany
– 20.9 kHz signal broadcast from Rosnay, France.

The analysis of daily dynamics of received VLF radio signal strength samples contributes to the understanding of near-real time local ionospheric status, especially of lower ionospheric layers (Rice et al, 2009). A characteristic observed daily dynamics of received Rosnay VLF radio signal strength, as seen at the GNSS Space Weather Laboratory, Faculty of Maritime Studies, University of Rijeka, Croatia (in the rest of the paper: Rijeka, Croatia monitoring station) during the undisturbed ionospheric conditions, is presented on Figure 2.

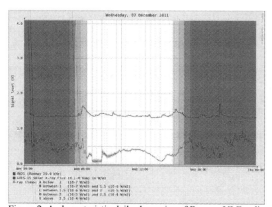

Figure 2. A characteristic daily dynamics of Rosnay VLF radio signal strength during the period of quiet ionospheric condition, as observed at Rijeka, Croatia monitoring station.

Two dedicated SID monitors are deployed along with the stationary reference single-frequency GPS receiver at Rijeka, Croatia monitoring station for continuous monitoring of two VLF signals from Rhauderfehn, Germany and Rosnay, France.

4 EXPERIMENTAL DATA ANALYSIS

The assessment's results generally fall into two categories:
– ionospheric dynamics, resulting from the geomagnetic storm in effect
– GPS positioning error dynamics, resulting from the ionospheric disturbances.

4.1 Ionospheric dynamics

Although announced as a potentially significant event, the geomagnetic storm of 7 March, 2012 turned to be a moderate storm. Caused by substantial Coronal Mass Ejection (CME) from the Sun, the geomagnetic storm followed a usual pattern, reaching the moderate activity level and dying out by the end of the day. Although its activities grew the following days, the storm remained at the moderate levels. Considering that the storm was preceded by a time interval of a low solar and ionospheric activity, the case of the geomagnetic storm on 7 March, 2012 provided a good environment for studying the GPS positioning performance during the moderately disturbed ionospheric and geomagnetic conditions, without the presence of remaining effects of previous ionospheric events.

General space weather conditions are summarised in Figure 1, provided by NOAA. It is evident from the figure that the space weather disturbance hit the Earth's environment in the early hours of 7 March, 2012, imminently causing geomagnetic disturbance with two slightly emphasised local maxima (around

0900 and 1500, both times UTC). The geomagnetic storm died out towards the dusk and the end of the day.

The effects of the global geomagnetic storm were conveyed to the ionosphere, as evident from SID observations of VLF signals shown in Figures 3 (Rhauderfehn station signal) and 4 (Rosnay station signal), respectively. Compared with the common daily variations of received signal strengths (Figure 2), observations of the received VLF signal strengths can indirectly point to the status of the low ionospheric dynamics. The comparison of Rhauderfehn signal observations on 7 March, 2012 (Figure 3) with a common daily variations of VLF signal strengths reveals intensive low-ionosphere activities around 0900, 1500, 1800 and 2100 (all times UTC). The analysis of Rosnay signal observations on 7 March, 2012 clearly shows increased low-ionosphere dynamics around 0600, 0900 and 1500 (all times UTC), while the increased night-time signal levels cannot be clearly related to the effects of the ionospheric storm.

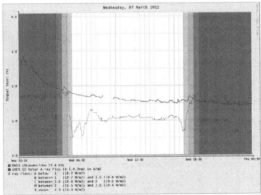

Figure 3. VLF Rhauderfehn, Germany 23.4 kHz signal received by SID monitor at Rijeka, Croatia monitoring station on 7 March, 2012.

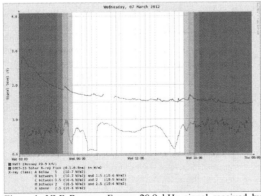

Figure 4. VLF Rosnay, France 20.9 kHz signal received by SID monitor at Rijeka, Croatia monitoring station on 7 March, 2012.

4.2 GPS positioning error dynamics

Two independent sets of single-frequency GPS positioning estimates time series were used in research:

- RINEX observables taken at IGS Medicina, Italy GNSS reference station, and
- NMEA observables collected by stationary reference single-frequency GPS receiver at GNSS laboratory, Faculty of Maritime Studies, University of Rijeka, Croatia

The selection of the sources of experimental observables was carried out with the aim to provide the coverage of the Northern Adriatic area, with intensive commercial, recreational and military maritime traffic. Corrective measures were in operation at both sites to minimise the effects of multipath, and space and control segment.

The Medicina RINEX observables at 30 s sampling rate, taken from the IGS internet archive, was used for simulated single-frequency GPS positioning estimates, from which the GPS positioning error response to disturbed ionospheric conditions was derived (Figure 5). Medicina GPS error components time series dynamics on 7 March, 2012 are summarised in Table 1.

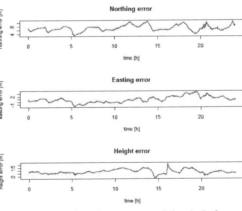

Figure 5. Time series of components of the single-frequency GPS position estimate errors, taken at Medicina, Italy GNSS reference station on 7 March, 2012 (30 s-sampling rate).

Table 1. Medicina (RINEX) GPS positioning error dynamics on 7 March, 2012.

Error component	Dynamics description (all times UTC)
Northing	Error dynamics noticeable, local maxima around 1200, 1400, 1730, 2100, plateau between 0700 and 0900
Easting	Negligible effects of the ionospheric storm, values slightly increased after 1600
Height	Daily variation pattern slightly disturbed, with local maxima around 1600 and 1800

168

The Rijeka NMEA observables at 2 s sampling rate were used as a source of receiver-based estimates of components of the GPS positioning error (Figure 6). The effects of the stochastic component were removed by utilisation of a simple Moving Average filter of the order of 120 (s).

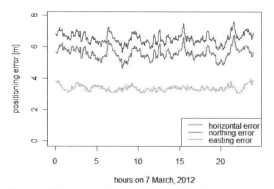

hours on 7 March, 2012

Figure 6. Time series of components of the single-frequency GPS position estimate errors, taken at Rijeka, Croatia monitoring station on 7 March, 2012 (2 s-sampling rate, simple moving average filter of order 120 s applied).

Table 2. Rijeka (NMEA) GPS positioning error dynamics on 7 March, 2012.

Error component	Dynamics description (all times UTC)
Northing	Noticeable effects of the ionospheric storm, local maxima around 0600, 0900, 1500, 1800
Easting	Negligible effects of the ionospheric storm, northing error consisting of DC bias and a minor stochastic component
Horizontal	Daily variation governed by easting positioning error dynamics, local maxima around 0600, 0900, 1500, 1800, 2100

5 DISCUSSION

The analysis of Medicina (RINEX) and Rijeka (NMEA) data sets revealed the impact of the minor ionospheric storm on 7 March, 2012 on GPS horizontal positioning error. Prominent effects were observed on the northing positioning error component both at Medicina, Italy, and Rijeka, Croatia. In both cases, the amounts of error exceed the common values at the times of local maxima. Prominent ionospheric storm effects on northing positioning error component, with the easting positioning error component experiencing only negligible effects, were observed in other case-studies in the region (Filjar, 2008, Filjar, Kos & Kos, 2007, Filjar, Kos & Kos, 2006).

Local maxima of horizontal GPS positioning errors at both Medicina, Italy and Rijeka, Croatia were well correlated with the local maxima of received VLF signals from both Rhauderfehn,

Germany and Rosnay, France, with the only exception of 1200 UTC maximum of GPS positioning error not correlated with VLF signal strength observables. This leads to the conclusion that the minor ionospheric storm on 7 March, 2012 created a regional ionospheric event that affected in the same manner the areas of both Medicina, Italy and Rijeka, Croatia. In due course, low ionosphere observations taken in Rijeka can be applied regionally for description of local low ionosphere dynamics, at least for the case in consideration.

The additional analysis of Rijeka GPS positioning errors dynamics was performed, with the outcome as follows. Histogram of Rijeka horizontal GPS positioning error estimates (Figure 7) shows a very good fit with the Gaussian (normal) distribution. This allows for establishment of simple statistical model of the process of the horizontal GPS positioning error generation during the ionospheric storm on 7 March, 2012.

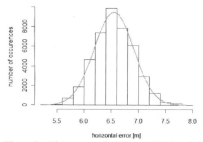

horizontal error [m]

Figure 7. Histogram of GPS positioning horizontal errors observed on 7 March, 2012 (curved-orange line represents the equivalent normal distribution).

Finally, a simple non-linear analysis (Hilborn, 2000, Ott, 2002) applied to GPS positioning time series did not reveal any embedded complex dynamics of horizontal GPS positioning error time series, as seen in two-dimensional state space diagrams of northing and easting GPS positioning errors (Figures 8 and 9, respectively).

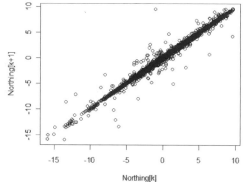

Northing[k]

Figure 8. Two-dimensional phase space diagram of northing GPS positioning errors.

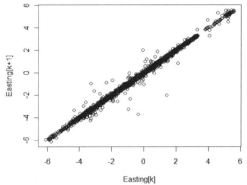

Figure 9. Two-dimensional phase space diagram of easting GPS positioning errors.

6 CONCLUSION

The results of the assessment of a minor ionospheric storm effects on a single-frequency GPS positioning performance are presented in this paper, with an emphasis given on the effects of the low-ionosphere dynamics. Time series of the actual GPS positioning error observables taken at Medicina, Italy and Rijeka, Croatia were correlated with common space weather indices, as well as with the indirect observables of low-ionosphere dynamics, taken at Rijeka, Croatia.

Our analysis show that the minor ionospheric storm of 7 March, 2012 created noticeable effects to GPS horizontal positioning error, but without embedded complexity. The effects of the ionospheric storm created similar satellite positioning conditions on both sides of the Adriatic Sea, suggesting an opportunity for a co-ordinated approach to identification and mitigation of the approaching ionospheric effects on GNSS PNT performance and operation.

The local maxima of horizontal GPS ionospheric errors were well correlated with VLF signal strengths recorded at Rijeka, Croatia, suggesting that the low-cost low-ionosphere activity monitoring may provide a valuable means for regional identification of approaching GPS positioning performance disturbances.

This paper reports the outcome of a case-study analysis which is to be explored and verified further in analysis of similar events taking place in the region in the future.

ACKNOWLEDGEMENTS

Research activities presented in this paper were conducted under the research project *Research into the correlation of maritime-transport elements in marine traffic*, supported by the Ministry of Science, Education and Sports, Republic of Croatia.

REFERENCES

American Meteorological Society. 2011. *Understanding Vulnerability & Building Resilience*. AMS Washington, DC. Available at: http://bit.ly/rnkRQu, accessed on 13 August, 2012.

Davis, K. 1990. *Ionospheric radio*. Peter Peregrinus Ltd. London, UK.

Filjar, R & Huljenić, D. 2012. The importance of mitigation of GNSS vulnerabilities and risks. *Coordinates*, 8(5): 14-16.

Filjar, R., Kos, T. & Kos, S. 2010. Low-cost space weather sensors for identification and estimation of GNSS performance spae weather effects. *Proc of NAV10 Conference*. Westminster, London, UK.

Filjar, R., Kos, T. & Kos, S. 2009. Klobuchar-like Local Model of Quiet Space Weather GPS Ionospheric Delay for Northern Adriatic. *Journal of Navigation*, 62(3): 543-554.

Filjar, R., Kos, T. & Gold, H. 2009. Availability of the international GNSS Service observables for GNSS performance studies in Croatia. *Proc of 2nd GNSS Vulnerabilities and Solutions Conference*. Baška, Croatia.

Filjar, R. 2008. A Study of Direct Severe Space Weather Effects on GPS Ionospheric Delay. *Journal of Navigation*, 61(1): 115-128.

Filjar, R., Kos, S. & Kos, T. 2007. GPS Positioning Accuracy in Severe Space Weather Conditions in Croatia: A 2003 Halloween Event Case Study. *Proc of NAV07 Conference*. Westminster, London, UK.

Filjar, R. & Kos, T. 2006. GPS Positioning Accuracy in Croatia During the Extreme Space Weather Conditions in September 2005. *Proc of the European Navigation Conference ENC 2006*. Manchester, UK.

Hapgood, M. & Thomson, A. 2010. Space Weather: Its impact on Earth and implications for business. *Lloyd's 360° Risk Insight*. Lloyd's, London, UK. Available at: http://bit.ly/9Pjk9R, accessed on 13 August, 2012.

Helliwell, R. A. 2006. *Whistlers and Related Ionospheric Phenomena*. Dover Publications,Mineola, New York, USA.

Hilborn, R. 2000. *Chaos and Nonlinear Dynamics: An Introduction for Scientists and Engineers*. Oxford University Press, Oxford, UK.

Hughes, I. G. and Hase, T. P. A. 2010. *Measurements and their Unscertainties: A Practical Guide to Modern Error Analysis*. Oxford University Press, Inc. New York, USA.

IGS: International GNSS Service. 2012. Available at: http://bit.ly/9K6tRn, accessed on 11 August, 2012.

NOAA: NOAA Space Weather Prediction Center. 2012. Available at: http://www.swpc.noaa.gov, accessed on 13 August, 2012.

Ott, E. 2002. Chaos in Dynamical System, 2nd ed. Cambridge University Press, Cambridge, UK.

Parkinson, B. W. & Spilker, Jr. J. J. (eds). 1996. Global Positioning System: Theory and Practice, Vol. I. AIAA, Washington, DC, USA.

R Development Core Team. 2010. R: A language and environment for statistical computing. R Foundation for Statistical Computing, ISBN 3-900051-07-0, Vienna, Austria. Available at: http://www.R-project.org, accessed on 11 August, 2012.

Rice, D. D. et al. 2009. Characterizing the Lower Ionosphere with a Space-Weather-Aware Receiver Matrix. URSI Radio Science Bulletin, 328:20-32. Available at: http://bit.ly/Ik0nyf, accessed on 11 August, 2012.

Stanford Solar Center. 2012. The SID Monitor website. Available at: http://bit.ly/h6e1lc, accessed on 11 August, 2012.

Thomas, M. et al. 2011. Global Navigation Space Systems: Reliance and Vulnerabilities. The Royal Academy of Engineering, London, UK. Available at: http://bit.ly/feFB2i, accessed on 09 August, 2012.

Global Navigation Satellite Systems (GNSS)
Advances in Marine Navigation – Marine Navigation and Safety of Sea Transportation – Weintrit (ed.)

Accuracy of the GPS Position Indicated by Different Maritime Receivers

J. Januszewski

Gdynia Maritime University, Gdynia, Poland

ABSTRACT: Nowadays on the ship's bridge two or even more GPS receivers are installed. As in the major cases the coordinates of the position obtained from these receivers differ the following questions can be posed – what is the cause of this divergence, which receiver in the first must be taken into account etc. The measurements of GPS position based on the four different stationary GPS receivers were realized in the laboratory of Gdynia Maritime University in Poland in the summer 2012. The coordinates of the position of all these receivers were registered at the same time. The measurements were made for different input data, the same for all receivers. The distances between the individual unit's antenna were considered also. Next measurements were realized on two ships in different European ports. The results showed that the GPS position accuracy depends on the type of the receiver and its technical parameters particularly.

1 INTRODUCTION

An uninterrupted information about the ship's position is one of the most important elements of the safety of navigation in the sea transport in restricted and coastal areas, recommended by International Maritime Organization – IMO (imo.org]. The information about user's position is obtained generally from specialized electronic position–fixing systems, in particular, satellite navigation systems (SNS) as the American GPS system (Januszewski J. 2010, Hofmann-Wellenhof B. et al. 2008, Kaplan, E.D. & Hegarty, C.J. 2006). Nowadays (January 2013) this system is fully operational with very robust constellation 31 satellites (www.gpsworld.com). On each ship's bridge one GPS stationary receiver is installed at least but on many ships there are two or even more GPS receivers. As in the major cases the coordinates of the position obtained from these receivers differ the following questions can be posed:
– what is the cause of this divergence?
– which receiver in the first must be taken into account and why?
– which input data has the biggest influence on position's accuracy?

2 GPS POSITION ACCURACY

The accuracy of the user's position solution determined by SNS is ultimately expressed as the product of a geometry factor and a pseudorange error factor (Kaplan, E.D. & Hegarty, C.J. 2006):

(error in SNS solution) =
(geometry factor) x (pseudorange error factor) (1)

As the error solution can be expressed by M – the standard deviation of the positioning accuracy, geometry factor by the dilution of precision (DOP) coefficient and pseudorange error factor by σ_{UERE} (UERE – User Equivalent Range Error), these relation can be defined as:

$$M = DOP \cdot \sigma_{UERE} \qquad (2)$$

If we can obtain all four coordinates of the observer's position (latitude, longitude, altitude above given ellipsoid, time – φ, λ, h, t), factor DOP is expressed by GDOP (Geometric Dilution of Precision) coefficient, if we want obtain horizontal coordinates, only, geometry factor DOP is expressed by HDOP (Horizontal Dilution of Precision) coefficient. In this situation the horizontal accuracy with 95% confidence level $M_{\varphi,\lambda}^{95\%}$ can be approximated by:

$$M_{\varphi,\lambda}^{95\%} \approx 2 \cdot HDOP \cdot \sigma_{UERE} \qquad (3)$$

Specializing the equation giving the functional relationship between the errors in the pseudorange values and the induced errors in the computed position for the horizontal dimension we can say that the 95% point for the distribution of the horizontal error (2 σ) can be estimated by the doubled product of HDOP coefficient and the user equiva-lent range error (σ_{UERE}).

Global GPS civil service performance commitment met continuously since December 1993. Root Mean Square (RMS) Signal-in-Space (SiS) User Range Error (URE) was equal 1.6 metres in 2001, 1.1 in 2006 and finally 0.9 in 2011 (Gruber, B. 2012), St. Marie T. 2012). Signal in Space User Range Error is the difference between a GPS satellite's navigation data (position and clock) and the truth projected on the line of sight to the user.

The HDOP coefficient value depends on the SNS geometry, the user's coordinates, time and the number of satellites fully operational in given moment, in particular. At the beginning of XXI century for GPS system this value was about 1.5, at present (January 2013) is about 1.0 because very robust GPS constellation consists of 31 space vehicles currently in operation (www.navcen.uscg.gov).

3 THE MEASUREMENTS OF GPS POSITION

The measurements of GPS position based on stationary receivers were realized in the laboratory of University and on two ships in different ports in Europe.

3.1 *The measurements in the laboratory*

The measurements of GPS positions based on four different stationary receivers were realized in the laboratory of Gdynia Maritime University in Gdynia in Poland in the summer 2012. These receivers were:
– MX 200 Professional Navigator, Magnavox, cal-led later MX 200,
– ap Mk10 Professional, Leica, called later MK10,
– MX 512 Simrad Navigation System, called later Simrad,
– NR–N124 Marine Navigator, MAN Technologies called later MAN.

The three coordinates of positions (latitude, longitude, height above ellipsoid WGS–84) of all four receivers, the positions of antennas of theses receivers in reality, obtained from geodetic measurements (Real Time Kinematic – RTK) and production year are presented in the table 1. The antennas of these receivers were installed on the masts on the roof of the university building. All positions were measured and presented in WGS–84 datum.

The measurements were made for different input data, the same for all four receivers:
– two different masking elevation angles H_{min}, 5^0 and 25^0, the most frequently used angle in the receivers in open area (e.g. ocean navigation) and typical angle for restricted area (e.g. coastal navigation), respectively,
– two different datums, WGS–84 and Timbolaia 1948, datum officially applied for GPS system and the datum for which the position offset relative to WGS–84 is significant, respectively. For Timbolaia datum (ellipsoid Everest) this offset is S $0^0 00.2696$' ; W $0^0 00.81$'.

The days of measurements, the registration duration and the number of positions registered by each receiver in mode 3D for different datums and angles H_{min} are given in the table 2. In each series during about 24 hours selected data:
– three coordinates of position: latitude, longitude and height (above selected ellipsoid),
– Horizontal Dilution of Precision HDOP coefficient,
– the number of satellites ls used in the receiver in position calculation

were registered by PC with sampling interval of 10 second (Mk10, Simrad and MAN) and 2 or 3 second (MX 200). If any data has been incomplete, this measurement was rejected. That's why for the given period of the measurements the number of positions obtained from different receivers is not the same.

In the case of all receivers if we change the masking angle the coordinates of the position determined for the new value of this angle signaled on the screen and registered by PC are the same. Meanwhile if we change the datum from WGS–84 for Timbolaia the coordinates on the screen can differ from coordinated registered. In the case of MX 200, MK10 and MAN receivers the coordinates signaled on the screen are determined in "the new" Timbolaia datum while registered by PC are still in "the old" WGS–84 datum. Only in Simrad receiver all coordinates are in Timbolaia datum in each case.

In MX 200, Mk10 and Simrad receivers the vertical separation between the geoid and reference ellipsoid WGS–84, called geoid undulation, was + 34.3 m (the value for Gdansk Bay). In these three receivers in mode 3D the current value of the geoidal height (the receiver's antenna height above geoide), and no ellipsoidal height, is signaled on the screen, while in MAN receiver the ellipsoidal height is signalized only.

The total number of positions registered by all four receivers was 192314: MX 200 – 114968, Mk10 – 25577, Simrad – 25 840, MAN –25929.

For each series of measurements and for each receiver were calculated:
– three coordinates (latitude, longitude and height above given ellipsoid) of mean position,

Table 1. The measurements in the laboratory – the coordinates of the GPS receivers installed in Gdynia Maritime University in Poland (RTK position), datum WGS–84.

Receiver		Coordinates		
Model	Production year	Latitude	Longitude	Height above ellipsoid WGS–84 [m]
MX 200	1991	$54^O 31'5.04922"$ N	$018^O 33'16.36049"$ E	57.48
Mk10	1995	$54^O 31'5.04918"$ N	$018^O 33'16.39215"$ E	57.48
Simrad	2011	$54^O 31'5.18485"$ N	$018^O 33'16.50982"$ E	54.21
MAN	2001	$54^O 31'5.07664"$ N	$018^O 33'16.38120"$ E	54.65

Table 2. The measurements in the laboratory – the days of measurements, the registration duration and the number of positions for different receivers, datums and angles H_{min}.

H_{min} [^{O}]	Datum	Day	Receiver	Start of measurements	Duration	Number of positions
5	WGS 84	12.07.2012	MX 200	10:03:03	24 h 36 min	39 513
			Mk10	10:02:29	24 h 19 min	8 686
			Simrad	10:04:26	24 h 34 min	8 851
			MAN	10:01:40	24 h 37 min	8 868
25	WGS 84	31.07.2012	MX 200	11:50:23	23 h 02 min	36 879
			Mk10	11:52:27	23 h 01 min	8 247
			Simrad	11:51:16	23 h 24 min	8 429
			MAN	11:51:01	23 h 41 min	8 506
5	Timbolaia 1948	01.08.2012	MX 200	11:33:56	24 h 05 min	38 576
			Mk10	11:35:17	24 h 01 min	8 644
			Simrad	11:50:56	23 h 46 min	8 560
			MAN	11:52:51	23 h 43 min	8 545

Table 3. The measurements in the laboratory – the coordinates of the mean position, the mean ellipsoidal height and the difference ΔH between mean height and RTK height for different receivers, angles H_{min} and datums.

Receiver	Datum	H_{min} [^{O}]	Mean position		Mean height [m]	ΔH [m]
			Latitude	Longitude		
MX 200	WGS 84	5	$54^O 31'5.07213"$ N	$018^O 33'16.39668"$ E	54.84	− 2.64
		25	$54^O 31'5.06263"$ N	$018^O 33'16.40463"$ E	57.28	− 0.20
	Timbolaia 1948	5	$54^O 30'48.88686"$ N	$018^O 32'27.80670"$ E	56.62	− 0.86
Mk10	WGS 84	5	$54^O 31'5.10862"$ N	$018^O 33'16.39753"$ E	60.36	+ 2.88
		25	$54^O 31'5.13099"$ N	$018^O 33'16.40908"$ E	62.23	+ 4.75
	Timbolaia 1948	5	$54^O 30'48.94974"$ N	$018^O 32'27.75545"$ E	61.92	+ 4.44
Simrad	WGS 84	5	$54^O 31'5.21343"$ N	$018^O 33'16.52619"$ E	53.75	− 0.46
		25	$54^O 31'5.21831"$ N	$018^O 33'16.52328"$ E	54.31	+ 0.10
	Timbolaia 1948	5	$54^O 30'49.03898"$ N	$018^O 32'27.92994"$ E	54.21	0
MAN	WGS 84	5	$54^O 31'5.08449"$ N	$018^O 33'16.41743"$ E	52.09	− 2.56
		25	$54^O 31'5.10673"$ N	$018^O 33'16.42334"$ E	54.12	− 0.53
	Timbolaia 1948	5	$54^O 30'48.91694"$ N	$018^O 32'27.82025"$ E	53.95	− 0.70

Table 4. The measurements in the laboratory – the distances [m] between RTK positions and between receiver's antennas for all 6 pairs of receivers.

Receiver	MX		Mk10		Simrad		MAN	
	RTK	antenna	RTK	antenna	RTK	antenna	RTK	antenna
MX	–	–	0.57	0.50	4.98	4.79	0.93	0.90
Mk10	0.57	0.50	–	–	4.70	4.92	0.87	0.85
Simrad	4.98	4.79	4.70	4.92	–	–	3.96	4.13
MAN	0.93	0.90	0.87	0.85	3.96	4.13	–	–

Table 5. The measurements in the laboratory – the distance d between RTK position and mean position of the receiver and the bearing α between them from RTK position for different receivers, angles H_{min} and datums.

Datum	H_{min} [^{O}]	MX 200		MK10		Simrad		MAN	
		d [m]	α [^{O}]	d [m]	α [^{O}]	d [m]	α [^{O}]	d [m]	α [^{O}]
WGS 84	5	0.86	42	1.84	3	0.92	16	0.694	69
	25	0.89	62	2.55	83	1.06	77	0.862	61
Timbolaia 1948	5	1004.56	240	1004.89	240	1004.72	240	1003.46	240

- 2D – two dimensional (horizontal) distance from the known position of the receiver (RTK) to the mean receiver position,
- 3D – three dimensional distance from the known position of the receiver (RTK) to the mean receiver position,
- latitude error σ_φ and longitude error σ_λ,
- height error σ_H,
- horizontal position error $\sigma_{\varphi,\lambda}$, (σ_{2D} with confidence level 95%),
- three dimensional position error $\sigma_{\varphi,\lambda,H}$ (σ_{3D} with confidence level 95%),
- mean value of HDOP coefficient.

Position fix in mode "3D" can be calculated only from these satellites, which elevation angle in observer's receiver at the moment of measurement is higher than masking elevation angle H_{min}. At any moment the user's receiver needs to see at least four satellites.

3.2 *The measurements on the ships*

The measurements were realized on the ship DP3 Edda Fides (the first floating hotel and service vessel to be built exclusively for the offshore oil and gas industry (www.marinetraffic.com) equipped with:
- two identical SAAB R4 GPS/DGPS Navigation System (www.saabgroup.com) receivers, called later SAAB A and B, port Haugesund in Norway, November 2011, a dozen or so one-hour sessions, This receiver L1, C/A consists of 12 channels (2 dedicated to SBAS and DGPS by SBAS or externally RTCM corrections). SAAB R4 is receiver approved for SOLAS and any other precision navigation application. The actual distance between receiver's antennas was 1.5 m;
- two identical Kongsberg DPS 232 (GPS L1/L2 + GLONASS L1/L2 + SBAS) receivers, called later Konsgberg A and B, in the region of the port Castellon in Spain, November and December 2011, a dozen or so one-hour sessions. DPS 232

is an all-in-one DP new generation GNSS-based position reference system, which takes positioning to the next level for secure and robust solutions exerting GPS and GLONASS. SBAS or DGPS horizontal user's position accuracy is less than 1 m (95%) (www.km.kongberg.com). The actual distance between receiver's antennas was 7.0 m.

In each series of the measurements the geographic coordinates of two receivers were registered at the same time with sampling interval of 60 second, mode 3D, datum WGS–84 for different input data introduced in receiver A or in receiver B or both receivers.

The next measurements of positions based on two GPS receivers installed on the general cargo ship SMT Bontrup (length overall 200.5 m) were realized in three European ports in 2012 year, in each port two one hour sessions. In each case as the real position it was assumed the position coordinates red from the chart which datum was WGS–84. These receivers were Furuno GPS Navigator GP–90, called later Furuno and SAAB R4 GPS/DGPS Navigation System, used as GPS only, called later SAAB. The coordinates of the receivers were registered at the same time with sampling interval of 60 second, mode was 3D, datum WGS–84 and masking angle $H_{min} = 5^O$.

Table 6. The measurements in the laboratory – minimal ls_{min} and maximal ls_{max} number of satellites used in the position calculation for different angles H_{min} and datums and for different receivers.

Receiver	WGS 84				Timbolaia 1948	
	$H_{min} = 5^O$		$H_{min} = 25^O$		$H_{min} = 5^O$	
	ls_{min}	ls_{max}	ls_{min}	ls_{max}	ls_{min}	ls_{max}
MX 200	5	6	3	6	5	6
Mk10	4	6	3	6	4	6
Simrad	8	10	4	8	6	10
MAN	7	12	4	8	8	12

Table 7. The measurements in the laboratory – the errors σ_φ, σ_λ, σ_{2D}, σ_H, σ_{3D} and HDOP coefficient value for different receivers, angles H_{min} and datums.

H_{min} [^{O}]	Datum	Receiver	σ_φ [m]	σ_λ [m]	σ_{2D} [m]	σ_{2D} (95%) [m]	σ_H [m]	σ_{3D} (95%) [m]	HDOP
5	WGS 84	MX 200	1.74	1.07	2.04	4.08	3.43	7.98	1.74
		Mk10	2.51	1.88	3.14	6.28	3.51	9.42	1.27
		Simrad	0.60	0.45	0.75	1.50	0.81	2.43	0.92
		MAN	0.71	0.44	0.83	1.66	2.00	4.33	0.89
25	WGS 84	MX 200	4.03	1.20	4.20	8.40	9.23	20.29	2.04
		Mk10	3.43	2.36	4.16	8.32	8.09	18.20	2.01
		Simrad	1.71	0.74	1.86	3.72	3.52	7.96	2.00
		MAN	1.66	0.70	1.80	3.60	3.22	7.38	2.06
5	Timbolaia 1948	MX 200	1.87	1.34	2.30	4.60	3.81	8.90	1.73
		Mk10	3.25	2.71	4.23	8.46	5.52	13.91	1.31
		Simrad	0.59	0.52	0.79	1.58	1.21	2.89	0.92
		MAN	0.72	0.68	0.99	1.98	2.03	4.52	0.89

Table 8. The measurements on the ship DP3 Edda Fides – latitude error σ_φ, longitude error σ_λ, horizontal position error σ_{2D} (95%), distance D_{AB} between mean positions of two SAAB R4 GPS/DGPS receivers for different input data.

No	Input data Receiver A	Receiver B	Receiver A σ_φ [m]	σ_λ [m]	σ_{2D} [m] (95%)	Receiver B σ_φ [m]	σ_λ [m]	σ_{2D} [m] (95%)	D_{AB} [m]
1	GPS/SBAS, $H_{min} = 5^O$	GPS/SBAS, $H_{min} = 5^O$	0.17	0.19	0.51	0.22	0.14	0.52	0.75
2	GPS/SBAS, $H_{min} = 5^O$	23 h 56 min later than the measurements number 1, the same satellite constellation	0.17	0.19	0.51	0.17	0.14	0.44	0.60
3	GPS/SBAS, $H_{min} = 5^O$	GPS/SBAS, $H_{min} = 20^O$	0.18	0.15	0.47	1.72	0.62	3.66	1.02
4	GPS/SBAS, $H_{min} = 5^O$, different GPS satellite constellation for A and B		0.52	0.48	1.42	1.48	0.60	3.19	4.64
5	GPS/SBAS, $H_{min} = 5^O$	GPS/DGPS, $H_{min} = 5^O$	0.15	0.32	0.71	0.27	0.16	0.63	1.20
6	GPS/SBAS, $H_{min} = 5^O$	GPS only, $H_{min} = 5^O$	0.21	0.27	0.68	1.45	1.38	4.00	1.88

Table 9. The measurements on the ship DP3 Edda Fides – latitude error σ_φ, longitude error σ_λ, horizontal position error σ_{2D} (95%), distance D_{AB} between mean positions of two Kongsberg DPS 232 receivers for different daytimes and the measurements conditions.

No	Measurements conditions Receiver A	Receiver B	Receiver A σ_φ [m]	σ_λ [m]	σ_{2D} [m] (95%)	Receiver B σ_φ [m]	σ_λ [m]	σ_{2D} [m] (95%)	D_{AB} [m]
1	GPS/GLONASS, sunset		0.25	0.17	0.60	0.33	0.21	0.78	0.16
2	GPS/GLONASS, sunrise		0.37	0.31	0.97	0.24	0.48	1.07	0.56
3	GPS/GLONASS, trans-shipment		0.30	0.19	0.71	0.24	0.30	0.77	2.21
4	change of GPS and GLONASS satellites used in both receivers in position calculation		2.16	2.97	7.34	0.37	3.23	6.55	1.69

Table 10. The measurements on the ship SMT Bontrup – latitude error σ_φ, longitude error σ_λ, horizontal position error σ_{2D} (95%), distance D between mean position and real position, maximal distance between GPS position and real position for different receivers in different days and ports.

Port	Day	Receiver	σ_φ [m]	σ_λ [m]	σ_{2D} [m] (95%)	D [m]	D_{max} [m]
Amsterdam (Netherlands)	22.02.2012	Furuno	1.25	2.12	4.90	8.17	13.05
		Saab	1.48	2.36	5.58	7.50	13.00
	23.02.2012	Furuno	1.37	1.79	4.50	6.78	11.34
		Saab	1.60	1.96	4.92	8.85	16.67
Antwerp (Belgium)	21.01.2012	Furuno	1.98	1.84	5.40	5.60	10.98
		Saab	1.63	1.46	4.36	7.88	11.58
	02.03.2012	Furuno	1.63	2.44	5.86	4.18	9.23
		Saab	1.28	1.87	4.54	4.98	9.83
Bremanger (Norway)	01.01.2012	Furuno	1.80	2.54	6.22	7.19	15.17
		Saab	1.35	1.60	4.16	7.30	13.14
	18.02.2012	Furuno	1.89	2.30	5.96	5.53	13.53
		Saab	1.35	2.26	5.26	7.66	13.68

4 THE RESULTS

The distances between the two positions on the ellipsoid WGS–84 were calculated from the relation (Admiralty Manual of Navigation. 2008):

$$1' = 1852.22 - 9.32 \cos 2\varphi_m \quad (4)$$

where φ_m = middle latitude of two positions.

4.1 The measurements in the laboratory

The coordinates of the mean position, the mean ellipsoidal height and the difference ΔH between the mean height and RTK height for all four receivers, different angles H_{min} and datums are presented in the table 3. For Mk 10 the mean height is greater than RTK height considerably (almost 5 m), for MX 200 and MAN the mean height is less (– 2.78 m) than RTK height for all H_{min} and datums, and for Simrad the difference ΔH is the least, its absolute value doesn't exceed 0.5 m.

The distances between RTK positions and between receiver's antennas for all 6 pairs of receivers are showed in the table 4. After the comparison of RTK distance with the actual antennas distance for each pair we can say that the smallest and the greatest difference ΔD of these distances is in the case of pair Mk10 & MAN receivers (2 cm) and of the pair MK10 & Simrad receivers (22 cm) respectively. The difference ΔD doesn't exceed 13% of the bigger distance that time.

The distance d between RTK position and mean position of the receiver and the bearing α between them from RTK position for different receivers, angles H_{min} and datums are presented in the table 5. For datum WGS–84 and for all receivers because of smaller number of satellites used in position calculation the distance d is for $H_{min} = 25^O$ greater than for $H_{min} = 5^O$ but this increase is little. For datum Timbolaia for all receivers distance d increases considerably, until almost 1005 m. It's because the position offset relative is for this datum significant.

The minimal ls_{min} and maximal ls_{max} number of satellites used in the position calculation for different angles H_{min}, datums and receivers are showed in the table 6. For all receivers the numbers ls_{min} and ls_{max} depend on the number of receiver's channels lc. In the case MX 200 and Mk10 receivers as lc = 6 the number ls_{max} is for $H_{min} = 5^O$ and independently of datum equal 6 only, it means that the number ls_{max} is less than the number of satellites visible by the antenna considerably. For the same receivers if $H_{min} = 25^O$ the number ls_{max} is the same (6) but ls_{min} decreases up to 3, it means that the position was determined in mode 2D. For Simrad and MAN receivers the number ls_{max} is for $H_{min} = 5^O$ equal the number lc, 10 and 12 respectively. For these receivers the number ls_{min}, 8 and 7, respectively, is greater than the number ls_{max} (6) for MX 200 and Mk 10 receivers.

The errors σ_φ, σ_λ, σ_{2D}, σ_H, σ_{3D} and HDOP coefficient value for different receivers, angles H_{min} and datums are showed in the table 7. We can say that:

– for each receiver and independently of datum the errors σ_φ, σ_λ, σ_H are for $H_{min} = 25^O$ greater than for $H_{min} = 5^O$,
– for all 3 series of measurements the errors σ_φ, σ_λ, σ_H are for Simrad and MAN receivers smaller than for MX 200 and MK10 receivers considerably, twice or even more,
– for all 3 series of measurements the error σ_H is for all receivers greater than σ_φ as well as σ_λ. This difference is particularly evident if $H_{min} = 25^O$, at least twice. It's because the number of satellites used in position calculation is that time less considerably,
– the error σ_{2D} (95%) is for all receivers less than 10 m, also for angle $H_{min} = 25^O$,
– if $H_{min} = 5^O$, independently of datum, the error σ_{3D} (95%) is the smallest for Simrad receiver, if $H_{min} = 25^O$ this error is the smallest for MAN receiver. It's because of the greater number of satellites used in position calculation, MAN – 12, Simrad – 10 only,
– if $H_{min} = 5^O$, independently of datum, HDOP coefficient value is for Simrad and MAN receivers less than for MX 200 and MK 10 receivers, if $H_{min} = 25^O$ this coefficient is almost the same for all receivers. It's because for this

angle the number of satellites which can be used for calculate HDOP is for all receivers almost the same (6 or 8) while for $H_{min} = 5^O$ this number is for MAN (12) and Simrad (10) greater than for two others (6) considerably.

4.2 The measurements on the ships

The latitude error σ_φ, longitude error σ_λ, horizontal position error σ_{2D} (95%) and the distance D_{AB} between mean positions of two SAAB receivers for different input data (6 one-hour sessions) and two Kongsberg receivers for different daytimes and the measurements conditions (4 one-hour sessions) are showed in the table 8 and table 9 respectively. We can conclude that:

– in the case of two identical receivers, error σ_{2D} is not the same, but the difference is few per cent only,
– the error σ_{2D} of position determined 23 h 56 min later than the first position with the same satellite constellation is not the same because the measurements conditions (signal in space in particular) change with time,
– for different daytimes and the measurements conditions the position's accuracy is almost the same, about 1 m or less,
– if masking angle of the receiver increases few times, all errors increases considerably also,
– the distance between mean positions of two identical SAAB receivers and two identical Kongsberg receivers (D_{AB}) is not greater than 4.7 m and 2.3 m, respectively.

In the case of the measurement based on Furuno and Saab receivers on the ship SMT Bontrup we can say that the errors σ_φ, σ_λ and σ_{2D} are almost the same for these receivers, error σ_{2D} is about 4.1 ÷ 6.2 m (table 10). For both Furuno and Saab the distance between mean position and real position is from interval 4.1÷8.9 m while maximal distance between GPS position and real position from interval 9.2÷16.7 m. These values are typical for GPS accuracy in real conditions.

5 CONCLUSIONS

– the choice of the SNS receiver and the mode of its use depend on the type of the ship and its region of navigation. The accuracy of the position GPS stand-alone (e.g. SAAB GPS only) for the cargo ship during ocean and coastal navigation is sufficient while the same receiver SAAB on specialized ship where high accuracy is needed determines the position using SBAS or DGPS augmentation.
– according to Federal Administration and maritime receivers producers the GPS system makes possible the determination of horizontal

user's position (95% confidence level) with the accuracy few metres. The measurements based on different stationary GPS stand-alone receivers realized in the laboratory and on the ships confirm it. With augmentation DGPS or SBAS this accuracy increases to 1 m and less.

– the accuracy of the user's position obtained from the GPS system depends on the number of channels (lc) of user's receiver and the number of the GPS satellites visible at given moment above masking angle. That's why the knowledge of technical performances of the receiver and the total number of the GPS satellites fully operational (ls) is very important for the users. There is no direct relation between the number lc, the number ls and the position error M, but we can say the following "when lc and ls greater, M is less" and inversely "when lc and ls is, M is greater".

– the accuracy of SNS position indicated by different maritime receivers differ because these units use different methods and algorithms which enable to change the results of pseudorange measurements and information obtained from the navigation messages into the user's coordinates. It concerns the identical models also because in each unit the local oscillator, fundamental element in all radio receivers, is different.

– the accuracy of the position obtained from professional GPS receiver with augmentation DGPS, SBAS or from GPS/GLONASS receiver, error σ_{2D} (95%) less than 1 m, is greater than from GPS stand-alone considerably.

– in the case of maritime GPS receiver the biggest influence on the accuracy of its position has the masking elevation angle and sudden changes in satellite constellation, in integrated GPS/GLO-NASS receiver, in particular.

– if on the ship's bridge two or more GPS receivers are installed, the ship's position must be obtained from professional receiver or from "younger" receiver which will ensure the greater accuracy.

REFERENCES

Admiralty Manual of Navigation. 2008. The Principle of Navigation, vol. 1, The Nautical Institute, London.

Gruber B. 2012. Status and Modernization of the US Global Positioning System, Munich Satellite Navigation Summit, Munich.

Hofmann-Wellenhof B. et al. 2008. *GNSS–Global Navigation Satellite Systems GPS, GLONASS, Galileo & more.* Wien: SpringerWienNewYork,.

Januszewski J. 2010. *Systemy satelitarne GSP, Galileo i inne.* Warszawa: PWN, (in polish).

Kaplan, E.D. & Hegarty, C.J. 2006. *Understanding GPS Principles and Applications.* Boston/London: Artech House,.

St. Marie T. 2012. GPS Constellation Status and Performance, 52nd Meeting CGSIC, Nashville.

www.gpsworld.com
www.imo.org
www.km.kongsberg.com
www.marinetraffic.com
www.navcen.uscg.gov
www.saabgroup.com

Cost-efficient, Subscription-based, and Secure Transmission of GNSS Data for Differential Augmentation via TV Satellite Links

R. Mielniczuk
University of Applied Sciences, Konstanz, Germany
AGH Univ. of Science and Technology, Kraków, Poland

H. Gebhard
University of Applied Sciences, Konstanz, Germany

Z. Papir
AGH University of Science and Technology, Kraków, Poland

ABSTRACT: In this paper, a system for broadcasting GNSS data for differential augmentation such as Differential GPS (DGPS) and Real-Time Kinematic (RTK) via TV satellite links is presented. With the omnipresence of "free-to-air" TV satellite signals the proposed system constitutes a cost-efficient solution for versatile worldwide dissemination of GNSS data to a large number of receivers, especially in areas not covered by mobile IP networks. Its compliance with the Networked Transport of RTCM via Internet Protocol (NTRIP), which is widely used for GNSS data dissemination via IP networks, makes the system compatible with existing GNSS data networks and client infrastructures. What is more, this system enables the usage of conditional access to broadcast content and verifies the data origin through the use of digital signatures. Additionally, the satellite capacity costs can be kept low through use of spare broadcast protocol and dedicated lossless data compression.

1 INTRODUCTION

Differential augmentation techniques for GNSS, such as DGPS and RTK, ensure improved positioning of a mobile receiver in real time when provided with a radio link. The radio data link is required to deliver additional information, known as GNSS data, continuously to the mobile receiver. This allows mitigation of the effects of error sources responsible for decreased positioning accuracy (Strang & Borre 1997).

Currently, GNSS data is widely available on the Internet, and it is possible for mobile GNSS receivers to access GNSS data on the Internet with the use of the NTRIP (Weber et al. 2005b). This solution requires an Internet connection for a receiver in the field to work and limits its usage to the areas with mobile IP network coverage, such as GSM, GPRS, UMTS, WiMaX, and LTE, or requires expensive worldwide satellite Internet connectivity.

To provide worldwide, cost-efficient access to the GNSS data available on the Internet in areas not covered by mobile IP networks, a system using TV satellite links has been developed and implemented (Mielniczuk & Gebhard 2011a,b). This solution provides an arbitrary number of mobile satellite receivers with cost-free reception to broadcast GNSS data. For commercial services an embedded conditional access solution is used to charge clients who access the protected content. In this system, the GNSS data already available on the Internet is first delivered with the use of the NTRIP to a broadcast server. Next, the GNSS streams are sent via an IP channel to a teleport that sends the data via an uplink to a satellite. Finally, the data is broadcast from the satellite to mobile satellite receivers, which can provide the GNSS data through the use of the NTRIP via a local radio link, e.g. Wi-Fi, to handheld NTRIP clients.

The transmission from the broadcast server to the mobile satellite receivers in the field is realized using a spare multiplex-based protocol. This lowers satellite link costs, which are significantly more expensive than mobile Internet connections of the same speed. This protocol ensures also an introduced latency limit for delay-sensible transmission of GNSS data and is used for transmission of metadata information, conditional access solutions data, as well as messages to users.

Another option for reducing transmission costs is data compression. For that purpose, a realtime and lossless compression method for one of most popular GNSS data formats, i.e. RTCM SC-104, has been developed. This method benefits from both the statistical properties of RTCM data and the linearity of particular corrections in RTCM streams. Its construction works with RTCM streams composed of arbitrary types of messages, but the resulting space savings depends on statistics of the messages and their content.

The remainder of this paper is structured as follows. First, some techniques of GNSS data transmission, abandoned as well as currently operating, are briefly described. Next, the NTRIP, which is used in GNSS data networks and client infrastructure, and its work are described. After that, motivation for this work and the cost-effective system that can provide versatile worldwide GNSS data dissemination to a large number of receivers are explained. Further, construction of server components targeted for high service availability and low latency is presented. Next, the spare protocol used for broadcast communication with mobile satellite receivers is described. Later, a lossless RTCM SC-104 data format compression working in real time is presented. Finally, security concerns in relation to broadcast satellite connectivity are addressed.

2 TRANSMISSION OF GNSS DATA

Currently, a plethora of systems and used data link technologies exist for GNSS data dissemination for various differential augmentation services. These services can be distinguished by parameters such as achieved positioning accuracy, supported area, initialization time, and GNSS receiver requirements, or whether they provide information on integrity, continuity, accuracy, and availability. The important parameters for a utilized data link are costs, range and coverage, transmission's throughput and latency, and whether error checking or correction is realized. Another matter is whether the data link disturbs GNSS signal reception. Thus other frequencies or PRN codes have to be used. The existing solutions for GNSS data distribution can be divided into three categories according to connectivity: ground, satellite and IP based.

2.1 Ground-Based Augmentation Systems (GBAS)

GBAS have been used primarily for GNSS data distribution. Their performance in terms of range depends on the frequency used. A short overview and some examples are given. The ultra-low frequency (ULF) band was used by the Accurate Positioning by Low Frequency (ALF) (Dittrich & Kuhmstedt 2005) system, providing service with a 600 km range over the entire territory of Germany. The low (LF) and medium (MF) frequency bands are used by the US Coast Guard (Ketchum et al. 1997) and Canadian Coast Guard (Canadian Coast Guard 2007) for distribution of corrections for their coasts. The practical range can reach even 480 km over the sea and dozens of kilometers over the land. The MF band is also used in Germany by the International Association of Lighthouse Authorities' (IALA) DGPS stations for covering seaside and inland territory of the country with the transmission range of 250 km (Hoppe et al. 2006). Communications in the high frequency (HF) band are based primarily on reflections from the ionosphere, resulting in communication ranges of up 700 km and more, and some two-way communication tests in this band between a ship and a ground station were performed (Vetter & Sellers 1998). Generally, communication in the LF, MF, and HF bands is affected by interferences in crowded bands and changing propagation conditions, causing signal fading. Moreover, the narrow band links constitute a bandwidth limit for the transmitted corrections.

In the very high frequency (VHF) band the Radio Aided Satellite Navigation Technique (RASANT) (Raven et al. 1996) was used in Germany with the FM Radio Data System (RDS) and a set of radio stations to cover the whole country. Another way to use the free part of FM radio bands for GNSS data transmission is with Data Radio Channel (DARC), which is used, for example, in Korea (Park et al. 2001). In the VHF band works also Local Area Augmentation System (LAAS), a civil all-weather aircraft landing system based on DGPS technique using reference receivers located around the airport. The Joint Precision Approach and Landing System (JPALS) is a military landing system that works in the ultra-high frequency (UHF) band. Communication in the VHF and UHF bands is considered to be limited to the line of sight.

2.2 Satellite-Based Systems

For nationwide and continental-wide augmentation systems, geostationary satellites allow broadcast over a large area. Currently operating Regional Satellite Augmentation Systems (RSAS) are the Wide Area Augmentation System (WAAS) in the US, the European Geostationary Navigation Overlay System (EGNOS) in Europe, and the Multi-functional Satellite Augmentation System (MSAS) in Japan, and a number of RSAS are currently under development. Satellite-based systems mainly transmit in the band of GNSS systems to a handheld receiver that can be embedded in a GNSS receiver and use the same antenna circuits. Global Satellite Augmentation Systems (GSAS) are operated by John Deere (Sharpe 2009), Furgo Group (Pflugmacher 2009), and LandStar (Fotopoulos et al. 1998) companies. These provide fee-based services using proprietary protocols and receivers.

2.3 IP-Based Systems

Currently, GNSS-capable devices are equipped with wireless IP connectivity. Such channels are typically more robust, faster, and more affordable than dedicated terrestrial or satellite links and are usually bidirectional. These characteristics make it possible to construct professional and interactive services for differential augmentation systems. The usage of IP protocol was a milestone, enabling cheap, worldwide, and real-time access possibilities to GNSS data wherever the receiver has Internet connectivity.

Following, two IP-based solutions for GNSS data dissemination are briefly enumerated, and a description of a third, the NTRIP system, is provided in the next section. The first IP-based solution, the SISNeT platform (Mathur et al. 2006), is designed to deliver EGNOS correction data through the Internet to land mobile users in urban areas where the visibility of GEO satellites is frequently poor. The SISNeT service is free of charge and has operated since 2002. It uses TCP streaming for data delivery and supports only EGNOS messages. The second IP-based solution, the Lightweight GNSS Support Protocol (LGSP) (Tyson & Kopp 2007), is designed to provide an alternative, secure distribution channel for GNSS data without tight bandwidth restrictions. It may also help to limit receiver initialization delays in low signal availability areas and deliver high-rate GNSS data. In this system, the server architecture comprises master and mirror servers which provide two messaging mechanisms, request/response and multicast streaming. Other than the concept and its precise design, nothing is known about current LGSP development or deployment.

3 NTRIP SYSTEMS

Since the rise of the 1.0 version of the NTRIP (Weber et al. 2005a) in 2004, GNSS data has become available anywhere the mobile receiver can establish Internet access. Mobile IP networks have drawn attention for their usage because of the possibility they offer to integrate their receivers into portable GNSS equipment. The NTRIP has made it possible to provide interactive and lower priced GNSS data services in comparison with ground-based systems, such as ALF or RASANT, and satellite-based systems.

In the NTRIP system, the server component, which is called NTRIP caster (Fig. 1), services requests sent by clients' programs that upload the streamed content from NTRIP sources, called NTRIP servers, and download the streams, known as NTRIP clients. Such architecture allows both client programs to connect from local Internet access networks where Network Address Translation (NAT) is used and preserves peer isolation, which increases the security. What is more, the only component that has to be equipped with a high-speed Internet connection when a large number of clients download the same stream is the NTRIP caster and NTRIP servers can run on resource-limited devices.

In the 1.0 version the protocol uses an HTTP-like request-response mechanism. After the HTTP connection is successfully established a continuous GNSS data stream is sent to or from NTRIP caster. Such construction is necessary because it assures problem-free work over proxies accepting only HTTP traffic (Gebhard & Weber 2005).

Each NTRIP caster provides information on its current state and configuration in the form of plain text table, called a source-table, which can be sent to an NTRIP client on a request. The source-table contains information on available NTRIP sources, networks of NTRIP sources, and NTRIP casters in the form of various types of entries. The source entries are complemented with all necessary metadata information for geodetic and navigation purposes, e.g. the GNSS data format or the coordinates of the reference station. The entries for casters contain, among other data, the information on the fallback caster addresses, which can be use by NTRIP client programs in case of service fallout. The network entries are used for grouping NTRIP sources, e.g. according to the data provider.

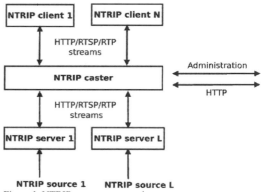
Figure 1. NTRIP system construction

The NTRIP was standardized by the Radio Technical Commission for Maritime Services (RTCM) and since 2009 the 2.0 version (Radio Technical Commission for Maritime Services 2009) has been in the operational state. In this version, two mechanisms of data transmission were added: the first one using RTSP for control and RTP for data transport and the second one using RTP only. Both allow the reduction of data latency over packet lossy links at the expense of data integrity. In addition, chunked transfer encoding for HTTP streaming, which allows the verification of the transfer completeness, was added. A further change was the add-on of source-table filtering requests, which allows NTRIP clients to gain more precise responses from the NTRIP caster, e.g. to find streams in the given country or coordinately restricted location, which could be more comfortably viewed on small displays of handheld devices.

Nowadays, NTRIP clients are installed or embedded in thousands of GNSS-like devices and used by the biggest GNSS data world networks (Söhne et al. 2009). Moreover, there are open-source applications available for all NTRIP system components.

4 MOTIVATION AND THE PROPOSED SYSTEM

Currently, economical usage of NTRIP systems is limited to populated regions where mobile Internet access networks are present. Satellite Internet connectivity makes it possible to use NTRIP in rural areas and in uninhabited places, such as deserts, seas, and mountains, if a sight line between a transceiver and a satellite can be established. Such worldwide Internet satellite access can be achieved by services like Iridium, Globalstar, and Inmarsat. However, the costs of the required equipment and continuous access to even a single GNSS data stream via these services are considerably more expensive than that of IP mobile networks.

A less expensive solution are broadband satellite Internet access services like Tooway, StarBand, or ASTRA2Connect with monthly fees comparable to terrestrial Digital Subscriber Line (DSL) Internet connections at the time of this writing. However, these services use proprietary receivers that cannot be mixed and work only on one network in a specific region. Additionally, if a broadband connection is used only to access a single GNSS stream with a speed of only a few kbps, such a solution is likely to be expensive.

A system of geostationary TV satellites that allow the transfer of IP packets, e.g. in DVB-S and DVB-S2 technologies, is a more economical and versatile alternative. The one-way, point-to-multipoint network offered by geostationary TV satellites is suitable for massive-scale applications, including GNSS data dissemination because data sent only once can be received by any number of recipients and only the data sender is charged for data transmission, regardless of the number of recipients. Moreover, the popularity of DVB solutions has made reception components available from many manufacturers at low prices. For example, a development set including a receiver, an embedded computer, and an antenna can be constructed for less than 250 euro.

Data broadcast on these IP downlinks can be received on Very Small Aperture Terminals (VSAT) from about seventy geostationary satellites located 2–3 degrees apart over the equator, using regional footprints for data broadcasts to cover most of earth's surface. The satellite capacity can be bought on MHz or kbps basis, which allows even small private companies to provide inexpensive GNSS data broadcast transmission to a large number of users in a wide region.

A DVB satellite receiver is capable of receiving IP data from any transponder, sending "free-to-air" DVB signals on any satellite in the reception range. Operation of such a receiver is not limited to a specific region. What is more, there exist mobile reception satellite antennas for vehicles on the move, such as cars, trains, boats, and airplanes.

Given reasons listed above, a cost-effective system for GNSS data dissemination (Mielniczuk & Gebhard 2011a,b) using satellite TV links is presented (Fig. 2). In this system, the GNSS data already available on the Internet, which originated from reference stations or network RTK products, is first delivered with the use of the NTRIP to a broadcast server. Next, the GNSS streams are sent via an IP channel to a teleport that sends the data via an uplink to a satellite. Finally, the data is broadcast from the satellite to mobile satellite receivers.

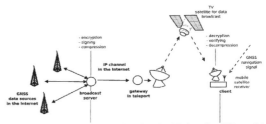

Figure 2. A system for broadcasting GNSS data for differential augmentation via TV satellites

A direct application of the presented architecture could include a vehicle equipped with a satellite receiver for a data link, a GNSS receiver, and reception-only antenna circuits installed on the roof. The received GNSS data would be used for the vehicle's precise guidance. For example, in maritime assignments, the augmented positioning and navigation of a vessel can be used in a few ways for providing safety and security enhancements. For inland waterways (Januszewski 2009), such as channels and rivers, and inshore navigation and harbor approach, the corrected position can be used for a more-dense arrangement of vessels (Ilcev 2011) and will allow for more precise maneuvers and navigation (Kujawa & Rogowski 2008). Differential augmentation also can be used for creation of precise bathymetric maps and information about under water obstacles (Popielarczyk & Oszczak 2007). Sailors equipped with such charts and positioning devices (augmented or not) can navigate safer on shallow waters with reefs, stones, and other underwater obstacles.

For other approaches, the client application, which runs on mobile satellite receivers, works in the proxy mode in which GNSS data can be delivered using a local radio data link in an ISM band, e.g. Wi-Fi, to handheld mobile GNSS receivers using the NTRIP, as presented in Figure 3. This option makes the system useful for surveying, retrieval of Earth observation data, precise vehicle guidance for off-shore explorations and mining vehicle operations, and automated farming in which a satellite receiver is installed a short way off and mobile GNSS receivers are used for tasks which require high precision positioning.

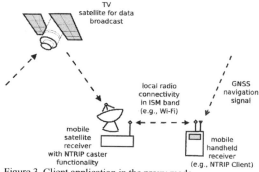

Figure 3. Client application in the proxy mode

The system uses an IP channel with the services of a teleport that ensures practical and inexpensive data delivery without requiring the satellite to be in the line of sight of the server component or an expensive uplink infrastructure for each satellite. The redundant architecture of the server component together with the possibility of a redundant downlink support to geographical areas ensures high service availability. Additionally, the system allows for optional data compression (Mielniczuk et al. 2012), which reduces satellite transmission costs.

In such architecture, two security issues arise. First, data sent with the "free-to-air" signal can be easily received, and that would not generate revenue for commercial GNSS data services. Therefore the system embeds a conditional access solution to protect the broadcast content. Second, the data streamed from the broadcast server travels via the Internet and the network of the satellite operator where the data could be modified. Similarly, the satellite's air interface could be spoofed. Therefore, digital signatures are used to verify data origin.

5 THE SERVER ARCHITECTURE

Based on the distributed architecture, the broadcast server comprises a number of broadcast nodes placed in data centers that are used for achieving server redundancy or hot-plug functionality, as presented in Figure 4. Each broadcast node is permanently connected via an IP channel to the teleports, allowing them to quickly perform a takeover in the case of an outage. Only one node is allowed to send data for a particular transport (destination IP address and UDP port on a given satellite transponder) at a given time. This ensures consistency of transport and, optionally, cryptographic and compression contexts and does not confuse the receiver application. The teleport services usually are protected by the use of redundant leased lines for Internet connectivity to avoid a single point of failure, which also increases the availability of our system.

Each broadcast node supports fully the NTRIP so that it can act as an NTRIP system component, i.e. it can connect to another NTRIP caster to pull (as an NTRIP client) or push (as an NTRIP server) a stream or receive a stream from a NTRIP source (as an NTRIP caster). However, to reduce the transmission latency, direct connections of GNSS data sources are recommended.

Ideally, a unit pushing GNSS data to a broadcast server uses the round-robin algorithm to choose a broadcast node to connect to from a dynamic list. This list can be updated using multiple results of DNS queries or the information on caster entries from the source-table on each broadcast node. However, configuration options of some GNSS sources can be limited. For example, it is possible to enter only one IP address of a caster or fixed firewall settings does not allow to connect to another address. For such sources it is possible to forward the data stream to the right broadcast node as soon as it connects, as shown in Figure 4. Similarly, an intentional disconnection of all streams from a node can force them to use another one while this node is maintained. For latency minimization, when a node is broadcasting data, all resource-intensive load tasks, such as software compilation, operating system updates, or file indexing services, are disabled.

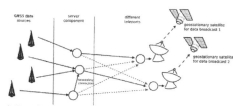

Figure 4. Broadcast core and satellite redundancy

The NTRIP standard allows for additional entries and complementary stream entries to the source-table. This option is used in the system to provide mobile satellite receivers with the satellite transponders settings, data and network layers, and transport protocols. For example, a satellite transponder entry contains not only basic information such as technology, modulation, and polarization but also a URL address to the footprint of a satellite transponder. By contrast, a transport entry contains information on the destination UDP port number, transport protocol, and, optionally, information for the conditional access solution. A mobile satellite receiver can download such an extended source-table when an Internet connection is available, or it can be provided in the form of a text file. Such plain text information can also be given by phone and entered on the mobile satellite receiver. This information constitutes the configuration used by the mobile satellite receiver and can be used for its automatized setup.

Broadcast nodes use a distributed database that includes system settings and user privileges for encrypted content. This database is also used by the user's WWW interface, which can be utilized by data providers and clients. For example, the data providers are able to choose the most appropriate geographical area from those available in the system to which to send content and choose a set of authorized devices capable of gaining access to the protected content. In contrast, registered users who want to gain access to the protected content can do this before by web interface or by phone and make the payment later.

Each node supports also a TLS protocol that can be used to secure an HTTP-like NTRIP communication among data nodes and data nodes and data sources. This ensures data encryption and peer authentication. In TLS protocol, the GNSS data provider authenticates the broadcast node by checking its certificate. The GNSS data provider is verified on the application layer; the user name and password, which are given only to the data provider, is sent to a broadcast node encrypted with TLS protocol. The security concerns in relation to broadcast communication are described in section 7.

6 THE BROADCAST PROTOCOL

For GNSS data transmission from the broadcast server to the mobile satellite receivers in the field, a novel broadcast protocol was developed. It is based on a spare multiplex protocol (Mielniczuk et al. 2013), which ensures an introduced latency limit for every GNSS data stream and significantly reduces the packetization overhead. For example, in some configurations the multiplexing concept can reduce the overhead from about 58% to 8%. The multiplex works with a wide range of input stream numbers and stream speeds. It does not fragment data chunks delivered to the broadcast server as long as they are smaller than the MTU of the broadcast link. This process reduces the impact of lost packets if complete entries of GNSS data format were carried in the input chunk boundaries.

The multiplexing scheme permits the spare transmission of information to the clients, e.g. metadata, conditional access scheme data, and messages. Encrypted and plain text messages can be sent to, a subgroup, or individual users, and their validity can be limited to a specific time range. These messages can contain information shown on a display, changes of satellite transponder settings, or expected dysfunctions. In this protocol, textual data are compressed to limit the traffic. For this purpose, zlib compression with an initial dictionary is used.

The described protocol was implemented and tested, and its security is currently being investigated. Current implementation of server and client components also supports two simple transports: UDP and RTP. Because of their simplicity and robustness, they were referenced during transmission latency and integrity measurements and helped debug and identify construction and implementation errors.

7 RTCM SC-104 STREAM COMPRESSION

Nonproprietary GNSS data formats found in the literature, such as RTCM SC-104 (Radio Technical Commission for Maritime Services 2001, Radio Technical Commission for Maritime Services 2006), CMR (Talbot 2003), ATOM (Artushkin et al. 2008), and mRTK (Alanen et al. 2006), are binary formats, which means that a stream of messages in those formats is already compact. Thus, universal compression methods may require the buffering of data to save space, the length of which can be unacceptable for low-latency data transmission. For compression, one of the most popular formats, i.e. RTCM SC-104, has been selected, and the compression involves data stream decoding and encoding.

The developed compression works with a large number of streams, and with a single RTCM SC-104 stream (Mielniczuk et al. 2012), in real-time without buffering delay for compression and decompression procedures. Its construction works with RTCM streams composed of arbitrary type of messages, but the resulting space savings depends on the statistics of the messages and their content. An analysis of the RTCM SC-104 format usage by the users has shown that a few types of messages constitute about 83% of sent messages. As a result, only these types of RTCM SC-104 messages are compressed while the others are simply forwarded. The compression involves the removal of information, such as parity or framing bits that can be reconstructed during the decompression process.

This process separately compresses each supported message in a stream. In such an operation, binary data fields of the same type in a message are compressed using one of the described methods. The first method uses a lesser number of bits than originally used if possible. The second method saves the differences between values instead of the absolute values. In the third method, the modal value is removed from the array, and the remaining array and the modal value are sent with occurrence indexes. In addition, some methods for specific data fields were located. This type of compression achieves about 58% and 19% of space saving for the most frequently sent RTCM SC-104 messages in this format in versions 2 and 3, respectively.

The compression can also work on a sequence of consecutive messages of one type. The first message is compressed as described above, and instead of sending the next messages, only differences to the last compressed message are sent. For such a differential encoding, two methods are used. The first, which is used mainly for short data fields, sends only values and the indexes of data fields that have changed. The second method, which is used mainly for long data fields, uses a linear error predictor with outlier detection. It benefits from the linear dependency found in some data fields. Only the difference between the value estimated in the receiver and the original value is sent. This type of compression improves the above-reported results to about 90% and 53% of space savings, respectively, for sequences of length from 5 to 11 messages.

The compression of a message's sequence determines its performance but it forces the receiver to wait on the start frame of its differential encoding before decompression. In addition, lost packets cause the receiver to be unable to decompress the next received data until the next start frame arrives.

8 BROADCAST LINK CONCERNS

Conditional access solutions protect the distributed content from unauthorized access. Because in the introduced system only a one-way connection to the mobile satellite receiver is available, the choice has to be limited to Broadcast Encryption (BE) (Horwitz 2003) schemes. BE schemes are the basis of the key management systems for broadcast. They equip users with a device that contains secret information protected from access. A BE protocol delivers all users the same information (broadcast channel) but only the authorized set of devices is capable of getting the decryption key for the broadcast content. When the set of authorized devices changes, e.g. a device is revoked (removed from the authorized set), the next data sent by the BE protocol will be unusable on revoked device.

There are many methods for gaining access to secret information held in electronic circuits, and new methods are constantly being developed (Bar-El

2003, Kömmerling & Kuhn 1999). However, designing and programming electronic circuits to be resistant against such attacks is costly, difficult, and time consuming. Therefore, relaying system security based on the secrecy of information held in an electronic device is risky. The Conditional Access (CA) system used in DVB technology is such an example. Security is based on the secrecy of a key on a smart card. If this key is revealed, the given CA system implementation is broken, and the smart card can be cloned. The only way to restore system security is to change and replace all of the smart cards. If a broken or cloned device can be traced and identified and a BE scheme provides revocation, this device is simply removed from the privileged set and the system remains secure.

Until now, for the conditional access solution the Subset Cover framework (Naor et al. 2001), which provides revocation and tracing, has been utilized. What is more, it is used in Blu-ray and HD-DVD technologies (Henry et al. 2007) for digital content protection. The Subset Cover framework does not define encryption algorithms used in it and allows the key length to be chosen freely. In the current implementation, the AES encryption with a key length of 128 bits is used.

For data origin authentication, digital signatures are used. The asymmetric cryptography of public and private keys ensures that even non-revoked users are not capable of properly signing faked data (Ferguson & Schneier 2003). For realization the cryptography of Elliptic Curves (EC) has been used because of the much shorter signatures for broadcast messages and an even greater level of security in comparison to RSA cryptography, e.g. 160 bits for EC vs. 1024 bits for RSA (Bos et al. 2009).

For the realization of the device, which contains the secret information that is given to users, a smart card microcontroller with USB interface support is preferred because of its universality and delivery of power supply to the connected device. If it is assumed that the resources of the satellite receiver will be very limited, data decryption and verification of packets can be performed on and prepared for this smart card microcontroller.

9 CONCLUSIONS AND CURRENT STATUS

The construction of the presented system provides professional and secure worldwide GNSS data service at a lower cost. The solution is not limited to satellites using DVB technology because all communication is IP based. With a little effort, such as changes in satellite receiver type and antenna circuits, the solution can be used in other technologies.

System implementation was not profiled to introduce considerable buffering or processing latency and to fulfill the low-latency transport restrictions. As a result, the end-to-end latency of a data stream consists mainly of the time required for Internet data transport and propagation and serialization delay on the broadcast link. The security of the broadcast protocol is currently being reviewed. A test system installation with a DVB-S2 beam over Europe is being performed, and usage experiences are being collected.

REFERENCES

Alanen, K., Wirola, L., Kappi, J. & Syrjarinne, J. (2006). Inertial Sensor Enhanced Mobile RTK Solution Using Low-Cost Assisted GPS Receivers and Internet-Enabled Cellular Phones. Position, Location, and Navigation Symposium, 2006 IEEE/ION: 920–926.

Artushkin, I., Boriskin, A. & Kozlov, D. (2008). ATOM: Super Compact and Flexible Format to Store and Transmit GNSS Data.

Bar-El, H. (2003). Known Attacks Against Smartcards.

Bos, J. W., Kaihara, M. E., Kleinjung, T., Lenstra, A. K. & Montgomery, P. L. (2009). On the Security of 1024-bit RSA and 160-bit Elliptic Curve Cryptography.

Canadian Coast Guard (2007). Canadian Marine Differential Global Positioning System (DGPS) Broadcast Standard.

Dittrich, J. & Kuhmstedt, E. (2005). Accurate Positioning by Low Frequency (ALF) Eine Chronologie und Reminiszenz zu Entwicklung und Betrieb des ersten flachendeckenden DGPS-/RTCM-Rund- strahldienstes in Deutschland .

Ferguson, N. & Schneier, B. (2003). Practical Cryptography. Wiley Publishing, Inc.

Fotopoulos, G., Cannon, M.E., Bogle, A. & Johnston, G. (1998). Testing LandStar's Performance Under Operational Conditions.

Gebhard, H. & Weber, G. (2005). Using the Internet for streaming differential GNSS data to mobile devices.

Henry, K., Sui, J. & Zhong, G. (2007). An Overview of the Advanced Access Content System (AACS).

Horwitz, J. (2003). A survey of broadcast encryption.

Ilcev, S. D. (2011). Maritime Communication, Navigation and Surveillance (CNS). International Journal on Marine Navigation and Safety of Sea Transportation 5(1):39–50.

Januszewski, J. (2009). Satellite and Terrestrial Radionavigation Systems on European Inland Waterways. International Journal on Marine Navigation and Safety of Sea Transportation 3(2):121–126.

Ketchum, R.L. and Lemmon, J.J. and Hoffman, J.R. (1997). SITE SELECTION PLAN AND INSTALLATION GUIDELINES FOR A NATIONWIDE DIFFERENTIAL GPS SERVICE.

Kömmerling, O. & Kuhn, M. G. (1999). Design principles for tamper-resistant smartcard processors. Proceedings of the USENIX Workshop on Smartcard Technology on USENIX Workshop on Smartcard Technology.

Kujawa, L. & Rogowski, J. (2008). Possibility of Precise Positioning and Precise Inshore Navigation Using RTK and Internet. International Journal on Marine Navigation and Safety of Sea Transportation 2(1):31–35.

Mathur, A., Toran, F. & Ventura, J. (2006). SISNET User Interface Document.

Mielniczuk, R., Engstler, A. & Gebhard, H. (2012). Lossless RTCM-SC104 data compression for broadcasting via satellite links. Mobilkommunikation Technologien und Anwendungen - Vorträge der 17. ITG-Fachtagung vom 9. bis 10. Mai in Osnabrück: 105–110.

Mielniczuk, R. & Gebhard, H. (2011a). A PLATFORM FOR IP STREAMING OF DIFFERENTIAL GNSS DATA OVER SATELLITE INTERNET SERVICES. PRZEGLĄD TELEKOMUNIKACYJNY - WIADOMOŚCI TELEKOMUNIKACYJNE - CDROM: KKRRiT 2011 : Krajowa Konferencja Radiokomunikacji, Radiofonii i Telewizji : 8–10 czerwca 2011 r., Poznań

Mielniczuk, R. & Gebhard, H. (2011b). An IP Service for Secure GNSS Data Broadcast over Satellite Links. Mobilkommunikation Technologien und Anwendungen - Vorträge der 16. ITG-Fachtagung vom 18. Bis 19. Mai in Osnabrück: 96–99.

Mielniczuk, R., Gebhard, H. & Papir, Z. (2013). Multiplexing of GNSS Data for Differential Augmentation via Satellite IP Broadcast Links.

Naor, D., Naor, M. & Lotspiech, J. (2001). Revocation and Tracing Schemes for Stateless Receivers.

Park, J. & Joh, J. & Lim, J. & Park, P. & Lee, S. (2001). THE DEVELOPMENT OF DGPS SERVICE SYSTEM USING FM DARC (EYEDIO) IN KOREA.

Pflugmacher, A., Heister, H. & Heunecke, O. (2009). Global investigations of the satellite-based Fugro OmniSTAR HP service.

Popielarczyk, D. & Oszczak, S. (2007). Application of GNSS Integrated Technology to Safety of Inland Water Navigation. International Journal on Marine Navigation and Safety of Sea Transportation 1(2):129–135.

Radio Technical Commission for Maritime Services (2001). RTCM 10402.3 RECOMMENDED STANDARDS FOR DIFFERENTIAL GNSS (GLOBAL NAVIGATION SATELLITE SYSTEMS) SERVICE.

Radio Technical Commission for Maritime Services (2006). RTCM STANDARD 10403.1 FOR DIFFER-ENTIAL GNSS (GLOBAL NAVIGATION SATELLITE SYSTEMS) SERVICES VERSION 3.

Radio Technical Commission for Maritime Services (2009). RTCM STANDARD 10410.1 NETWORKED TRANSPORT OF RTCM via INTERNET PROTOCOL (Ntrip) - Version 2.0.

Raven, P., Sandmann, S. & Schoemackers, G. (1996). RASANT Radio Aided Satellite Navigation Technique.

Sharpe, T. & Hatch, R. F. (2009). John Deere's StarFire System: WADGPS for Precision Agriculture.

Söhne, W., Stürze, A. & Weber, G. (2009). Increasing the GNSS Stream Dissemination Capacity for IGS and EUREF.

Strang, G. & Borre, K. (1997). Linear Algebra, Geodesy, and GPS. Wellesley - Cambridge Press.

Talbot, C. (2003). Compact Data Transmission Standard for High-Precision GPS.

Tyson, M. & Kopp, C. (2007). The Lightweight Global Navigation Satellite System (GNSS) Support Protocol (LGSP).

Vetter, J. R. & Sellers, W. A. (1998). Differential Global Positioning System Navigation Using High-Frequency Ground Wave Transmissions JOHNS HOPKINS APL TECHNICAL DIGEST 19(3):340–350.

Weber, G., Dettmering, D. & Gebhard, H. (2005a). Networked Transport of RTCM via Internet Protocol (NTRIP). In S. Verlag (ed.), Sanso F. (Ed.): A Window on the Future, Proceedings of the IAG General Assembly, Sapporo, Japan, 2003, Volume 128 of Symposia Series: 60–64.

Weber, G., Dettmering, D., Gebhard, H. & Kalafus, R. (2005b). Networked Transport of RTCM via Internet Protocol (Ntrip) - IP-Streaming for Real-Time GNSS Applications. Proceedings of the 18th International Technical Meeting of the Satellite Division of the Institute of Navigation (ION GNSS 2005).

Global Navigation Satellite Systems (GNSS)
Advances in Marine Navigation – Marine Navigation and Safety of Sea Transportation – Weintrit (ed.)

Method of Improving EGNOS Service in Local Conditions

A. Felski
Polish Naval Academy, Gdynia, Poland

A. Nowak
Technical University of Gdansk, Poland

ABSTRACT: Since EGNOS Service is declared fully operational appeared the question of limitations in the use of the system to specific operations. Defined horizontal accuracy better than 3 m (95%) and very high integrity suggest for example its usefulness' in harbour operations. However some reports suggests influence of local environmental conditions onto behavior of system aside of RIMS stations net. In addition some harbour tasks have more restricted demands than offered by EGNOS. Thus, it seems to be reasonable to ask the question about the real system performance in Poland. It is especially important if we consider RIMS distribution. Because of this stations location Poland is divided into two parts: inside and outside the 99% APV-I availability coverage. Unfortunately, the boarder runs near the polish biggest harbours - Gdansk and Gdynia, so lower availability has to be expected on polish coastal waters. Authors' analyses of the accuracy and availability of EGNOS service in Gdynia leads to advisability to monitoring of EGNOS behavior in local conditions. For this purpose dedicated software has been made by authors. The results and dedicated software for EGNOS service monitoring will be presented in the paper.

1 INTRODUCTION

The European Geostationary Navigation Overlay Service (EGNOS) is the system for augmenting the US GPS satellite navigation system and makes it suitable for safety critical applications. Its foundations are modeled in essential degree on Wide Area Augmentation System (WAAS) – the aid developed by US Federal Aviation Administration with the goal of improving GPS accuracy, integrity, and availability. Essentially, EGNOS, as well as WAAS is intended to enable aircraft to rely on GPS for all phases of flight, including precision approaches to any airport within its coverage area. His properties and the general accessibility suggest considering of the usefulness also in the maritime navigation, especially when navigating through narrow channels. EGNOS consists of three geostationary satellites and a network of ground stations for monitoring GPS constellation and GPS Signal-in-Space, and achieves its aim by transmitting a signal containing corrections to measured variations of pseudo-ranges and information on the reliability and accuracy of the positioning signals sent out by GPS. EGNOS was initiated by European Commission, but now is operated by – European Satellite Services Provider

(ESSP), founded on 2001 and in 2008 transformed into company of limited liability with shareholders of 7 key European Air Navigation Service Providers. ESSP declares that the system allows users in Europe to determine their position with accuracy (horizontal, 95%) of 3 meters [Web-1].

At the time of this writing EGNOS offers two of five planned services: Open Service and Safety of Live Service. Since October 1, 2009 when the EGNOS Open Service has been available EGNOS positioning data are freely available in Europe through satellite signals to anyone equipped with an EGNOS-enabled GPS receiver. The EGNOS Safety of Live service has been officially declared available for aviation on March 02, 2011. Space-based navigation signals have become usable for the safety-critical task of guiding aircraft - vertically as well as horizontally - during landing approaches [Web-1].

Besides European EGNOS there are two similar SBAS operational worldwides [Web-2]:
– The US has the Wide Area Augmentation System (WAAS), developed and operated by the Federal Aviation Administration (FAA), with an extension over Canada called CWAAS (Canadian WAAS),

- Japan has the Multi-functional Satellite Augmentation System (MSAS), developed and operated by Japan's Civil Aviation Bureau.

Two more systems are being developed for future certification by the International Civil Aviation Authority: Russia's System of Differential Correction and Monitoring (SDCM), under development by Roscosmos, and India's GPS and Geo-Augmented Navigation (GAGAN) system, under development by Indian Civil Aviation and India's ISRO space agency [Web-2]. All of them are developed on the basis of the identical standards, so the same receiver can be use all over the world.

2 DECLARED EGNOS PERFORMANCE

As it was mentioned before, two different levels of EGNOS performance can be observed in Poland depending on which part of Poland we are talking about. It depends of distribution of monitoring (RIMS) stations – generally speaking distribution of external stations determines limits of the coverage area of the system. Location of RIMS station (shown in fig. 1) clearly shows that in eastern part of Poland worse accuracy and availability of EGNOS services should be expected.

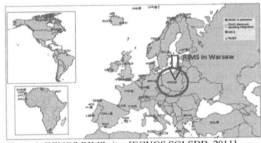

Figure 1. EGNOS RIMS sites [EGNOS SOLSDD, 2011]

EGNOS Safety of Life Service Definition Document 2011 directly shows distribution of different levels of accuracy in Poland (fig. 2) but charts of EGNOS APV-I availability, continuity and SoL service continuity, seems to clearly justify researches concerning EGNOS performance in context of the system usefulness in polish harbour operations.

Table 1. contains EGNOS Safety of Life performance values and fig. 2-4 show problem of different EGNOS services performance in Poland.

On the basis of this drawings and lists we can consider, that EGNOS can potentially be applied as the navigational service of port-operations, however not always will assure suitable exactitudes. This leads for the conclusion about the need of the monitoring of the property of the system in local circumstances.

Table 1 EGNOS Safety of Life Service performance values (EGNOS SOLSDD, 2011)

Parameter		Performance
Accuracy	Vertical	4 m (95%)
	Horizontal	3 m (95%)
Integrity	Integrity Risk	2×10^{-7}/approach
	Time To Alert	Less than 6 s
Availability		99.9% for NPA[1] in all the ECAC[3]
		99% for APV-I[2] in most ECAC
Continuity		For NPA:
		$<2 \times 10^{-4}$ per hour in most of ECAC
		$<2 \times 10^{-3}$ per hour in other areas
		For APV-I:
		$<1 \times 10^{-4}$ per 15 s in the core of ECAC
		5×10^{-4} per 15 s in most of ECAC
		$<10 \times 10^{-3}$ per 15 s in other areas

[1] NPA – Non-Precision Approach
[2] APV – Approach with Vertical Guidance
[3] ECAC - European Civil Aviation Conference

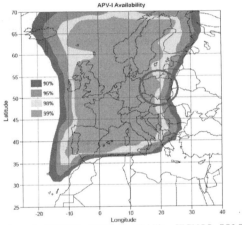

Figure 2. EGNOS APV-I Availability [EGNOS SOLSDD, 2011]

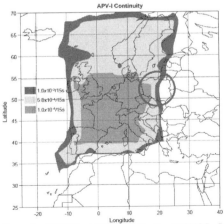

Figure 3. EGNOS APV-I Continuity [EGNOS SOLSDD, 2011]

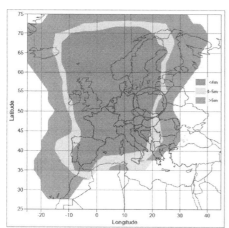

Figure 4. APV-I accuracy expected from EGNOS [EGNOS SOLSDD, 2011]

3 DEDICATED AUTHOR'S SOFTWARE TO EGNOS SERVICES MONITORING

Dedicated author's software is called "EGNOS Forecaster". The software was built in Delphi XE as multithread application for 32-bit platform of Windows. It is based on single dialog form with fixed size (no resizable). The main task of application is monitoring EGNOS services current work and on this basis is forecasting reliability characteristics of its. Appropriate work of the software demands two weeks observations at least. Shorter period could cause inadequate results of forecasting however monitoring function works correctly since first turn on. The main window of the software is shown in fig.5. The software could

cooperate with every EGNOS receiver on condition that it is able to transmit at least GGA, ZDA and GSV messages in NMEA 0183 standard.

Forecast is calculated on the base of past EGNOS behavior observations where MTBF – mean time between failures (mean time in state of work) and MTTR – mean time to repair (mean time in state of failure) are taken into consideration. State of work is defined as:

The current horizontal fix errors are lower than HAL (Horizontal Alert Limit) and current vertical error is lower than VAL (Vertical Alert Limit) and EGNOS corrections are available.

In the other case, the system is in state of failure.

Horizontal and vertical errors are measured as distance between current position fixed by EGNOS receiver and true position (reference position) entered into the system. It is extremely important to enter accurate reference position. Otherwise the software will work incorrectly and the forecasts could significantly differ from reality. According to the algorithm only after two weeks of constant observations user obtains adequate forecast concerning estimated EGNOS availability and estimated mean time between failures (MTBF) for chosen values of HAL and VAL. The forecast is valid for next fifteen minutes. Longer period of observation causes more accurate forecasting but the last two preceding weeks are the most important (have the higher weight for the calculations). If user changes value of HAL or VAL the forecast will be recalculated. It allows to fit forecasting process to current navigational task.

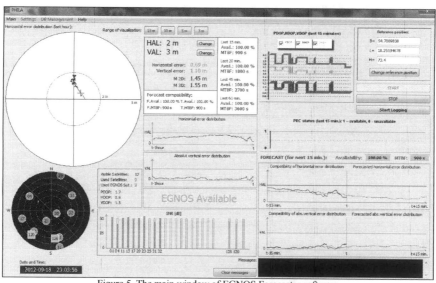

Figure 5. The main window of EGNOS Forecaster software

Current observations are stored in data base and are used to forecast calculation and analyze EGNOS behavior. User obtains information about system availability and MTBF in every 10 s. It contains calculated values for last hour (in fifteen minutes period) and forecast compatibility for last fifteen minutes (in numerical and graphical forms). It allows evaluating service quality if process of forecasting is going correctly.

Figure 6. Information about EGNOS availability, MTBF and forecast compatibility

Apart from forecasting functions the software realizes standard system monitoring tasks like:
- monitoring of current accuracy of service and presenting distribution of horizontal and vertical fix errors,
- monitoring number of visible and used satellites (including geostationary satellites transmitting EGNOS corrections) and presenting sky-plot,
- presenting signal to noise values of visible satellites and information about availability of EGNOS corrections,
- monitoring and presenting values of Dilution of Precision (DOP) coefficients,
- allows to record current observations (coordinates of fixed position, DOP values, number of visible and used satellites, etc.),
- etc.

Local monitoring of EGNOS services allows to determine the system usefulness in specified harbor operations. By changing HAL value (as far as we talk about maritime navigation only horizontal accuracy can be taken in account) operator in monitoring station is able to judge (on the base of forecasted availability, MTBF and MTTR) if EGNOS can be used as primary positioning system to navigation and transmit this information to operating vessels. In author's opinion it is interesting direction for future works on usage of EGNOS in navigational task which have strong accuracy, availability and continuity demands.

4 EXEMPLARY RESULTS OF OBSERVATIONS AND CONCLUSIONS

The observations were done for several weeks in monitoring station put out in harbor of Gdynia, located in Polish Naval Academy (see fig. 7). The EGNOS Forecaster software cooperate with Hemisphere R110 GNSS Receiver.

Because of limited size of the paper exemplary results of observations are presented as general conclusions. They are:
- Observed horizontal and vertical accuracy of the EGNOS was better than declared in EGNOS Safety of Life Service Definition Document 2011 (respectively 3 and 4 m for 95% level),
- Observed SoL service availability was near 100% (only few gaps in EGNOS transmission shorter than 2 s),
- Horizontal and vertical accuracy are timely better than declared in EGNOS SOLSDD 2011, thus it seems to possible using EGNOS to harbour operations which have stronger accuracy demands. For example observed horizontal accuracy for 95% level for 1 hour periods never exceeded 2.3 m and many times was better than 1.2 m,
- Observed mean values of availability and MTBF for 1 hour periods for chosen thresholds of HAL were:
  - for HAL = 2.0 m : Availability – 94.86%, MTBF – 876 s,
  - for HAL = 1.5 m : Availability – 79.51%, MTBF – 298 s,
  - for HAL = 1.0 m : Availability – 53.53%, MTBF – 91 s.

In author's opinion the presented results of observation are optimistic and open doors to further researches concerning usage EGNOS for precise navigational tasks. They show, that in spite of long term horizontal accuracy of the system in on the 3m (95%) level, timely is better than 1 m, so precise forecasting of the system behavior could significantly extend EGNOS applications.

REFERENCES

EGNOS SOLSDD (2011), EGNOS Safety of Life Service Definition Document, European Commission, Directorate-General for Enterprise and Industry.

EGNOS SDDOS (2009), EGNOS Service Definition Document Open Service, European Commission Directorate-General for Energy and Transport.

EGNOS – A Cornerstone of Galileo (2007), SP-1303. ESA

Federal Radionavigation Plan (2008), National Technical Information Service, Springfield. DOT-VNTSC-RITA-08-02/DoD-4650.5.

Felski A., Nowak A., Woźniak T. (2011), Accuracy and Availability of EGNOS – Results of Observations. Artificial Satellites vol. 46 no. 3/2011 pp. 111-118.

Felski A., Nowak A. (2012), Local Monitoring of EGNOS Services. Annual of Navigation, No 19/2012 part 1, pp. 25-34.

Mięsikowski M., Nowak A., Oszczak B., Specht C. (2006), EGNOS – Accuracy Performance in Poland, Annual of Navigation, No 11/2006, pp. 63-72.

Web-1: http://www.esa.int/esaNA/egnos.html, (consulted September 1, 2012).

Web-2: http://www.esa.int/esaNA/SEMNX8NXDXG_egnos_html, (consulted September 1, 2012).

Study of the RF Front-end of the Multi-Constellation GNSS Receiver

F. Vejražka, J. Svatoň, J. Pop & P. Kovář
Czech Technical University in Prague, Czech Republic

ABSTRACT: One of the main problems of the multi-constellation GNSS receiver consists in the amplifying and splitting of signals of partial systems working in different bandwidths received by a wideband antenna. We require a low noise figure of the frontend of such a receiver and resistance to strong jamming signals (mobile phones etc.). These problems are rising up in connection with the requirements of small dimensions and low power consumption as is usually needed for handheld receivers. We will study a variety of possibilities of the design of the multi-constellation GNSS receiver frontend and will present our theoretical and experimental results together with a philosophy of the multi-constellation GNSS receiver.

1 BASIC CLASSIFICATION OF GNSS RECEIVERS

We will be able to receive more than 60 navigation satellites by the end of the next decade. The GPS is widespread at present, the GLONASS is quickly developing in Russia, the COMPASS has entered the application phase in the Eastern Asia region and will be matured as a global system soon. We can expect that the GALILEO will be put into operation in Europe quickly. Originally military applications have expanded into the civil sector. Some of them are critical from the safety and lives rescue point of view. Great demands are made on precision, reliability, accessibility and integrity of such applications (O'Keefe 2001).

Simultaneous operation of all four systems is interesting for users from point of view reliability and navigation integrity, especially taking into consideration the fact that they are independent. Moreover the precision of navigation is better; the number of satellites is greater, because the root mean square of position error is

$$\sigma_P = \sigma_D DOP \qquad (1)$$

where σ_D is the rms of distance error and DOP depends on the number of satellites and their distribution in the sky. The receiver which is able to receive signals of satellites of several different navigation systems is called *the multi-constellation* or *multi-system receiver*.

The multi-constellation receiver is the subject of activity of the Centre of Competence at the Czech Technical University in Prague. A purpose of the Centre consists in providing the industry with assistance with establishing innovations and an attainment of competitiveness in the field of satellite navigation receivers and their applications. It means looking for algorithms of signal processing of the GNSS satellites which would
- be very effective with smaller requirements for the receiver processor and memory power
- provide the user with high precision of position determination or lead to a significant improvement of the parameters
- be easily implementable into some applications in the Czech industry conditions
- be so revealing that they would constitute a potential for competitive receiver development

From the point of view of applications we can classify the GNSS receivers in outline as follows:
- Receivers situated on a fixed or movable stand. Requirements for their size and input power are not substantial. As an example we can mention marine receivers, railway receivers, receivers of ionosphere stations. Although such receivers can determine position in movement, we will designate them as stationary. A *stationary* receiver can serve also as a development tool for signal processing algorithms validation.
- Receivers of small size, in most cases hand-held. Small size and low power consumption are substantial for the given application. Although

such receivers can be operated on a fixed place under certain conditions we will call them as *mobile*.

2 THE SOFTWARE RECEIVER OF THE GPS C/A SIGNALS ON L1 FREQUENCY

2.1 *The receiver structure*

One of the possible arrangements of the main blocks of the C/A code receiver for L1 frequency is in Figure 1. Signals from the antenna output are amplified by the low noise amplifier (LNA). The following filter passes signals from the GPS frequency band centered on L1=1,575.42 MHz. The frequency band of the filter is usually B=2.046 MHz for simpler receivers, i.e. the main lobe of spectra carrying most of the signal power is processed. Top quality receivers of high precision use usually of the frequency band width ±12 MHz i.e. the whole C/A spectrum transmitted by satellite.

The mixer transposes the signal on a convenient intermediate frequency, or to the base band. On the mixer output there can be another filter and an amplifier. The processing of signals is analogue up to this place and we call this receiver part a high frequency (HF) or radio frequency (RF) part or block.

The RF part is followed by an A/D converter and further signal processing takes place in the digital part of the receiver.

The RF part of the receiver must not distort the signal, to increase additive noise content, and produce phase fluctuations (phase noise, jitter). It has to mitigate interfering signals.

2.2 *Antenna*

The NOVATEL 704 X antenna (Novatel 2012) was used in the first stage of the stationary receiver design, also in consideration of the multi-constellation receiver development. It is a wide band antenna for 1.1 – 1.6 GHz band without an amplifier. It is designed as the so called PinWheel antenna with a slot array. Its phase center doesn't move in the whole frequency range. The radiation characteristic is nearly semi-spherical and suppresses reception from elevations lower than 5 deg, where Earth temperature noise is significant. The gain in zenith in L1 GPS band is 6.8 dBic and the axial ratio (ratio of polarization ellipse semi axes) is 1 – 2 dB for elevation 5 to 90 deg. The axial ratio characterizes the capability of suppression of reflections and should not exceed 3 dB value.

We expect our own antenna design to be developed in the future especially for a mobile receiver. It is a question whether to use the only one multi frequency antenna or two or more antennas for a single frequency. A solution is seen in patch or helix antennas on substrates with a high permeability. The antenna designed on the basis of metamaterials could be another solution.

2.3 *Antenna preamplifier*

The receiver arrangement according to Figure 1 can be advantageously used in stationary receivers where the antenna is connected with the following receiver blocks by cable. The LNA, and filter sometimes, are placed near the antenna, usually outdoors. We call them often as antenna preamplifier. The signal is led by a cable to the indoor part of the receiver, where it is amplified by the next amplifier.

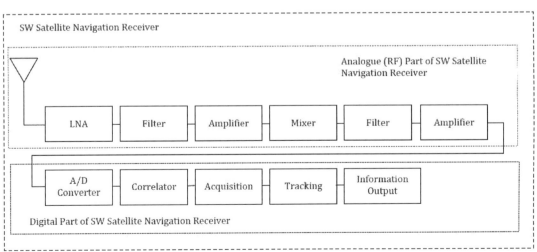

Figure 1. SW satellite navigation receiver structure

This arrangement advantage consists in the low noise figure of the circuits before the mixer. However the interfering signals with high level (signals of base stations of mobile telephony, e.g.) come on the LNA input. Therefore the LNA must have high dynamics and low intermodulation distortion (high the IP3 value). To meet this requirement we have to apply a low noise *power* transistor in the LNA which is resulting in large power consumption of the LNA (not acceptable in the mobile receiver).

In mobile receivers the antenna is usually placed in the immediate vicinity of the other blocks. We can omit the LNA and place the filter directly behind the antenna. Its pass band is typically 4 MHz (only when we use the main lob of the radiation pattern, however it is possible to use a wider band up to 24 MHz), the attenuation is typically 1 dB, and suppression outside the frequency pass band is 30-35 dB. An amplifier with the low noise figure follows the filter. The advantage of this arrangement lies in suppression of high level interfering signals by an input filter; therefore demands on dynamics and suppression of the intermodulation distortion are not as high as in the previous case. We do not have to use the power element in the amplifier, which is resulting in substantially lower power consumption.

The OEM RF preamplifier.

For experiments with design of the stationary receiver we have used the amplifier Mini-Circuits ZX60-P162LN1+. On frequency 1.6 GHz it has the following parameters according to the catalogue (Mini-Circuits 2012):
- gain 18.53 dB
- noise figure F=0.7 dB
- dynamics (output power for 1 dB compression of output signal) 19.5 dB
- IP3 = 29.6 dBm
- power supply 4 V, 52 mA

The amplifier is provided with the SMA connectors (Fig. 2).

Figure 2. Antenna preamplifier Mini-Circuits ZX60-P162LN1+ (Mini-Circuits 2012)

Design and realization of antenna preamplifier 2009.

When designing the preamplifier we are looking in particular for the noise adaptation, not the impedance or power one. We use s-parameters at different working points. We are designing noise adaptation circuits and observing the results in the Smith diagram. This process can be significantly accelerated by the simulation software use (AWR Design Environment (Popp 2013), Ansoft Serenade or Ansoft Designer (Rohde 2000), e.g.).

We designed and realized the antenna preamplifier (Fig. 3) in 2009 by the principles mentioned above (Kovar 2009). Both by simulation and by measurements the noise figure was determined as F=0.86 with power consumption 4 V/135 mA.

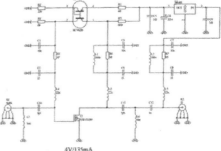

4V/135mA

Figure 3. Antenna preamplifier 2009 (Kovář 2009)

Design and realization of antenna preamplifier 2013.

Currently the preamplifier with a more up-to-date transistor is being designed. Its noise figure should be better than 0.4 and power consumption should be also lower. An example of simulation results is in Figure 4.

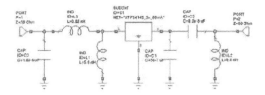

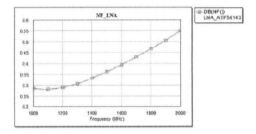

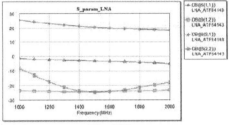

Figure 4. Antenna preamplifier 2013.

2.4 *The mixer*

It is necessary to digitize the signal for futher processing, it means we have to sample and quantize it. We can do it

– on a carrier frequency
– on an intermediate frequency
– in the base band

We will choose the type of conversion by the type of digitalization.

Receiver with direct amplification and direct sampling.

The signal on the output of the RF preamplifier is digitalized. Because the carrier frequency is high we have to use the A/D converter with the wide frequency band. The sampling frequency should be high to comply with the Shannon-Kotelnikov theorem. A possible sampling frequency jitter will deteriorate the signal to noise ratio on the converter output. A signal processor of the digital part of the receiver has to operate on a high frequency and will be very complex. The simple RF part of the receiver is the advantage of this solution. The disadvantage consists in high power consumption of A/D converter and signal processor. The ADC12D1800RF (Texas Instruments 2012) can be the example of this solution.

Superhet.

The signal is shifted to a lower intermediate frequency in the mixer. It can be then more easily filtered, amplified and digitalized. The design requires to optimize the frequency plan and level budget of the receiver. This method makes it possible to realize receivers with the best functional parameters. Difficulties of it consist in

– reception of image frequencies and other side signals caused by intermodulation distortion of the mixer; suppression of 20 dB and more is required
– phase noise of the local oscillator

In the best quality receivers transversal SAW filters are used behind the mixer. Because of effortlessness analogue filtering by the intermediate filter it is not necessary to perform digital filtering in the processor and therefore we need not to quantize into many levels. Two levels (one bit quantization) can be enough for inexpensive receivers.

The receiver with conversion to the base band.

The conversion into the base band can be done by multiplication of RF signal by a local oscillator signal. After filtering we obtain components of the complex envelope of the modulation signal (see Figure 5) which are digitalized after that.

The disadvantage of this method is the strict requirement of symmetry of multiplication elements. If it is not satisfied, a DC offset appears on the output.

It is necessary to suppress interference from the local oscillator which is oscillating on a carrier frequency.

All these drawbacks can be mitigated when we implement the mixer by the monolithic technology. An example of such realization is the circuit MAX2120 (Maxim 2010). Large power consumption is its main deficiency and we can use it in a stationary receiver.

For digitalization of orthogonal components in the base band we use the two-channel A/D converter. Digitalization and the further signal processing in the signal processor are relatively simple. There are not any special requirements for the A/D converter; a small number of quantization levels is enough for simple receivers.

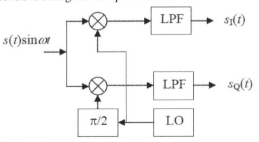

Figure 5. Conversion into base band

Hybrid receiver.

It is a receiver in which we shift the signal on lower frequency and then convert it into the base band, or we use the direct quantization, which we may call a hybrid receiver. Good selectivity as a result of easy filtering on a low intermediate frequency is its advantage.

2.5 Power of signal and noise in GPS

The power led to the antenna of the GPS satellite is 27 W (Misra 2006). A sufficient angle for irradiation of the Earth is ±13.88 deg, however satellites are transmitting into angle ±21.3 deg. The corresponding gain of the satellite antenna is 14.66 dB and EIRP 29 dBW, i.e. 794 W. The gain of the satellite antenna is dependent on the angle of irradiation, which belongs to the given elevation angle. Misra (Misra 2006) has assessed the gain GT of the transmitting antenna 12.1 dB, 12.9 dB and 10.2 dB for elevations 0 deg, 40 deg and 90 deg.

We obtain power density on the Earth surface when the satellite is in zenith as

$$P = \frac{P_T G_T}{4\pi R^2 L_A} \tag{2}$$

where R is the distance of the satellite from the ground (20 190 km) and LA expresses losses of propagation in the atmosphere (0.5 dB). When we rewrite (2) in decibels, we obtain

$$P_{dB} = P_{TdB} + G_{TdB} - 10\log(4\pi) - 20\log R - L_{AdB} =$$
$$= 14.3 + 10.2 - 11 - 146.1 - 0.5 = -133.1 \,\text{dBW/m}^2 \tag{3}$$

The power on the output of receiving antenna is given by its effective area

$$A_E = G_R \frac{\lambda^2}{4\pi} \tag{4}$$

The gain is $G_R=1$ for an isotropic antenna. Its effective area is $2.885625.10^{-3}$ m^2; which is equal to -25.4 dBm^2. In applications the patch antenna is usually used which has (Misra 2006) the gain $G_R=+4$ dB for 90 deg elevation. Then the power on the receiving antenna output is

$$P_R = P \frac{\lambda^2}{4\pi} G_R \tag{5}$$

and in decibels we obtain

$$P_R = P_{dB} + A_{EdB} = -133.1 + (-25.4 + 4) =$$
$$= -154.5 \,\text{dBW} \tag{6}$$

When we presume the antenna noise temperature $T_A=100$ K, the input filter attenuation $G_1=1$ dB, the gain of the LNA 20 dB, its noise figure $F_2=3$ dB and the equivalent temperature 290 K and finally the attenuation of the cable to the indoor part of the receiver 10 dB then we obtain for the spectral noise power density (Misra 2006)

$$N_0 \approx 10\log k \left[T_A + 290 \left(\frac{F_2}{G_1} - 1 \right) \right] =$$
$$= -201.3 \,\text{dBW/Hz} \tag{7}$$

where $k=1.38 \cdot 10^{-23}$ J/K is Boltzmann's constant.
Then the signal to the noise ratio is

$$C/N_0 = -154.5 + 201.3 = 46.8 \,\text{dB-Hz} \tag{8}$$

We have to realize that the root mean square of error of the measured distance depends on the effective width β of the frequency band

$$\sigma_D = \frac{c^2}{(2E/N_0)\beta^2}$$

When we use the frequency band width of B = 20 MHz, the ratio of the signal power C and the noise power P_N on the preamplifier output is

$$C/P_N = \frac{C}{BN_0} = -154.5 - (-201.3 + 10\log(20.10^6)) = \tag{10}$$
$$= -26.2 \,\text{dB}$$

A narrower width of the frequency band, 2 MHz e.g., brings us a 10 dB improvement of that ratio, however the precision of position determination is worse (see (9)). In Table 1 there are different values of these parameters for antenna gains and the LNA noise figures that we used during our analysis of the RF part of the GNSS receiver.

Table 1. Parameters of RF part of GPS receiver

G_R [dB]	F_2 [dB]	N_0 [dBW/Hz]	C/N_0 [dB-Hz]	C/P_N [dB]
4.0	3.00	-201.3	46.8	-26.2
4.0	0.86	-204.5	50.0	-23.0
4.0	0.50	-205.2	50.7	-22.3
6.8	0.86	-204.5	52.8	-20.2
6.8	0.50	-205.2	53.5	-19.5

3 THE MULTI-CONSTELLATION GNSS RECEIVER

3.1 Processed signals

We suppose that the multi-constellation receiver will process the following signals:
- GPS (L1, L2, L5)
- Galileo (E1, E5a, E5b)
- GLONASS (L1(G1), L2(G2))
- COMPASS (E1(B1), E2(B3))

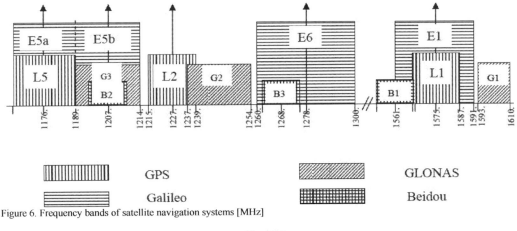

Figure 6. Frequency bands of satellite navigation systems [MHz]

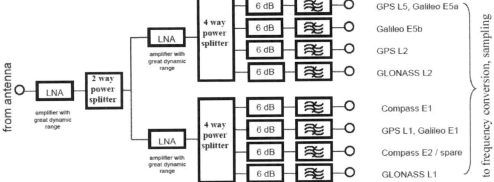

Figure 7. Navigation systems signal splitting in RF receiver part

The position of these signals on the frequency axis is outlined in Figure 6. It is evident, that the signals
- GPS L1 and Galileo E1
- GPS L5 and Galileo E5a

occupy the same frequency bands.

Currently, with our technological possibilities we are only able to design a stationary multi-constellation receiver. We will describe roughly its RF part hereinafter.

3.2 The RF part of the stationary multi-constellation GNSS receiver

Considering the demands on the RF part of a receiver of navigation satellites signals as mentioned above, the RF part of the multi-constellation receiver will correspond to the structure in Figure 1. Its design will agree with the principles aforesaid, too.

Power levels of the mentioned systems are in range from -155 dBW to -163 dBW on typical antenna output on the Earth surface as we can see in Table 2. GPS L2 signals are weaker and L5 stronger

than L1 signals. The GLONASS signals are the weakest ones.

The first LNA (outdoor) is placed by the Novatel 704 X antenna and is connected with an indoor part by cable. In the indoor part, of which the RF part is arranged according to Figure 7, the particular signals are separated. Most of the components are from Mini-Circuits with a design similar to that in Figure 2.

Table 2. Power level of the signals on antenna output on the Earth (BeiDou 2012, European Union 2010, GPS Directorate 2012, Glonass 2008)

System	Power level [dBW]
GPS L1	-155.5
GLONASS L1	-158.0
Galileo E5	-152.0
Compass B1	-160.0

We suppose that the filters will be ceramic resonators with the band width (3 dB) 20 MHz for the first four channels and 10 – 12 MHz for GLONASS and Compass channels.

3.3 The mobile multi-constellation receiver

The method of GNSS signals reception described above is suitable for stationary receivers only because of the receiver dimensions and power consumption. Despite that the stationary receiver can be used in many applications like railway and bus transportation. It is an ideal tool for the algorithms design, too.

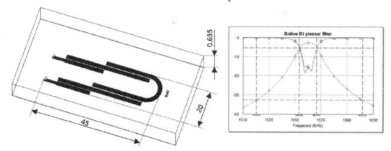

Figure 8. Microstrip filter for Beidou/COMPASS B1 band (Svatoň 2013)

For a mobile receiver, the only solution possible is an integrated circuit. We are currently working on this solution.

3.4 The microstrip filters

Filters are a fundamental but considerably bulky element of the RF part of the GNSS multi-constellation receiver. We have experimented with many types, among them with planar (microstrip) ones (Svatoň 2013). The first experience was obtained with classical microwave substrates for printed boards. The results for B1 frequency band were relatively successful as it is possible to see in Figure 8. Presently experiments with corundum substrate with higher permittivity are being prepared. Precision of filter motif realization is a particular problem; our experiments lead to silk screen printing of thick layers.

4 CONCLUSIONS

According to the principles described above, the stationary receiver was designed and implemented in the past four years. It makes it possible to receive signals of all four satellite navigation systems and to master separately their processing (Kovář 2012).

For a mobile multi-constellation receiver it is necessary to use integrated technologies which allow to achieve not only a small size of the product but mainly its small power consumption.

It is difficult to draw up a multi-constellation receiver; the fulfillment of many requirements is often contradictory. It is the reason why we are considering a design of a simulator which would allow the optimal choice of the receiver parameters.

ACKNOWLEDGEMENT

The submitted paper has arisen during work on the project which has been financially supported by the Technology Agency of the Czech Republic.

REFERENCES

BeiDou Navigation Satellite System 2012. Open Service Signal B1I (Version 1.0). China Satellite Navigation Office.
European Union 2010. European GNSS Open Service. Signal in Space Interface Control Document.
GPS Directorate 2012. Navstar GPS Space Segment and Navigation User Segment interfaces.
Glonass 2008. Interface Control Document. Ed. 5.1. Moscow: Russian Institute of Space Device Engineering.
Kovář, P. et al. 2009. Universal front end for software GNSS receiver. In Proceedings of 13th IAIN World Congress [CD-ROM]. Bergen: Nordic Institute of Navigation.
Kovář, P. 2012. Witch Navigator. www.witchnav.cz
Maxim Integrated Products: Max2120. Complete, Direct-Conversion Tuner for DVB-S and Free-to-Air Applications. Sunnyvale: Maxim.
Mini-Circuits 2012. http://217.34.103.131/pdfs/ZX60-P162 LN1+.pdf
Misra, P. & Enge, P. 2006. Global positioning system. Lincoln: Ganga-Jamuna Press.
Novatel 2012. http://www.novatel.com/assets/Documents/ Papers/GPS-704xWhitePaper.pdf
O'Keefe, K. 2001. Availability and reliability advantages of GPS/Galileo integration. In Proceedings of ION GPS 2001. Salt Lake City.
Popp, J. 2013. Anténní předzesilovač pro přijímač systému GNSS. (Antenna preamplifier for GNSS receiver). Bachelor thesis. Prague: Czech Technical University in Prague.
Rohde, U.L. & Newkirk, D.P 2000. Microwave circuit design for wireless applications. Hoboken: John Wiley and Sons.
Svatoň, J. 2013. Vstupní obvody multikonstelačních přijímačů pro družicovou navigaci. (Multiconstellation Satellite Navigation Receivers Frontends). Master degree thesis. Prague: Czech Technical University in Prague.
Texas Instruments: ADC12Dxx00RF Direct RF-Sampling ADC Family. http://www.ti.com/lit/ml/snwt003/ snwt003.pdf

A New User Integrity Monitoring for Multiple Ramp Failures

H. Yun & C. Kee

School of Mechanical and Aerospace Engineering and the Institute of Advanced Aerospace Technology, Seoul National University, Republic of Korea

ABSTRACT: This paper develops and analyzes a new RAIM algorithm as a candidate of future architecture of RAIM algorithm which can treat not only a single failure but also simultaneous multiple failures. A proposed algorithm uses measurements residuals and satellite observation matrices of several consecutive epochs for multiple Failures Detection and Exclusion (FDE). It can detect multiple failures without limitation of number of faulty measurements. In this paper, we give detailed explanation of the FDE algorithms with rigorous mathematical expression. Simulation results show that proposed algorithm can detect and exclude the multiple failures of tens of meters.

1 INTRODUCTION

In accordance with the GPS modernization, the emergence of Galileo and redevelopment of GLONASS, users are expected to be able to take advantage of the large number of visible satellites. These new or renewed GNSS satellites provide users with improved signal quality. As a result, users are expected to have improved navigation performance thanks to the increased number of visible satellites and the improved signal quality. This has raised the possibility of using Receiver Autonomous Integrity Monitoring (RAIM) for much more demanding phases of flight such as LPV-200 or Cat-I. However, achieving LPV-200 or Cat-I, conventional RAIM algorithm alone has limitations. Conventional RAIM algorithms were designed to assure that integrity requirements are met under the assumption that a signal fault exists on only a single satellite at any given time. Such an assumption is valid for these applications, which can tolerate horizontal position errors up to a Horizontal Alert Limit (HAL) of 556 meters or greater, and which do not use vertical guidance from GNSS. Simultaneous Range errors that could cause a hazard for vertical guidance might occur on more than one satellite at the same time. Especially in multi-constellation environment, the number of visible satellites will be increased and possibility of simultaneous multiple failures will be higher.

In these circumstances, a series of new developments has taken place in the field of Receiver Autonomous Integrity Monitoring (RAIM). Of particular interest were the topics of multi-constellation RAIM and analyzing the impact of simultaneous multiple failures. [1-3] developed multiple hypothesis RAIM algorithms modifying a single Hypothesis Solution Separation algorithm. Theses algorithms use range residual as a test statistic of Fault Detection and Exclusion (FDE). Range residual is the vector, which is projected component of range error vector, and it cannot be used as an absolute indication of a range error. Its detection capability can be degraded when the error vector and satellite line of sight vectors are in bad geometry. [4] proposed a Sequential Multiple Hypothesis RAIM (SMHR) algorithm which can handle all the case of multiple failures including the failures which cannot be detected by conventional least squares algorithm or solution separation algorithm. This algorithm reconstructs the error vector itself using sequential scheme, and uses it as a test statistic. Test statistic is estimate of an error vector itself not a projected vector (i.e. range residual vector), its detection capability is less affected by satellite geometry. The SMHR algorithm reconstructs the error vector under the assumption that the error vector is constant in several consecutive epochs. Therefore it has a weakness that it cannot detect the failures which is varying with time. This paper modifies and improves the algorithm of [4] and proposes a new multiple hypothesis RAIM algorithm which can detect any types of failures.

2 SMHR ALGORITHM ANALYSIS AND CHARACTERIZATION

Most of RAIM algorithms use the residual vectors to determine whether the current measurement is normal or not. However range residual cannot be used as an absolute indication of a range error, because it is a projection of the error vector not itself. And some situations, the detection performance can be degraded, especially in cases of multiple failures. To overcome this limitation [4] proposed a new FDE algorithm that monitoring the error vector itself, not monitoring the projection of error vector. Figure 1 shows the concept of this approach in 2-dimensional case.

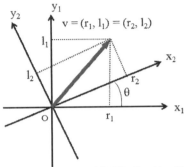

Figure 1. Conventional SMHR Algorithm (2D example)

This technique, for example, can be used to find the error vector $\mathbf{v}$. In this problem, $\mathbf{r}$ is available at each epoch, but $\mathbf{l}$ is not. The residual vector at each epoch is equivalent to $\mathbf{r}_1$ and $\mathbf{r}_2$, and the lost components are equivalent to $\mathbf{l}_1$ and $\mathbf{l}_2$. The single measurement of $\mathbf{r}_1$ is not sufficient to estimate the vector $\mathbf{v}$, because $\mathbf{r}_1$ has only one degree of information while $\mathbf{v}$ requires two degrees of information. Therefore, two independent observations are necessary to estimate the magnitude of $\mathbf{v}$. $\mathbf{v}$ can be estimated by collecting information from two consecutives epochs and measuring the change in coordinate frame θ. This problem can be expressed in terms of the following mathematical expression:

$$\mathbf{v} = (\mathbf{r}_1, \mathbf{l}_1) = (\mathbf{r}_2, \mathbf{l}_2) \tag{1}$$

$$\begin{cases} \mathbf{r}_2 = \mathbf{r}_1 \cdot \cos\alpha + \mathbf{l}_1 \cdot \sin\theta \\ \mathbf{l}_2 = -\mathbf{r}_1 \cdot \sin\alpha + \mathbf{l}_1 \cdot \cos\theta \end{cases} \tag{2}$$

where the unknowns are $\mathbf{v}$, $\mathbf{l}_1$, and $\mathbf{l}_2$, and the observables are $\mathbf{r}_1$, $\mathbf{r}_2$, and θ. The next step is to reconstruct the vector $(\mathbf{r}_1, \mathbf{l}_1)$ with known $\mathbf{r}_1$, $\mathbf{r}_2$, and θ values. Assuming that $\mathbf{v}$ is constant for two consecutive epochs, $\mathbf{l}_1$ can be calculated as follows:

$$\hat{\mathbf{l}}_1 = -\frac{\cos\theta}{\sin\theta} \cdot \mathbf{r}_1 + \frac{1}{\sin\theta} \cdot \mathbf{r}_2 \tag{3}$$

In this problem, the two reference systems ($x_1 O y_1$ and $x_2 O y_2$) must be linearly independent (i.e. $\theta \neq 0$).

Finally the estimate of an error vector can be calculated as Eq. (7).

$$\hat{\mathbf{v}} = \left(\mathbf{r}_1, \hat{\mathbf{l}}_1 \right) \tag{4}$$

Expanding this 2D problem to a navigation case, the error vector $\mathbf{v}$ has N degrees of information, and its projection $\mathbf{r}$ has N − 4 degrees of information. This approach tries to reconstruct the error vector, collecting projections on the consecutive epochs, and taking advantage of the change of subspaces. The satellites and the receiver move relatively and the mutual geometry changes. This correspond to changing the point of view for observing the error vector $\mathbf{v}$. This allows retrieving the missing information. While conventional RAIM algorithms use only $\mathbf{r}_1$ for FDE, SMHR utilizes $\mathbf{r}_1$ and $\mathbf{r}_2$ which are projected on different subspaces each other. This corresponds to changing the point of view for observing the error vector. Therefore the detection capability of SMHR is independent on the direction of an error vector or the direction of coordinate frame. However this approach has two limitations for applying to real world.

– Ill-conditioned system

In Eq. (3), θ means the difference in subspaces of observation matrix of two consecutive epochs. In this case, changes of Line Of Sight (LOS) vector are very small, because of large distances between receiver and satellites. Therefore θ has a very small value, and system becomes ill-conditioned. As a result this approach has a weakness that it is very sensitive to measurement noise. According to [4], this approach can detect a range bias that has a magnitude of about 5km.

– Constant bias assumption

This Approach is valid for estimating the constant bias range error only, because it assumed that the error vector is constant in two consecutive epochs. When ramp error occurred, this assumption cannot be used anymore, and the error vector estimation may have invalid values.

3 A NEW SMHR ALGORITHM FOR MULTIPLE RAMP FAILURES DETECTION

Conventional SMHR algorithm cannot be used in real world because of two limitations which has been described above. Nevertheless it has the advantage of detecting the range errors using more information than conventional RAIM algorithms. The proposed algorithm has been developed to make the best use of strength of the Conventional SMHR and to minimize the weakness of the Conventional SMHR algorithm. It minimizes the standard deviation of estimate by using precise carrier phase measurement. And it can handle the ramp error also assuming that the error vector can vary with the time.

3.1 Adoption of RRAIM concept

Due to an ill-conditioned problem, the SMHR algorithm responds to measurement noise very sensitively. As getting lower noise level of measurement, the algorithm can obtain robust error vector estimation. For lower measurement noise level, the proposed algorithm uses carrier phase measurement as well as pseudorange measurement. In general case, in order to use a carrier phase as a range measurement, we have to solve an integer ambiguity. Resolving integer ambiguity can be done with the help of ground facility, and the users would be heavily dependent on ground facility. This causes a latency of processing information, therefore it can be a hazard for satisfying a Time To Alert (TTA) requirement. For using precise carrier phase measurement without resolving and integer ambiguity, the proposed algorithm adopted navigation equation and error models of Relative RAIM (RRAIM) algorithm [5,6].

In RRAIM approach, the receiver uses carrier-smoothed psudorange measurements, and propagates these measurements by compensating for the difference between current and past carrier phase measurement.

$$\hat{\rho}_t = \rho_{t-T} + \Delta\phi_{t,t-T} \qquad (5)$$

In Eq. (5), $\Delta\phi_{t,t-T}$ is the difference in carrier phase measurements between time t and t-T, i.e. $\phi_t - \phi_{t-T}$, and these are expressed in distance rather than angular units. The propagated range measurement $\hat{\rho}_t$ is related to the true range r_t by

$$\hat{\rho}_t = r_t + B_t + \delta\rho_{t-T} + \delta\Delta\phi_{t,t-T} \qquad (6)$$

where B_t is receiver clock bias, $\delta\rho_{t-T}$ is error in ρ_{t-T}, and $\delta\Delta\phi_{t,t-T}$ is error in $\Delta\phi_{t,t-T}$. The methods for specifying each errors in Eq. (6) are well described in [5].

3.2 Ramp error assumption

In order to solve the problem sequentially, previous work assumes that the error vector is constant for several consecutive epochs. However there are a lot of ramp errors in real GNSS failures such as satellite clock failures. Therefore this approach has a limitation on handling this type of error. The proposed algorithm assumes that each component of

the error vector increases or decreases for several consecutive epochs.

$$\mathbf{v}_k = \mathbf{v}_{k-1} + \boldsymbol{\alpha} \qquad (7)$$

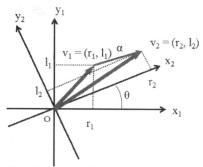

Figure 2. Concept of the proposed algorithm (2D example)

In Eq. (7), $\boldsymbol{\alpha}$ is an n x 1 constant drift vector. Under this assumption, for 2-dimensional case, estimating the error vector using range residual vectors can be described as follows.

In Figure 2, 1-dimensional range residual vectors $\mathbf{r}_1$ and $\mathbf{r}_2$ are observables at each epoch. In nominal condition, bias component of $\mathbf{v}_1$ and $\mathbf{v}_2$ is near 0, and we can assume that there is only noise component in $\mathbf{v}_1$ and $\mathbf{v}_2$. In this condition, $E(\boldsymbol{\alpha}) = \mathbf{0}$ and $E(\mathbf{v}) = \mathbf{0}$. When fault is occurred in certain measurements, corresponding components of $\boldsymbol{\alpha}$ have a non-zero value. By estimating $\boldsymbol{\alpha}$ and integrating this value, we can get an estimate of the error vector. In Figure 2, residual vector at first epoch $\mathbf{r}_1$ is a projection on x-axis of x_1Oy_1 frame, and the lost component $\mathbf{l}_1$ is projection on y-axis.

$$\begin{bmatrix} \mathbf{r}_1 \\ \mathbf{l}_1 \end{bmatrix} = \begin{bmatrix} 1 & 0 \\ 0 & 1 \end{bmatrix} \mathbf{v}_1 \qquad (8)$$

Residual vector and lost component at second epoch are projections of error vector $\mathbf{v}_2$ onto x and y axis of x_2Oy_2 frame which rotated θ from x_1Oy_1 frame. And according to the assumption, $\mathbf{v}_2$ can be expressed as a sum of the previous error vector $\mathbf{v}_1$ and drift vector $\boldsymbol{\alpha}$.

$$\begin{bmatrix} \mathbf{r}_2 \\ \mathbf{l}_2 \end{bmatrix} = \begin{bmatrix} \cos\theta & \sin\theta \\ -\sin\theta & \cos\theta \end{bmatrix} \mathbf{v}_2 = \begin{bmatrix} \cos\theta & \sin\theta \\ -\sin\theta & \cos\theta \end{bmatrix} (\mathbf{v}_1 + \boldsymbol{\alpha}) \quad (9)$$

Eq. (10) which is for getting unknown vector $\boldsymbol{\alpha}$ is derived from combining Eq. (8) and (9).

$$\begin{bmatrix} 0 & -\sin\theta \\ -1 & \cos\theta \end{bmatrix} -\mathbf{I}_2 \begin{bmatrix} \mathbf{l}_1 \\ \mathbf{l}_2 \\ \boldsymbol{\alpha} \end{bmatrix} = \begin{bmatrix} 1 & -\cos\theta \\ 0 & -\sin\theta \end{bmatrix} \begin{bmatrix} r_1 \\ r_2 \end{bmatrix} \qquad (10)$$

The matrix in left-hand side of Eq. (10) has a dimension of 2 x 4 and it is not invertible. While

SMHR approach which assumes constant bias error need measurement of only 2 epochs in 2-dimensional case, the proposed algorithm needs more measurements. By stacking 2 more epochs the matrix has a dimension of 6 x 6 and becomes invertible.

$$
\begin{bmatrix}
0 & -\sin\theta_{1,2} & 0 & 0 & \\
-1 & \cos\theta_{1,2} & 0 & 0 & -\mathbf{I}_{2\times2} \\
0 & 0 & -\sin\theta_{1,3} & 0 & \\
-1 & 0 & \cos\theta_{1,3} & 0 & -2\mathbf{I}_{2\times2} \\
0 & 0 & 0 & -\sin\theta_{1,4} & \\
-1 & 0 & 0 & \cos\theta_{1,4} & -3\mathbf{I}_{2\times2}
\end{bmatrix}
\begin{bmatrix} l_1 \\ l_2 \\ l_3 \\ l_4 \\ \boldsymbol{\alpha} \end{bmatrix}
$$

$$
=
\begin{bmatrix}
1 & -\cos\theta_{1,2} & 0 & 0 \\
0 & -\sin\theta_{1,2} & 0 & 0 \\
1 & 0 & -\cos\theta_{1,2} & 0 \\
0 & 0 & -\sin\theta_{1,2} & 0 \\
1 & 0 & 0 & -\cos\theta_{1,2} \\
0 & 0 & 0 & -\sin\theta_{1,2}
\end{bmatrix}
\begin{bmatrix} r_1 \\ r_2 \\ r_3 \\ r_4 \end{bmatrix}
\qquad (11)
$$

In Eq. (11), $\theta_{1,2}$ means the angle between the coordinate frame x_1Oy_1 and x_2Oy_2. Solving Eq. (11), the estimate of lost component vectors and the drift vector is obtained.

Expanding this 2-dimensional problem to a navigation case, the $\mathbf{v}_k$ and $\boldsymbol{\alpha}$ has a dimension of N each, and the $\mathbf{r}_k$ has a dimension of $N-4$.

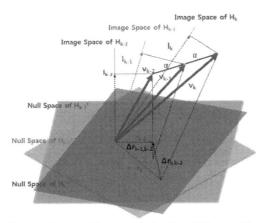

Figure 3. Concept of the proposed algorithm (N-D example)

The observable $\mathbf{r}_k$ is the error vector's projection on the null space of $\mathbf{H}_k^T$, and the lost component $\mathbf{l}_k$ is the projection on the image space of $\mathbf{H}_k$. Relationship between each component in Figure 3 can be expressed as follow.

$$
\mathbf{v}_k = \mathbf{r}_k + \mathbf{l}_k = \left(\mathbf{I} - \mathbf{H}_k\mathbf{K}_k\right)\mathbf{v}_k + \mathbf{H}_k\mathbf{K}_k\mathbf{v}_k
\qquad (12)
$$

Similarly to the 2-dimensional case, in order to estimate valid lost component, the subspaces of every epoch should be linearly independent. However, the image space of each observation matrix shares the same subspace, which spans the 4th column of H matrix. Therefore, common component exists in every epoch and the system is linearly dependent. This problem can be solved by estimating common component independently and eliminating the common component from the range residual and observation matrix. Then Eq. (12) can be re-written as Eq. (13).

$$
\mathbf{v}_k = \mathbf{r}_k + \tilde{\mathbf{H}}_k\tilde{\mathbf{K}}_k\mathbf{v}_k
\qquad (13)
$$

where $\tilde{\mathbf{r}}_k = \mathbf{z}_k - \tilde{\mathbf{H}}_k\hat{\mathbf{x}}'_k$, $\tilde{\mathbf{H}}_k$ is an N x 3 observation matrix, which consists of the first three columns of $\mathbf{H}_k$, $\tilde{\mathbf{K}}_k$ is a pseudo inverse matrix of a $\tilde{\mathbf{H}}_k$ and $\hat{\mathbf{x}}'_k$ is composed of a 3×1 user position estimation vector and 1 receiver clock estimation, which is calculated from the independent clock estimation filter

Because $\mathbf{v}_k = \mathbf{v}_{k-1} + \boldsymbol{\alpha} = \mathbf{v}_{k-2} + 2\boldsymbol{\alpha}$, the differences between the common component eliminated residual vector at each epochs, i.e. $\Delta\tilde{\mathbf{r}}_{k,k-2}$ and $\Delta\tilde{\mathbf{r}}_{k-1,k-2}$ can be expressed as in Eq. (14) and (15).

$$
\Delta\mathbf{r}_{k-1,k-2} = \mathbf{H}_{k-2}\mathbf{K}_{k-2}\mathbf{v}_{k-2} - \mathbf{H}_{k-1}\mathbf{K}_{k-1}\mathbf{v}_{k-2}
$$
$$
+\left(\mathbf{I} - \mathbf{H}_{k-1}\mathbf{K}_{k-1}\right)\cdot\boldsymbol{\alpha}
\qquad (14)
$$

$$
\Delta\mathbf{r}_{k,k-2} = \mathbf{H}_{k-2}\mathbf{K}_{k-2}\mathbf{v}_{k-2} - \mathbf{H}_k\mathbf{K}_k\mathbf{v}_{k-2}
$$
$$
+\left(\mathbf{I} - \mathbf{H}_k\mathbf{K}_k\right)\cdot 2\boldsymbol{\alpha}
\qquad (15)
$$

Eq. (14) and (15) can be expressed together in matrix form such as Eq. (16).

$$
\begin{bmatrix} \Delta\mathbf{r}_{k-1,k-2} \\ \Delta\mathbf{r}_{k,k-2} \end{bmatrix}
=
\begin{bmatrix}
\mathbf{H}_{k-2} & -\mathbf{H}_{k-1} & \mathbf{O} & \mathbf{I} \\
\mathbf{H}_{k-2} & \mathbf{O} & -\mathbf{H}_k & 2\mathbf{I}
\end{bmatrix}
\begin{bmatrix}
\mathbf{K}_{k-2}\mathbf{v}_{k-2} \\
\mathbf{K}_{k-1}\left(\mathbf{v}_{k-2}+\boldsymbol{\alpha}\right) \\
\mathbf{K}_k\left(\mathbf{v}_{k-2}+2\boldsymbol{\alpha}\right) \\
\boldsymbol{\alpha}
\end{bmatrix}
\quad (16)
$$

In Eq. (16), the number of unknowns is $3\cdot(m-1) +$ N and the number of observables is 2N, where m is the number of used epochs constructing the system matrix. Therefore, for solving Eq. (16), minimum number of necessary epochs m varies depending on the number of visible satellite. Table 1 gives for each number of visible satellites (N) the number of epochs (m) to be collected to estimate $\mathbf{v}$ and the estimation delay = m − 1.

Table 1. Estimation delay of the proposed algorithm

Number of visible satellites (N)	Number of minimum necessary epochs (m)	Estimation Delay
5	5	4 sec
6~8	4	3 sec
9 or more	3	2 sec

In multiple constellation circumstance, users would have 13 visible satellites at least. In that case the estimation delay is only 2 seconds. The estimation delay is not critic because it allows meeting the requirement of TTA smaller than 6 seconds.

4 SIMULATION RESULTS

Simulations are conducted to verify the feasibility of the proposed algorithm. The satellite orbit is generated using real GPS ephemeris data from the RINEX navigation files. In this simulation, it is assumed that the GIC provides users with correction messages and integrity messages every 30 seconds. The user can obtain integrity-assured measurements by applying the information from the GIC. It is also assumed that ionospheric delay, tropospheric delay, satellite orbit and clock errors are eliminated with the help of the GIC information. Under these assumptions, for each line of sight, residual error in the GIC-generated range correction and a random noise with a standard deviation modeled as a function of satellite elevation angle are added to each geometrical range. In addition to this, various combinations of range errors are inserted at t = 100 to 150 seconds. The inserted range errors are listed below.

Table 2. Magnitude of inserted errors

	Constant Bias (m)	Ramp Error (m/s)
Magnitude of Error	10	0.5
	13	0.75
	15	1
	20	2

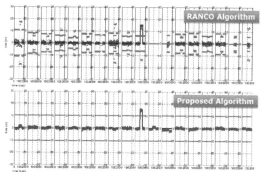

Figure 4. Results of single constant bias error case

Figure 4 shows the results of the RANCO algorithm and the proposed algorithm under a single failure condition with 15 meters of biases inserted. In this case, detection rate of each algorithm are 70% and 100%. In the RANCO algorithm, there is some points that have a value of zero. Because these points were used for calculating a subset solution, so

they have no residual values and thresholds. RANCO algorithm uses N − 4 observables for detecting N range errors, and detection capability is affected by projection subspaces which are determined from subset selection and satellite geometry. When the bias is similar in magnitude with detection threshold, the RANCO algorithm might select ill-conditioned satellites as a subset in order to maximize the number of inliers. This results in the increase of detection threshold and degradation of detection capability.

On the other hand, the proposed algorithm uses measurements of several consecutive epochs, and estimates the error vector itself, its detection capability is independent on constellation geometry.

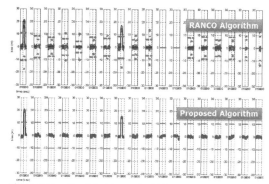

Figure 5. Results of double constant bias errors case

Figure 5 shows the results of double failures which have a magnitude of 15 and 20 meters. In this case, the detection rate of each algorithm is 54% and 100%. Although the magnitude of bias was increased, detection rate of the RANCO algorithm was decreased. In cases of multiple failures, direction of error vector can be similar with the direction of an image space of observation matrix. In this case, magnitude of lost component increases, and detection rate of the RANCO algorithm which uses only range residuals can be degraded. The proposed algorithm utilizes not only a range residual vector but also a lost component, its detection capability is not dependent on the number of failures. It is observed that analogous results are also valid in cases involving ramp failures.

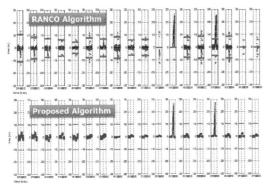

Figure 6. Results of double ramp errors case

Figure 6 shows the results when ramp failures which have a speed of 0.5 m/s occurred in the 12th and 15th satellites.

For ramp failures, it is very hard to detect the occurrence of failure at a time, because magnitude of the error is small at the beginning. In case of RANCO algorithm, undetected errors in subset solution directly affect all the other measurement, and cause an increase of range residual of normal satellite. This can be a potential threat of false alarm. On the other hand, in case of the proposed algorithm, the effect of undetected errors is negligible. The part that failures occurred was magnified for more detailed analysis.

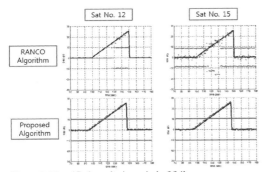

Figure 7. Magnified results in period of failures

In Figure 7, upper two figures are results of the RANCO algorithm and lower two figures are results of the proposed algorithm. RANCO algorithm detected an occurrence of failures about 25 to 30 seconds after the beginning of the errors. On the other hand, for the proposed algorithm it took only about 20 to 25 seconds. Because the RANCO algorithm tries to maximize the number of inliers with only partial information of the error, there is a threat for declaring a normal satellite as a faulty one. Therefore it cannot detect a failure until the range error grows sufficiently large. The proposed algorithm solved this problem by estimating the error vector itself with sequential approach,

therefore detection capability is uniform whether the range measurements contain undetected failures or not.

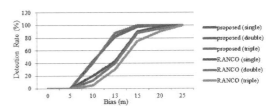

Figure 8. Detection rate of each algorithm

Figure 8 shows the detection rate of each algorithm in cases of single or multiple (double, triple) failures. Detection rate of the proposed algorithm is higher than that of the RANCO algorithm overall. And detection capability of the proposed algorithm is uniform no matter the number of faulty measurements, while that of the RANCO algorithm is degraded as the number of faulty satellites increased.

5 CONCLUSIONS

This paper proposed a new sequential multiple hypothesis RAIM algorithm which can estimate an error vectors whether they are constant bias or ramp errors. Conventional SMHR algorithm assumes only constant bias cases and uses only pseudorange measurement. Therefore it can detect and exclude only constant bias failures with magnitude of several kilometers and it cannot be used in general case. The proposed algorithm overcome these limitations by using precise carrier phase measurements and assuming magnitude of range errors can be varied with time. As a result, the proposed algorithm can detect failures which have magnitude of several tens of meters, while conventional SMHR algorithm can only detect it of several kilometers. Because the proposed algorithm detects and identifies failures by estimating the error vector itself, its detection capability is unaffected by satellite geometry or combination of faulty satellites, and besides its detection capability is unaffected by types of errors whether they are single or multiple failures. Furthermore the proposed algorithm is computationally efficient because it identifies faulty measurement by monitoring the estimated error vector directly while other conventional multiple hypothesis RAIM algorithm uses iterative method to identify faulty measurements. The simulation results show that detection rate of the proposed algorithm is maximum 50 % higher than that of conventional RANCO algorithm, and computation time is about 1/50 of the conventional RANCO algorithm.

This paper focused on fault detection and exclusion algorithm, and further researches about protection level and availability analysis are under investigation.

ACKNOWLEDGEMENT

This research was supported by a grant from "Development of Wide Area Differential GNSS," which is funded by Ministry of Land, Transport and Maritime Affairs of Korean government, contracted through SNU-IAMD at Seoul National University.

REFERENCES

[1] B. J. Ene A., Powell J. D., "Fault Detection and Elimination for Galileo-GPS Vertical Guidance," in Intitute of Navigation's National Technical Meeting, San Diego, CA, 2007.

[2] B. Pervan, Pullen, S., Christie, J. , "A Multiple Hypothesis Approach to Satellite Navigation Integrity," Navigation, vol. 45, 1998.

[3] G. Schroth, Rippl,M., Ene,A., Blanch,J., Belabbas,B., Walter,T., Enge,P., Meurer,M., "Enhancements of the Range Consensus Algorithm (RANCO)," in ION GNSS 21st. International Technical Meeting, Savannah, GA, US, 2008.

[4] I. Martini, Wolf, R., Hein, G. W. , "Receiver Integrity Monitoring in Case of Multiple Failures," in International Technical Meeting of the Satellite Division of the Institute of Navigation, Fort Worth, Tx, 2006, pp. 2608 - 2620.

[5] GEAS. (2008). Phase I - Panel Report. Available: http://www.faa.gov/about/office_org/headquarters_offices/ ato/service_units/techops/navservices/gnss/library/documen ts/media/GEAS_PhaseI_report_FINAL_15Feb08.pdf

[6] S. Hwang, "Relative GPS Carrier-Phase positioning using particle filters with position samples," in American Control Conference, 2009.

Chapter 7

Radar, ARPA and Anti-Collision

Were Improper Lights a Confusing and Fatal Factor in the BLUEBIRD/DEBONAIR Collision in Dublin Bay?

E. Doyle
Cork, Ireland

ABSTRACT: In clear visibility during the early dark hours of 20 May 2001, a collision occurred in the outer fairway channel to Dublin Port, between the motor vessel BLUEBIRD and the yacht DEBONAIR. The progress of the two vessels, along with other harbour traffic, was detected and monitored by the port's VTS system.

The report of the official investigation, while careful not to attribute blame or fault to any party, makes findings that are almost exclusively critical of DEBONAIR's actions. In doing so, the investigation overlooks a critical issue of artificial illumination and the impact of any such illumination on the collision outcome.

This paper examines how the impaired perception of BLUEBIRD's navigation and deck lights may have caused DEBONAIR's skipper to make a fatal turn to port in the mistaken belief that he was turning away from danger.

1 INTRODUCTION AND CONTEXT

1.1 Incident Overview

In the early hours of Sunday, 20 May 2001, a collision occurred in the fairway channel to Dublin Port, between the motor vessel BLUEBIRD and the sloop yacht DEBONAIR. BLUEBIRD was laden, inward bound with a cargo of wheat, and had embarked her pilot, while the yacht was outward bound towards Howth. The progress of the two vessels, along with other traffic in and around the harbour, was detected and monitored by the Dublin Port radar sensors and these images were displayed and recorded on the port's Vessel Traffic Services (VTS) system. The replay of this data enabled the Harbour Master, to establish the time of the collision at 0251 local time (UT+1). In the collision, DEBONAIR sustained massive hull damage from which she sank in a matter of seconds. Four persons, of the crew of five on board, lost their lives. BLUEBIRD was undamaged.

1.2 The Vessels Involved

BLUEBIRD was an unremarkable, coastal trading cargo vessel of 67 meters (LOA) and a service speed of about 11 knots. She was fully laden at the time of the incident, carrying approximately 1500 tonnes of wheat. The vessel was fitted, equipped and crewed in accordance with standard Convention requirements. There were seven persons on board, including the harbor pilot.

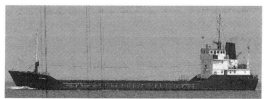
Figure 1. mv BLUEBIRD

Figure 2. Wreck of yacht DEBONAIR

DEBONAIR was a recreational cruiser/racer yacht, of a class known as the Club Shamrock 30 (LOA 9.2 meters). The boat was well maintained and equipped, and was capable of a speed of about 5 knots, under power. She was not under sail at the time of the accident. There were five persons on board.

1.3 The Collision Encounter

Having embarked her pilot at 0242 near No 1 Buoy, BLUEBIRD proceeded inwards along the fairway in the normal manner. She altered course to port at No's 3 and 4 Buoys (Dublin Bar), broadly as the axis of the channel required, and steadied on a course which was expected to afford a perfectly safe and normal "red to red" passing with the approaching DEBONAIR. When the vessels were about 150 metres apart, the yacht was seen to turn to port (showing her green starboard sidelight to BLUEBIRD). This inexplicable turn, shaping to cross BLUEBIRD's bows, brought the yacht on to an immediate collision course and extreme danger. The pilot had the ship's whistle sounded, and the master put his ship's engines "Full astern" at the same time or shortly thereafter. There was little or no effective time to avoid the collision, and there appeared to be little else that the master could have done, other than turning hard to starboard and accept the risk of running his ship aground. It is most probable that even this extreme option could not have prevented a collision.

Figure 3. Collision area at Dublin Bar

1.4 Investigation and Report by MCIB

The collision, and the circumstances surrounding the tragedy, was the subject of a prolonged investigation by the Marine Casualty Investigation Board (MCIB), who published their findings in a Report dated 11 February 2005. It is a matter of some regret that the investigation and report are marred by significant defects and shortcomings; fundamental questions are not addressed and, inexplicably, moonlight is offered as a source of ambient natural illumination at a time when the moon had not yet risen. Since the accident occurred during the early morning hours, at a time of near total darkness, the issue of illumination is critical to a proper understanding of the sequential events leading to the collision. That the immediate pre-collision scene was illuminated, in some way, will become clear in the following section.

1.5 Fundamental Questions Overlooked

By ignoring the issue of possible artificial illumination the MCIB investigation failed to address these fundamental questions:

1 Did BLUEBIRD exhibit the proper lights, and only those lights, prescribed by the COLREGS?
2 If, as the evidence inescapably suggests, 'lights, contrary to Rule 20' were a factor in the collision, could such lights have confused DEBONAIR's skipper when he made an *inexplicable(?)* and catastrophic last-minute turn to port?
3 Was the navigation of BLUEBIRD into Dublin Port conducted in accordance with the Dublin Pilotage District Bye-Laws then in force?

1.6 Witnesses

There were three material witnesses on board BLUEBIRD, all qualified and all directly concerned with the navigation or operation of the ship. On the bridge were the Master and the Dublin Port pilot, while on the foc'sle, the Chief Mate was making the usual arrival preparations.

The sole survivor of DEBONAIR's crew, while on deck at the time, was not involved with her outward navigation and he had no appreciation of the unfolding danger until seconds before the collision.

2 AMBIENT LIGHT AND ILLUMINATION

2.1 Level of Darkness

As the collision encounter unfolded at around 0251, each of the BLUEBIRD witnesses made observations of extraordinary visual detail that would normally be impossible in the ambient light, a condition of near total darkness. There is corroborative evidence, from the Coroner's Court, about the level of darkness; the crew of the pilot launch TOLKA passed close by DEBONAIR at about 0233 (in mid-channel, near No 9 Buoy) but they could not see or confirm if any of the yacht's crew were on deck because of the darkness. The collision occurred much further out in the Bay where the marginal effect of background shore lights would be even less significant. A further indication of the darkness level may be noted from the rescue operation mounted immediately after the collision;

when the first (and only) survivor had been rescued one or two other persons in the water were heard calling for help but they could not be seen in time to be rescued.

2.2 What Did the Witnesses See?

The details of what the BLUEBIRD witnesses saw are emphasized (author's emphasis) in bold italics in the following extracts from the MCIB Report:

"5.1.1 When about two ships lengths away from the yacht, the pilot and the Master of the "Bluebird" suddenly observed the sidelights of the "Debonair" changing from red to green. The moon was shining and *the pilot could actually see the hull outline alter course across the channel*."

"5.1.2 The Chief Officer, who was on the forecastle head of the "Bluebird", looked around when he heard the whistle sounding *and saw a yacht on the port side, altering course to port.* When the yacht started coming more quickly to port, the Chief Officer shouted at the yacht "what are you doing". *He recalls seeing the face of the helmsman and maybe two other persons.* When he saw the yacht's mast getting close to him he stepped back and heard a blow as the mast hit the bulwark…"

"5.1.3 On the bridge of the "Bluebird" they felt a bump and the yacht came out under the "Bluebird's" starboard bow. At this stage the Captain stopped the engine, in case there were any persons in the water. *The yacht was next seen on the starboard bow. However before the yacht reached midships it sank. The Captain saw one person come to the surface of the water.* Then the Captain released the starboard bridge wing lifebuoy (fitted with smoke and light). The crew also released a lifebuoy with a light attached. The Captain activated the GPS man overboard button. This was timed at 0252, the position readout was 53°20.584N 06°07.367W."

2.3 Accuracy of Chief Mate's Observation

The Chief Mate's recollection of *"…seeing the face of the helmsman and maybe two other persons."* is supported by the VTS audio tape record of the ensuing radio traffic; BLUEBIRD's pilot, in reporting the collision, stated that there were three persons on board the yacht. The actual number of crew on board was later confirmed as five persons, by the sole survivor, who also confirmed the Chief Mate's observation of three persons on deck at the moment of collision—the other two members of the yacht's crew had turned-in below deck.

2.4 Evidence at Coroner's Court

The accounts of these three witnesses were subsequently related to the Coroner's Court, in direct evidence by the pilot, and by way of statements from the Master and Chief Mate. (The Master's statement has a curious notation, apparently in his own hand, as follows: "This statement given to Capt. Wiltshire (Harbour Master) which is an abbriviated *(sic)* version of the fact".) None of these witnesses refer to the moon as the source of illumination that allowed them, individually and separately, to see so much detail with such confirmed accuracy.

2.5 Moonlight Not Source of Illumination

There is a very good reason why the three BLUEBIRD witnesses made no reference to the moon; the moon was not visible at the time of the collision, and did not rise at Dublin Bay until 0442 (0342 UT), almost two hours after the collision. The times of moonrise and moonset are readily computed with the aid of The Nautical Almanac 2001. Alternatively, a rapid on-line solution is publicly available on the website of the US Naval Observatory.

2.6 Perverse Outcome

The MCIB reference to a false moon, namely the statement at paragraph 5.1.1 that "The moon was shining…", perverts the proper findings and outcome of the Report because it deflects and distorts an objective investigation of the true source of the illumination that allowed the three witnesses to see the actual detail as reported.

2.7 Floodlights in Use

There is, on the public record, one unambiguous reference to a source of illumination that, in the ambient light, would have allowed the three witnesses to see the persons, objects and detail, which they claimed were visible. It is the Coroner's Court statement of the Chief Mate, headed "1735 hrs on Bluebird Berth No. 29, 20th May 2001", and in which, referring to his own ship, he states *"There were two floodlights on the ship and I could see the mast."* Floodlights were fitted at each bridge wing, directed forwards and slightly down and served to illuminate the main deck, including the pilot embarkation station. These lights would most definitely have illuminated the ship and her surrounds and allowed the witnesses to see considerable detail in that immediate area around and ahead of the ship. But their continued use in this fashion, by a ship under way, would have been quite improper.

2.8 COLREGS Rule 20(b)

However, not only does the MCIB investigation make no reference to any such improper lights, the investigators seem to have excluded the probability that such lights were in use. And the reality that floodlights were a factor in the collision compounds the shortcomings of the Report. Such lights, displayed in those particular circumstances, were a violation of the Collision Regulations, on the several grounds prescribed in Rule 20 (b), which states: *"...no other lights shall be exhibited, except such lights as cannot be mistaken for the lights specified in these Rules or do not impair their visibility or distinctive character, or interfere with the keeping of a proper look-out."*

2.9 Statutory Requirements of the Port

There is the further statutory duty in this regard, imposed on the pilot by the Dublin Pilotage District Bye-Laws, specifically Bye-Law 18, to *"...keep a constant lookout and use his utmost care and diligence to conduct such ship without damage, or doing any injury to others, duly observing the regulations for the time being in force for preventing collisions at sea."*

Other key documents in any serious attempt at resolving the issue of 'floodlights' must be the pilot's report, as required under Bye-Law 19, and the relevant logbook entry, as required under Bye-Law 21. The MCIB Report contains not a single reference to either of these statutory records, nor is there any indication of those records having been examined by the investigators.

3 OUTWARD PROGRESS OF DEBONAIR

3.1 Alcohol Consumption

DEBONAIR's crew had spent much of the day and evening prior to the collision in socializing with their sailing fraternity at the Poolbeg Yacht Club, where alcohol was consumed. Subsequently, at the Coroner's Court, the state pathologist detailed the blood alcohol levels found at post-mortem examination but could not say how intoxicated the victims might have been as she did not have an accurate estimate of the level at the time of death. She stated that people react to alcohol differently depending on how used they are to drinking, but above the road-driving level of 80 the victim may well have had some problem with co-ordination. DEBONAIR's helmsman and co-skipper, had approximately twice this level, but it is quite possible that fatigue rather than alcohol was a factor in the tragic events. The Report is silent on the question of fatigue.

3.2 Outward Track History Consistent with Alert Navigation

However, there is evidence, ignored in the MCIB Report, which supports the opinion that the skipper's co-ordination and spatial awareness were not adversely affected. It is this: the outward progress of DEBONAIR, as tracked on the Port radar and VTS, confirms that her helmsman/skipper was alert to the challenging nighttime navigational task. It would not have been possible to proceed safely downriver past several piers, jetties, channel buoys, fixed beacons/marks and lighthouse structures in the manner recorded on the track history without a proper lookout being maintained. Further, DEBONAIR had to navigate past these hazards without the benefit of radar. In considering the question of keeping a proper lookout, the Board's conclusions are defective in failing to recognize a feat of diligent nighttime navigation on the part of DEBONAIR's skipper.

3.3 DEBONAIR's Outward Track Influenced by Working Dredger KRANKALOON

Having left her berth, DEBONAIR's departure track history, as recorded on the port's VTS system, shows her tending to the north side of the channel close to North Bull Lighthouse and No. 5 Buoy, before regaining a more normal outbound position on the south side of the fairway channel. The MCIB characterizes these changes of course as 'unusual' and 'erratic'. They were not. It is almost certainly the case that notified activity by the trailing suction hopper dredger KRANKALOON, not properly addressed by the MCIB, was responsible for compelling DEBONAIR to seek the north side of the channel. A proper examination of the VTS archive data tracks KRANKALOON approaching the river mouth from a north-easterly direction, crossing the fairway to the south side of the channel and remaining there for some time, as she completes her dredging operation close to the Poolbeg Lighthouse. This operation, at the very entrance to the harbour, temporarily obstructs the channel to all traffic and particularly hindered outbound traffic, compelling any such vessels to seek clear passage on the north side of the channel. All this occurred shortly after DEBONAIR left her berth to proceed outward, and it explains why her skipper took an outward course to the north side of the channel. By the time she had reached the dredger's operating area near the Poolbeg Lighthouse the dredger had ceased work and was heading seaward for the spoil dumping ground (This was at 0240 as KRANKALOON approached No's 3/4 Buoys—she passed the inbound BLUEBIRD in the fairway at about 0245). DEBONAIR now altered course to starboard (just after No 5 Buoy), and regained a correct position on the south (stbd) side of the channel fairway. These alterations of course by DEBONAIR were actions on the part of her skipper that were logical, rational and seamanlike, and fully justified in the prevailing circumstances.

3.4 Contentious Conclusions of MCIB

The MCIB Report is defective in the conclusions presented in paragraph 10.5. The Board's assertion that "…It is unclear as to what took place concerning the navigation of the "Debonair" after leaving Poolbeg Boat and Yacht Club…" is flatly contradicted by the VTS record. The unfortunate fact that the helmsman did not survive the accident appears to be the basis upon which the investigators fail to critically examine why DEBONAIR's skipper made such a catastrophic alteration to port. Equally contentious is "…the Board's view that the helmsman on the "Debonair" was not monitoring the steering progress of the yacht." On the contrary, the VTS evidence points to a diametrically opposite view that DEBONAIR's skipper was acutely alert to his navigational responsibility. There is no disputing the Board's conclusion that "…Whatever occurred to the yacht would appear to have taken place suddenly and with little or no forewarning to the crew…", but hardly because the skipper was oblivious to the proximity of BLUEBIRD, as implied at paragraph 10.6.

4 BLUEBIRD'S LIGHTS

4.1 Most Probable Collision Scenario

There is no disagreement with the witness and VTS evidence that very shortly (possibly no more than 30 seconds) before the collision, DEBONAIR altered course to port, across BLUEBIRD's bow, as shown at Figure 4. BLUEBIRD's track is seen to be 277°, making 9.9 kn.

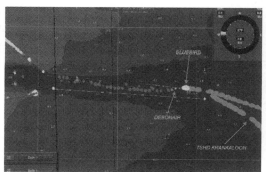

Figure 4. DEBONAIR turns across BLUEBIRD'S bow

A possible explanation for DEBONAIR's critical turn to port is her skipper's (erroneous and probably confused) assessment that the approaching BLUEBIRD had herself altered too much to port and now presented an imminent collision risk, compelling the yacht to turn to port in the fatally mistaken belief that this last-minute manoeuvre represented appropriate "evasive action" in the circumstances.

The VTS track history shows conclusively (Fig. 4) that BLUEBIRD did not make any such excessive alteration to port, but the fundamental point underpinning the hypothesis is illustrated in Figures 5-10. This series of photo images of BLUEBIRD was taken some years after the tragedy.

The images show BLUEBIRD approaching the observer on a 'port bow aspect', ie, replicating the aspect as presented to DEBONAIR in the moments immediately before the collision. She is exhibiting the normal navigation lights for a vessel of her size when underway— forward masthead light, second (higher) masthead light and port sidelight. The images also show two forward-facing floodlights (horizontally disposed) on the front of the ship's bridge, near the extremity of each bridge wing. Floodlights displayed in such manner are contrary to Rule 20 of the COLREGS.

Figure 5. mv BLUEBIRD, presenting a port bow aspect

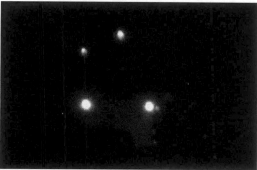

Figure 6. BLUEBIRD getting closer

As the distance between the two vessels closes, BLUEBIRD's forward masthead light (being closer to the observer) begins to appear higher in relation to the second masthead light.

Figure 7. Forward masthead light appears increasingly higher.

Figure 8. The masthead lights appear at same level

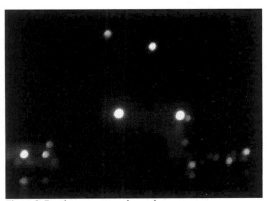

Figure 9. Port bow aspect unchanged

In Figure 8, the forward masthead light is apparently at the same level as the second masthead light, while the starboard floodlight is obscured by the ship's foc'sle head.

In Figure 9, as the distance continues to close, the forward masthead light is now apparently higher than the second (aftermast) light, while the starboard floodlight is no longer obscured by the focs'le head structure. This image presents an ever more confusing situation, as there is no such combination of lights within the COLREGS.

BLUEBIRD's "port bow aspect" remains unchanged throughout, ie, she has not altered course.

Figure 10. Increasingly confused disposition of lights

In Figure 10, the ship is ever closer but the confusion is unresolved. The increasing brightness of the floodlights makes it less easy to distinguish the sidelight, and, to an observer with low 'height of eye' such as from the deck of a yacht, perhaps impossible. The forward masthead light (to the left) appears to be considerably higher than the second masthead light. In these circumstances, where the distinctive character of the port sidelight has been impaired, it is possible to misconstrue the lights for those of an approaching vessel drawing to the right, in other words, a "starboard bow aspect".

4.2 Explaining the 'Inexplicable'

If, in the final moments or seconds before collision, DEBONAIR's skipper believed that the vessel on his port bow had herself turned so far to port as to be crossing his bow he must surely have perceived an imminent collision risk, and taken evasive action. This would explain his apparently inexplicable turn to port when the two vessels were just about 150 metres apart; he turned into the path of the oncoming BLUEBIRD when, in fact, he believed, mistakenly, that he was turning away from danger. Any suggestion that DEBONAIR had simply turned to port in blissful ignorance of the dangerous proximity of BLUEBIRD is stretching credulity beyond acceptable limits, if it is accepted that the approaching ship displayed two powerful floodlights in the general direction of the yacht —floodlights that gave so much illumination ahead of BLUEBIRD that the pilot could actually see the hull outline of DEBONAIR and the chief mate on the forecastle could see the face of the helmsman in her cockpit.

5 CONCLUSIONS

It is most unlikely that professional mariners, supported by modern bridge technology, could misread the masthead lights of another vessel when coming to such close quarters. But given a vessel with low freeboard and a matching 'height of eye' close to sea level, the risk of such dangerous misjudgment by small craft mariners or recreational sailors is very high. It may never be possible to establish the final tragic sequence of the DEBONAIR/BLUEBIRD collision with any certainty, but the scenario as outlined and illustrated in the images above is inescapable. In this instance, it is an unfortunate consequence of a clearly deficient investigation that profound lessons in the conduct of safe navigation, bought at the cost of four lives, have not been properly examined, addressed or promulgated.

REFERENCES

Coroner's Court Proceedings, Dublin
Dublin Port Company, *Dublin Pilotage District Bye-Laws.*
IMO, *International Regulations for Preventing Collisions at Sea, 1972 (COLREGs),*
Marine Casualty Investigation Board (MCIB), 2005, *Report of the Investigation into the Collision Between the Yacht "Debonair" and Cargo Vessel "Bluebird" in Dublin Bay on 20 May 2001,* Dublin, MCIB, http://www.mcib.ie
United States Naval Observatory, at: http://aa.usno.navy.mil/

Lesson Learned During the Realization of the Automated Radar Control System for Polish Sea-waters (ZSRN)

M. Fiorini
Selex ES –A Finmeccanica Company, Rome, Italy

S. Maciejewski

ABSTRACT: The historic decision in favour of Poland's accession to the EU expressed by the society in the accession referendum on 7–8 June 2003 sets de facto the basis for the "Zautomatyzowanego Systemu Radarowego Nadzoru (ZSRN) polskich obszarów morskich / Automatic National System of Radar Control for Maritime Areas of Poland" realized by the Finmeccanica group company Selex ES as representative of a temporary consortium formed together with a number of local companies. The paper will present the ZSRN system, the generic architecture and its main functionalities from technical and operational point of view trying to explain the challenges being faced and the lesson learned from the development of the project.

1 INTRODUCTION

ZSRN is being created because of the necessity to control sea-waters pursuant to domestic law regulations and the international obligations assumed by the Republic of Poland. The main mission of ZSRN is to develop and distribute a complex surface image covering the territorial sea and the internal sea-waters of the Republic of Poland. The information acquired and processed by the system will support the decision process and the management of marine activities at different responsibility levels of various institutions. Its start up will enable permanent and close cooperation between many departments in the discharge of their statutory duties related to the surveillance of sea-waters.

The user of the system is Border Guard (BG), [1]. The information processed in the system will be made available to the Maritime Administration and the Polish Navy according to competence.

The remain part of the paper is structured as follows. After describing the ZSRN generic architecture, the system structure and its main functionalities in paragraph 2, paragraph 3 describe the lesson learned from the development of the project. Conclusion and acknowledgements in paragraph 4 and 5 respectively, close the paper.

2 THE SYSTEM STRUCTURE AND ARCHITECTURE

The most important centers for ZSRN could be grouped into five categories as follows:
- Main Control Centre, CON;
- Reserve Main Control Centre, ZCON;
- Division Control Centres, DONs;
- Local Control Centres, LONs;
- Observation Posts, POs.

Figure 1. Distribution of ZSRN's centers along the Polish coastline on the Baltic sea and inland waterways.

They are distributed along the coast of Poland as indicated in Figure 1. Is part of the system one Mobile Radar Station (MRS) and three BG vessels too as well as the secure connections with external systems as sketched in Figure 2. All those centers form "The System" in accordance to the

International Council on Systems Engineering (INCOSE) definition: "a combination of interacting elements organized to achieve one or more stated purposes".

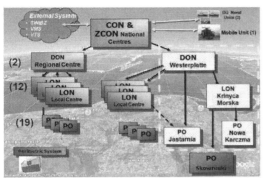

Figure 2. Gerarchical distribution of ZSRN's centers sketched on top of a Baltic sea map as background.

It is clearly indicate in Figure 2 the pilot session centres (in yellow) developed in a previous phase under a different contract and suppliers. The Pilot session is fully integrated with the rest of the System. ZSRN have also secure connections with external systems to exchange data, according to security policies and user privileges to the relevant Administrations.

All in all the borderline and areas under control of the Polish Border Guard are summarized in Table 1 and in Figure 3.

Table 1. Polish Maritime Borderline and Areas

The Polish Maritime Borderline	[km]
Length of Maritime Border	439,74
Length of Coastline	524,00
Length of Territorial Waters	395,31
The Polish Maritime Areas	[km²]
Maritime Internal Waters	1885
Territorial Waters	8681
Exclusive Economic Zone	26156
Total	36722

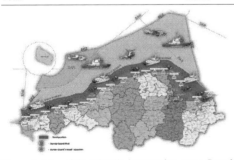

Figure 3. Map of the ZSRN's intervention areas. State border, Search And Rescue (SAR) and Economic Exclusive Zone (EEZ) areas are depicted as well as in land competence territory.

2.1 The main functions and the essential mission of the system

The system main functions are:
- sensors and communications interfaces;
- traffic picture handling;
- navigation management and traffic controls;
- decision aids;
- presentation manager;
- database management;
- recording, archiving and playback;
- simulation and training management;
- security management; and
- system management.

They are developed in order to support the duties of the end users. To this end the system is composed by its own sensors such as radar, radio (voice and data) and cameras (Low-Light-Level TV (LLLTV) and InfraRed (IR)), cf. Figure 4. Additional source of information are the external systems that provide valuable data such as Automatic Identification System (AIS) tracks, [2].

Figure 4. ZSRN radar and cameras sensors.

As well known, according to the Safety Of Life At Sea (SOLAS) convention - hold for the first time in London in 1914 after the Titanic's tragedy and then in 1929, 1948, 1960 and 1974 – modern large vessels have to be equipped with the AIS as follows:
- all ships of 300 gross tonnage and upwards engaged on international voyages (since 2004);
- cargo ships of 500 gross tonnage and upwards not engaged on international voyages (since 2008);
- passenger ships irrespective of size (since 2003).

On the contrary, radar are nowadays used for surveillance and tracking of any size of vessels at sea under the coverage area, cf Figure 5, [8]-[9].

Figure 5. Fishing vessels travelling parallel.

Surveillance and visual inspections are executed by means of camera too, cf. Figure 6.

Figure 6. Camera used for visual inspection of a buoy.

As mentioned above one Mobile Radar Station is part of ZSRN, cf. Figure 7.

Figure 7. The Mobile Radar Station (MRS) is equipped with Radar, Radio (data and voice) and Command & Control (C2) for two operators.

The essential mission of the ZSRN is to support:
- protection of the European Union external marine border;
- protection of the state's marine border against illegal immigration, narcotic and drug trafficking and illegal trade in goods;
- exercising of control over the usage of Polish sea-waters and the observation by vessels of the regulations in place;
- securing of Polish economic interests on Polish sea-waters;
- action against poaching and illegal fishing;
- control of suspicious vessels (fishing, sports and sailing boats);
- reconnaissance of detected unidentified surface and air objects;
- joint border actions conducted in cooperation with the border services of neighboring countries;
- use of BG vessels in activities aimed at protection of the country's marine border; and
- participation in Search And Rescue (SAR) operations.

The ZSRN mission is executed by BG personnel on duties at various level 24h per day, 365 days per year, cf. Figure 8.

Figure 8. Straż Graniczna (SG) Operators on duties at CON; photos courtesy of the Polish Border Guard.

3 LESSON LEARNED

Nearly two-thirds of software development projects in the US fail, due to either cancellation, overrunning of budgets, or delivery of software that is never put into production. Reasons for these failures should be identified in lack of stakeholders (e.g. users, customers) involvement, no clear statement of requirements, no project ownership, no clear vision and objectives and lack of planning, [5]. Fortunately the ZSRN is not an example of the two-thirds fails project but has been affected by some delay.

3.1 Engineering a System

It was recognized that the transition process of user needs to system requirements should be structured and traceable. This to mitigate the risk that a wish list turns into the requirement specification at a later stage of development or even after delivery to customer. This is where requirement engineering is needed, [6]. Requirements engineering is something very common for large and complex projects. To guarantee throughout the entire process that the (correct) user needs are realized.

Broadly speaking is possible to say that the Vee (V) Process Model has been widely used for requirements engineering of ZSRN. The model depicts a top-down development and bottom-up implementation approach, cf. Figure 9. On the left side, decomposition and definition descends while on the right side integration and verification ascends culminating at the system level.

The V-model is composed of three layers in increasing engineering details: *Customer's* perspective (it is the stakeholders view; who is interested in presenting requirements and receiving a finished product), *Systems Engineer's* perspective (it encompasses architectural details to address decomposition of the system-level specification into system design and subsystems' specifications and designs; build and tested subsystem and finally the tested system) and *Contractor's* perspective (it covers the implementation process —usually performed by contractors- and it is associated with component specifications and designs with fully tested components), [7].

To obtain a clear system requirements specification a process of interaction with stakeholder and users is necessary in order to have the needs of the different parties. The goal of the stakeholder requirement specification process is to formulate requirements for a system that can deliver the services as desired by users and stakeholders. When all user needs are analysed and transformed into requirements as appropriate, a requirements analyses process is needed to translate these requirements into a system requirement specification which is more of a technical nature.

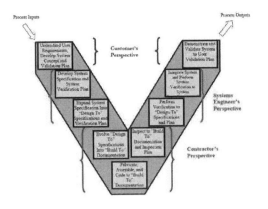

Figure 9. The Vee Process Model, [6].

The project was developed according to the modern methodologies of Systems Engineering nevertheless the "on-site" experience shown progress on the visibility of the project that have to be intended as execution and delivery of valuable results to the final users, that follows an hyperbole curve as reported in Figure 10. Such curve is known in literature as the hyperbole of technology in the supply chain of Gartner group research [3] and point out the perception of the adoption of technologies on to the four main phases of enthusiasm, disillusion, consolidation and diffusion. It result in something new from some (not all) of the people involved on the project when working to reach the complete delivery of the full system.

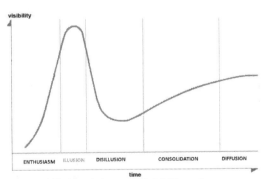

Figure 10. The hyperbole curve of visibility vs time.

It is also lightly indicated the illusion phase of the project; its correspond to 80% circa of the competition of the delivery of the equipments but is far away from the Operative use of the final system.

3.2 Budget allocation

Another important lesson learned is concerned with budget and it is sketched in Figure 11. In this context the word budget is synonymous with resources because it have to be indented as all the resources needed to complete each and every single piece of work (money, skilled people, etc.). The schematic representation means that technical or management decisions are too often driven by budget constraints instead of user needs and this result in the illusion to complete the project instead of the real completion of it, cf. Figure 10.

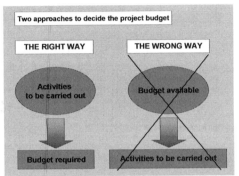

Figure 11. The right and wrong way to budget allocation, [4]. The concept should be extended to all resources required to complete each piece of work; budget is synonymous with resources.

3.3 Weather conditions and local habits

The Baltic sea is different from the Mediterranean sea and the differences became soon evident especially in winter months, cf. Figure 12. This affect the behaviors of persons but also machines if they are not well tuned to work in such extreme conditions. The experience demonstrate that these was the case of the radar extractor tracker for which more than one session of trials were necessary before to find the correct configuration and settings, cf. Figure 14.

Figure 12. The Baltic sea during "tipical" winter months.

It is relatively common in Poland (we learned by ourselves) to mark the path on a forest by means of red sign-marks on trees. We have to confess that it was challenging and a little tricky to find the way for all remote radar sites especially under snow storms when all trees looks very well covered by snow like the pub sign in Figure 13.

Figure 13. Sign-marks on trees and how a sign may look likes during winter months.

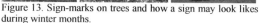
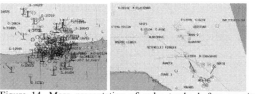

Figure 14. Mess presentation of radar tracks before on site trials (left) vs. clear presentation of radar tracks after proper settings (right).

4 CONCLUSION

Four systems engineering principal reported hereafter could be considered the lesson learned from the development of this project.

1 Stakeholder Involvement

Successful projects involve the customer, users, operators, and other stakeholders in the project development. Systems engineering is a systematic process that includes reviews and decision points intended to provide visibility into the process and encourage stakeholder involvement

2 Define the Problem Before Implementing the Solution

Very often, the develop team have a solution already in mind at the start of a project and may even find themselves "backing into" requirements to match their solution. Resist this temptation and instead use the systems engineering process to first define the problem. You'll find that there are actually multiple ways to solve the problem, and a good trade study will help to determine the best solution on the basis of a clear understanding of the stakeholder needs.

3 Divide and Conquer

A key systems engineering strategy is the decomposition large and complex systems into smaller subsystems and then of the subsystems into more manageable hardware and software components.

4 Traceability

As you move from one step to the next, it is important to be able to relate the items in one step with those in another. The requirements will be related to user needs as well as to the tests that will be used to verify them.

Last but not least, do not underestimate the local conditions and the habits and expectations of the locals !

ACKNOWLEDGEMENTS

The authors are grateful in debt with the Commanders of the Polish Border Guard (Straż Graniczna) for agreeing on the publication of this paper.

REFERENCES

[1] Polish Border Guard (Straż Graniczna) website, www.strazgraniczna.pl
[2] TU-R Recommendation M. 1371-1 "Technical characteristics of a universal shipborne automatic identification system using time-division multiple access in the maritime mobile band", 2001
[3] Hype Cycle for Supply Chain Management and Procurement, ID Number: G00149865, www.gartner.com
[4] Interact Point Qualification and Transfer: "Financial Management Handbook", 2006, p. 80
[5] I Grapham, Object-Oriented Methods: Principles and Practice, 3rd Ed., Addison-Wesley, 2001.
[6] P J W Hooijmans, M Fiorini, "Requirement Engineering Process for IALA e-Navigation", input document to the architecture WG5 e-Navigation of IALA, St-Germain-En-Laye (Paris), 19 May 2010
[7] ISO/IEC 15288
[8] M. Kwiatkowski, J. Popik, W. Buszka, R. Wawruch, "Integrated Vessel Traffic Control System", TransNav – Int. Journal on Marine Navigation and Safety of Sea Transportation, ISSN 2083-6473, ISSN 2083-6481 (electronic version), Vol. 6, No. 03 - 2012 pp. 323-327
[9] A. Krol, T. Stupak, R. Wawruch, M. Kwiatkowski, P. Paprocki, J. Popik, "Fusion of Data Received from AIS and FMCW and Pulse Radar - Results of Performance Tests Conducted Using Hydrographical Vessels "Tukana" and "Zodiak"" TransNav – Int. Journal on Marine Navigation and Safety of Sea Transportation, ISSN 2083-6473, ISSN 2083-6481 (electronic version), Vol. 5, No. 04 - 2011 pp. 463-469

Is ARPA Suitable for Automatic Assessment of AIS Targets?

F. Heymann, P. Banyś, T. Noack & E. Engler
Deutsches Zentrum für Luft und Raumfahrt eV., Neustrelitz, Germany

ABSTRACT: Collision avoidance is one of the high-level safety objectives and requires a complete and reliable description of maritime traffic situation. A combined use of data provided by independent data sources is an approach to improve the accuracy and integrity of traffic situation related information.
In this paper we study if the detection of real vessels via ARPA (Automatic Radar Plotting Aid) is suitable to combine and fuse these targets together with AIS (Automatic Identification System) targets for automatic assessment routines. The presented study is based on data recorded during a half day measurement campaign in the Baltic Sea and represents only a snapshot concerning available ARPA and AIS data and usable equipment. The results can help to find out, what the challenges of ARPA and AIS fusion process are. A method was developed to investigate the ARPA message quality utilizing dynamic AIS data. Insofar the positions provided by ARPA targets are checked against known positions of AIS targets in order to estimate distance between both vessel positions. Additional information is offered on the question if integrity information can be derived from fused ARPA and AIS data.
The work of this paper is integrated in the research and development activities of DLR Institute of Communications and Navigation dealing with the introduction of data and system integrity into the maritime traffic system. One of the aimed objectives is the automatic assessment of the traffic situation aboard a vessel.

1 INTRODUCTION

One of the important carriers of the worldwide economy is the transport of goods and persons realized by vessels. A rapid development of new technologies for the maritime traffic system occurred in the last decades to enable the handling of increased transport volume and to improve the safety. To harmonize the developments of electronic aids to navigation and dedicated systems and services aboard and ashore the International Maritime Organization (IMO) has initiated the e-Navigation strategy to integrate existing and new navigational tools, in particular electronic tools, in an all-embracing system.

The risk reduction of accidents between ships as well as ships and obstacles is the social goal associated to safe shipping from berth to berth. The technological goal covers the development of new tools and methods to support the ship-side and shore-side nautical staff during decision finding in complicated and complex navigational situations.

An error-free decision making to avoid collisions and groundings is only possible, if reliable and comprehensive information of the maritime traffic situation is available. Therefore only by complementary usage of all aboard and ashore data sources the necessary redundancy can be established to enable improvement of accuracy as well as introduction of data and system integrity.

Related to the Safety of Life at Sea Convention (SOLAS) the primary source for collision avoidance and traffic situation awareness is the radar system with the opportunity to detect and track objects by ARPA functionalities.

With the implementation of AIS in 2004 an additional important step was done to deploy a second measure for ship-side and shore-side vessel tracking. Like almost every technologies, neither ARPA nor AIS can be declared as an "altogether solution" and are subject to specific restrictions and limitations. Because of the cooperative character of AIS data (disengage able, dependent on the human initiated processes) and the dependency on other onboard devices (as for example the GPS receiver) there is still a margin for errors in the data. Insofar the possibility cannot be ruled out, that specific AIS data are wrong or not meaningful during important maneuvers of a vessel. An analysis of a comprehensive two month AIS data set (January and February 2010) describing the vessel traffic of the whole Baltic Sea (Banyś et al. 2012) came to conclusion, that specific parameters like Rate of Turn (ROT) as well as Heading (HDG) deliver significantly defective or implausible results.

The capability of radar sensors and APRA can be influenced by bad weather conditions with rain, possible shadows of shielding objects as well as artifacts from surrounding sources. Therefore the radar sensor has drawbacks in angular resolution (depends on the rotation rate) as well as in distance estimation (depending on the shape of the reflecting structure), whereby both can reduce the position accuracy of identified targets. A detailed ARPA sensitivity study was performed by (Lisowski 2012) using multi-stage positional non-cooperative and cooperative game and kinematics optimization control algorithms.

One approach to get an improved picture of the current traffic situation is the fusion of independent data sources – here of AIS and ARPA data (Neumann 2008, Krol et al. 2011 and Kwiatkowski et al. 2012). Aim of this paper is the analysis of ARPA echo. For this purpose AIS data sets of a test voyage are used to evaluate ARPA targets regarding position accuracy, number of detections and availability. The paper is structured in the following way: At first a short overview about ARPA and AIS technology is given in chapter 2. The logic of study is explained in chapter 3. Chapter 4 describes the data types used for the study. In chapter 5 the processes of data preparation and in chapter 6 of data fusion are outlined. Section 7 presents and interprets the analysis results before final conclusions are given.

2 GENERAL ASPECTS OF AIS AND ARPA

With the introduction of AIS in 2004 an important step in improving safety of sea was achieved. The AIS makes bridge watch keeping duties more comfortable and enhances vessel traffic management ashore. Its usage worldwide is widespread. As the Safety of Life at Sea Conventions state, all vessels of 300 gross tonnage and upwards engaged on international voyages, cargo vessels of 500 gross tonnage and upwards not engaged on international voyages and both passenger vessels and vessels carrying dangerous cargo irrespective of size shall be fitted with the AIS transponder. According to the Lloyd's List Intelligence, which is running the world's largest land and satellite based AIS monitoring network, there are currently about 72 000 vessels worldwide equipped with active AIS transponders. The main function of AIS is to broadcast dynamic and static navigational data of a vessel to other vessels located within a range of about 20 NM. The data are transmitted on two marine VHF channels reserved for AIS communication. Every message contains the information about the time of its next broadcast, too. On this way all vessels within a certain area can organize the AIS radio broadcasts themselves and share vital navigational data with one another.

The radar on the other side is an electromagnetic sensor used for object detection via reflected radio waves to determine the range, altitude, direction, or speed of objects. Maritime radar systems usually radiate pulses of almost rectangular shape in the frequency bands of either X-band (~9 GHz) or S-band (~3 GHz). If the maritime radar is installed with ARPA functionalities the opportunity is given to derive tracks based on radar targets. ARPA systems are able to calculate the course and speed of tracked objects as well as the closest point of approach (CPA) and time to closest point of approach (TCPA) in relation to the own vessel. The majority of ARPA systems integrate the ARPA features with the radar display.

Both systems use the standardized NMEA message protocol to make their outputs available to processing and monitoring systems via serial connection. The automatic echo extraction tool of the radar sends distance and true bearing information of all acquired targets inside the TTM message (NMEA 0183). The AIS transponder sends information with dynamic content like speed, position, course, etc. and static content like name, identification number with AIS messages defined in (Rec. ITU-R M.1371-3).

3 STUDY CONCEPT

Aim of the study is the investigation of automatic detected vessels by ARPA to describe their properties and to assess their applicability in a data fusion process.

The strategy of the study is illustrated in Figure 1 and covers 4 steps:

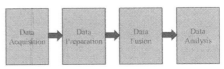

Figure 1. Strategy of study

In the first step of data acquisition we designed a measurement campaign in the Baltic Sea to record automatic mode ARPA data, position information of our own vessel from the PNT unit, and position data from surrounding vessels via AIS. The raw data were stored in a (SQL) database including sensor and exact time information (see section 4).

In the next step the data were prepared to make it usable for the fusion process. Initially the raw radar data, in form of distance and true bearing, has to be translated into position's longitude and latitude to be conform with position data of AIS messages. In addition all stored data has to be cleaned, ordered, characterized and time synchronized (see section 5).

The data fusion process was included to estimate the correlation between detected radar and AIS objects (see section 6). The following cases can be expected fusing ARPA and AIS data: ARPA targets are unavailable for an existing AIS target. A single AIS target has exactly one associated ARPA target. An AIS target has many associated ARPA targets. Last an ARPA target exists without associated AIS target.

As a last step we analyzed the properties of the ARPA extracted echoes concentrating on the separation between AIS and ARPA positions, the number of echoes and the availability of ARPA (see section 7).

4 DATA ACQUISITION

4.1 Data Types

The aimed study requires PNT data of the own vessel, radar data determined by the ship-side radar sensor, and AIS data provided by vessels operating in the surrounding area. A generic illustration of data acquisition process is shown in Figure 2.

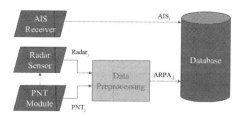

Figure 2. Block diagram of data acquisition

The PNT (Position, Navigation, and Time) data cover parameter describing the current 3-dimensional attitude and movement of a vessel in a defined coordinate and time reference system. Only the PNT data described in Table 1 are used for the study.

Table 1. PNT data of the own vessel

$PNT_j = tar[i, t_{PNT}(i), lat_{PNT}(i), lon_{PNT}(i)]$	
j	index of PNT data set
$t_{PNT}(i)$	reference time of PNT data set
$lat_{PNT}(i)$	latitude of own vessel position
$lon_{PNT}(i)$	longitude of own vessel position

The quality of PNT data depends on the used sensors and services on the one hand and on the

capability of PNT data processing algorithms on the other hand. At this point it should be mentioned that PNT data (e.g. HDG) are used by the radar sensor for automatic target detection.

Data obtained from radar sensors are the distance between the target and the own vessel and the true bearing of the target. Radar data given in Table 2 are used for our study.

Table 2. Radar data

$Radar_j = tar[j, t_{radar}(j), d_{radar}(j), b_{radar}(j)]$	
j	index of radar data set
$t_{radar}(j)$	reference time of radar data set
$d_{radar}(j)$	distance between own and detected vessel
$b_{radar}(j)$	true bearing angle

The radar based detection of targets depends on weather conditions, the justification of the radar system itself, and the environmental conditions.

The AIS enables the exchange of static, dynamic, and voyage related data between vessels. AIS data includes the MMSI (Maritime Mobile Service Identity) of vessel. The actual position of a vessel is part of the dynamic AIS data and can be considered as an independent data source to assess position data obtained by ARPA. For our study we use the AIS data given in Table 3. For analysis purposes 2 additional parameters are introduced to characterize each AIS target. The parameter $A(i)$ specifies the navigational status and area of the vessel regarding AIS target with index i. The parameter $N_{echo}(i)$ is set to zero and will count assignable radar targets during the data fusion (see section 5).

Table 3. AIS data and additional parameter

$AIS_i = tar[i, ID_{AIS}(i), t_{AIS}(i), lat_{AIS}(i), lon_{AIS}(i), A(i), N_{echo}(i)]$	
i	index of AIS data set
$ID_{AIS}(i)$	identification number of vessel
$t_{AIS}(i)$	reference time of AIS data
$lat_{AIS}(i)$	latitude of vessel position
$lon_{AIS}(i)$	longitude of vessel position
$A(i)$	characteristic of area and movement
$N_{echo}(i)$	number of assignable ARPA echoes

The quality of AIS position data depends either on ship-side PNT Module or single PNT sensors, which are included in or connected with the AIS transmitter.

ARPA data stands here for parameters, which describe the absolute position of a radar target on an automatic way. The position given in ARPA data is derived from the position of the own vessel (PNT data) and the aligned relative distance measurement of the radar target (radar data). Therefore the accuracy of ARPA position depends in equal measure on the accuracy of PNT and radar data as well as on the precision of data synchronization between both data sources. Table 4 specifies the ARPA data used for our study.

Table 4: .ARPA data

$ARPA_j = tar[j, t_{ARPA}(j), ID_{ARPA}(j), lat_{ARPA}(j), lon_{ARPA}(j)]$	
j	index of ARPA data set
$t_{ARPA}(j)$	reference time of ARPA data
$ID_{ARPA}(j)$	identification number of vessel
$lat_{ARPA}(j)$	latitude of vessel position
$lon_{ARPA}(j)$	longitude of vessel position

4.2 *Data Basis*

The data of this study were collected within a measurement campaign with the vessel "BALTIC TAUCHER II" as shown in Figure 3.

Figure 3. Measurement vessel "BALTIC TAUCHER II"

The voyage followed the trajectory shown in Figure 4. The vessel started in the fishery port of Rostock, passed the oversea port, crossed at the Baltic Sea, and returned to the anchorage. The measurement voyage took 12 h (from 0600 until 1800 UTC).

For our study data from following ship-side sensors are used:

– Position data came from the onboard GNSS receiver with an update of 1 Hz and are provided as GLL message in the NMEA 1083 format. In addition a geodetic GNSS receiver of type "Javad Delta G3T" with raw data output capability was used to achieve a second position determination with an update rate of 20 Hz.

– As the radar sensor the "Sperry Marine Visionmaster FT" operating in the X-band was used. This device provided approximately each 2.5 s a new radar image as well as a set of data describing distance and bearing angles per identified target. For the data provision the TTM message of the NMEA format was used.

– ARPA data were generated within a data preprocessing engine by determination of the absolute position of radar targets combining relative radar measurements with position data of the measurement vessel. Due to the high update rate of self-determined position data the residual time synchronization error is below 25ms.

– AIS data (dynamic and static) providing position data from vessel operating in the surrounded area were delivered by a class A AIS transponder of type "Furuno FA150". The AIS data are available as VDM message of the NMEA format. The AIS data per vessel were updated between 2.5s and 30s in dependence on navigational status and speed of each specific vessel.

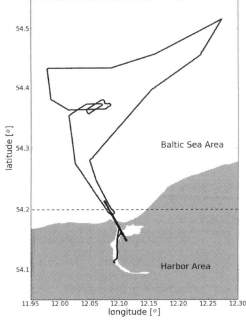

Figure 4 Trajectory performed during measuring voyage

During the measurement campaign around 70.000 AIS messages were received from more than 50 different vessels. More than 200.000 radar echoes were extracted and used to determine the position of radar targets. The own position was determined with a sampling frequency of 20 Hz without any interruptions during the measurement campaign.

5 DATA PREPARATION

Before AIS and ARPA targets can be fused, the data streams have to be prepared for the fusion process. The first 3 steps (a to c) comprise the preparation of AIS targets, whereby the 4th step (d) deals with the temporal association of AIS and ARPA targets.

1 In step 1 all AIS targets are divided in subsets of vessel specific AIS targets. This is achieved by sorting the AIS targets regarding MMSI of vessels.

2 AIS targets were proved regarding their consistency. For this purpose a consistency condition is specified taking into account the relation between 2 different AIS targets. If vessels' identification number in both AIS targets is different, than the distance between both targets should be greater than 300 m, if both targets are derived nearly for the same time (±5s). If a case would occur where the distance is lower than 300 m, both targets have been removed from the AIS data base. At this point it should be mentioned, that within the collected AIS data it was impossible to detect such a "specific" pair of AIS targets.

3 In step 3 the available AIS targets are characterized regarding vessel's navigational status and operational area. If the vessel operates or anchors in the harbor area, the spatial identifier is "H". If its operation area is the "Baltic Sea" the identifier is "B". The spatial separation between Baltic Sea and harbor is shown in Figure 4. A mooring vessel is characterized as "S=static" and a moving vessel as "D=dynamic". After the characterization the vessel ID is anonymized by counting vessels per specific characteristic (see Table 5).

4 In the last step the temporal association of AIS and ARPA targets is performed. For this purpose all ARPA targets j are associated to the AIS target i, if the difference between their time stamps is smaller than 5s. By this condition time slots are created with a length 10 s. A time slot of 10 s is long enough to allow more than one radar rotation (once every ~2.5s) and short enough to ensure an acceptable temporal and spatial resolution.

The process of data preparation is summarized in Figure 5.

6 DATA FUSION

Challenges on data fusion processes cover the spatial association as well as the temporal synchronization of different data sources. The time synchronization between AIS and ARPA targets was already realized during the data preparation process. In this context it was helpful, that the reference time points of different data sources are handled in the same time frame (GPS respectivly UTC). As diagrammed in Figure 5 the status of time synchronization between the targets ARPA(j) and AIS(i) is described by a matrix with the elements $assign_{i,j} \varepsilon$ {yes,no}.

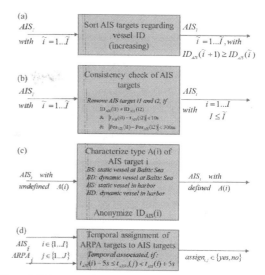

Figure 5. Data preparation

Consequently, the data fusion process is focused on the spatial association of targets coming from PNT, ARPA and AIS information. Figure 6 describes an example of AIS detected vessels (in red and green dots), ARPA detected objects (in yellow and grey dots) in relation to the position of vessel equipped with PNT, ARPA and AIS sensors (in blue dots).

The track of the measuring vessel (blue line) is given by a set of waypoints $Pos_{PNT}(t_i)$ with i=1…I. Assuming that the radio range of radar is 3 NM the orange circle(PNT_i) around $Pos_{PNT}(t_i)$ describes the area within ARPA and AIS targets can be compared. Due to movement of the measuring vessel a dynamic description of such areas is necessary. For example, the AIS target of vessel with ID=1 is within circle(PNT_i) but outside circle(PNT_{i+1}). For each AIS target which is assignable to a position of the measuring vessel the assignable ARPA targets are searched. An assignment is assumed, if the difference in time reference between ARPA and AIS target is smaller than 5s and if the distance between AIS and ARPA position is below 300 m. This mechanism around the AIS position $Pos_{AIS}(t_i,ID=7)$ is shown by the green circle. ARPA targets with the index j of 2, 5, 8, and 11 can be assigned to $Pos_{AIS}(t_i,ID=7)$. Due to their larger distance to $Pos_{AIS}(t_i,ID=7)$ the ARPA targets with the index j of 1, 3, and 4 cannot be assigned.

Related to the mathematical algorithm (see Figure 7) the fusion process is organized by 2 loops, the external loop counts up the AIS targets i and the internal loop counts up the ARPA targets j. That means, that for each AIS target the spatial association of all ARPA targets will be analyzed.

227

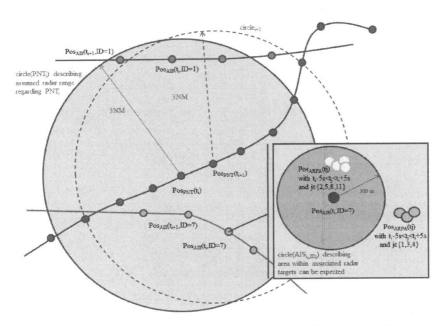

Figure 6. Example for the spatial assignment of PNT, APRA and AIS data within an area of 3 NM around a vessel

At the starting point of the AIS loop the AIS target counter (i=0), the vessel ID (iv=ID_{AIS}(i=1)), the total number of AIS targets per vessel ID ($Ntotal_{iv}$=0) and the total number of non-detections per vessel ID ($Nzero_{iv}$=0) are initialized. During this loop the first check is the change of the vessel ID in comparison to the last iteration. Depending on the result, we either reset the counters $Ntotal_{iv}$=0 and $Nzero_{iv}$=0, and update the vessel ID iv=ID_{AIS}(i) or we proceed the processing in the case of an identical ID. Because the database is sorted regarding increasing vessel ID it is guaranteed that we consider vessels providing AIS data. In the next step it is proved, if the considered AIS target is within the radar range (3NM) of the own vessel and is therefore qualified for the assessment of ARPA targets. For this comparison the distance between the AIS target and our own vessel is determined. Only for AIS targets within the radar range the counter $Ntotal_{iv}$ is increased by 1. Therefore at the end of processing the vector Ntotal describes the number of AIS targets per vessel utilized in the data analysis. For each AIS target qualified for the assessment of ARPA targets, the loop over all ARPA echoes with j=1…J is started. One result of the internal loop is the number of associated radar echos per AIS target. If this number is 0, than the counter $Nzero_{iv}$ is increased by 1.

The internal loop is shown in Figure 8 and starts with initialization of j=0 (the first ARPA echo) and of minimal distance Dmin(i) = 0 between ARPA j and AIS target i.

If $assign_{i,j}$ is "no", a temporal synchronization between AIS target i and ARPA target j does not exist. Than the assessment of the next ARPA target is started. If $assign_{i,j}$ is "yes", a temporal synchronization between AIS target i and ARPA target j is assumed. Than the assessment of the ARPA target j is started and the distance between ARPA and AIS based positions is calculated. If this distance is larger than 300 m the ARPA target j is considered as outside the ARPA accuracy. Then it is assumed, that the ARPA target j is not spatial associated with the AIS target i. Then the analysis of the next ARPA target starts. In the other case, if the distance is smaller or equal 300 m, the ARPA target is considered as echo associated to the AIS target. Than the counter Necho(i) is increased by 1. In addition we check if the distance D is smaller than the minimal distance Dmin(i) of all previously detected ARPA echoes associated with this AIS target. If this is the case the minimal distance is updated. The internal loop is finished, if all ARPA targets were analyzed. The counter Necho(i) describes at the end of each internal loop how often ARPA targets are assigned to the specific AIS target i. The analysis of this AIS targets ends when all ARPA echoes are checked. Then the internal loop is left.

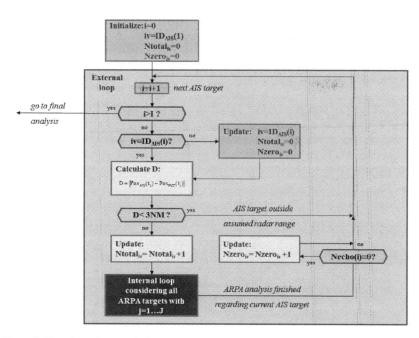

Figure 7. Flow chart of process fusing ARPA and AIS data under consideration of own position coming from PNT data

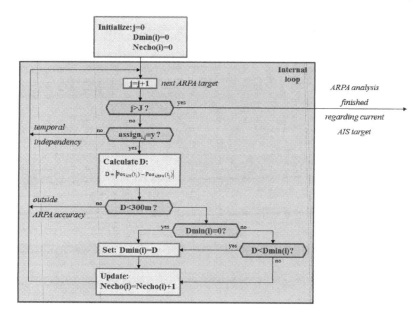

Figure 8. Flow chart of process fusing ARPA targets with a specific AIS target

The fusion process ends when all AIS targets are checked against all ARPA extractions. At the end of data fusion process following quality identifiers are determined:

- The number of AIS targets $Ntotal_{iv}$ per vessel, which are usable for the assessment of ARPA targets.
- The number of AIS targets $Nzero_{iv}$ per vessel, without associated ARPA targets.

- The number of ARPA targets Necho(i), which are spatial and temporal associated to the AIS target i.
- The minimum distance Dmin(i) between an AIS target i and associated ARPA targets j.

Furthermore the availability of ARPA targets can be determined by

$$Availability_{iv} = \frac{Ntotal_{iv} - Nzero_{iv}}{Ntotal_{iv}}$$

per vessel.

7 DATA ANALYSIS

This section discusses the results obtained from the fusion process described in section 6 and summarized in Table 5. In the radar range of the test vessel "BALTIC TAUCHER II" 23 dynamic vessels dedicated to the Open Sea Water (Baltic Sea) and 15 dynamic and 21 static vessels dedicated to the harbor area were identified.

For each vessel several AIS targets were identified, whereby for each AIS target the minimum distance between AIS and ARPA targets and the number of associated ARPA targets were determined. Table 5 lists the averaged values of minimum distance and Necho as well as the availability of ARPA targets per vessel.

In Figure 9 the mean values and the standard deviation of the parameters minimum distance, Necho, and availability is determined for a group of vessels with the same navigational status and operation area.

MINIMUM DISTANCE

The upper panel of Figure 9 illustrates the average minimum distance for the three analyzed regions. The statistical analysis leads to the following values of 154m ± 20m (mean value ± standard deviation) in the static harbor area, 142m ± 25m in the dynamic harbor area, and 140m ± 40m in the Baltic Sea area. The observed separation between AIS and ARPA targets can be induced by several reasons: One source might be the time resolution which contributes up to 50m (20kn over 5s ≈ 50m) to the total error budget in the distance estimation. An additional source for the observed separation can be the inaccurateness of the bearing angle induced by an error from the gyro compass which can be as large as 2 degrees (2° leads over 3NM to an error of around 200m). And last but not least the accuracy of the distance measurement, which depends mostly on the signal pulse length and can be as large as 75m for a 0.25μs pulse. Insofar the level of uncertainty is in line with the analyses results.

Table 5. Average of minimal distance between AIS and ARPA target, number of targets within a radius of 300m around the AIS target and the availability of ARPA signal given for vessels moving or mooring in harbor areas or operating at Baltic Sea

Dynamic vessels in harbor area				Static vessels in harbor area				Dynamic vessels at Baltic Sea			
Vessel ID	Distance [m]	Necho (r<300)	Availability [%]	Vessel ID	Distance [m]	Necho (r<300)	Availability [%]	Vessel ID	Distance [m]	Necho (r<300)	Availability [%]
HD01	169	2,8	39	HS01	167	2,5	36	BD01	86	9,4	95
HD02	130	3,5	41	HS02	112	2,7	66	BD02	101	1,5	75
HD03	155	1,2	20	HS03	116	2,2	39	BD03	148	1,7	46
HD04	157	1,3	28	HS04	169	2,7	37	BD04	98	1,9	70
HD05	139	1,5	47	HS05	148	2,4	60	BD05	108	3,6	79
HD06	149	1,8	64	HS06	132	2,6	39	BD06	140	2,7	70
HD07	122	2,1	47	HS07	202	2,1	29	BD07	135	1,3	55
HD08	183	1,6	44	HS08	121	2,6	41	BD08	152	2,3	53
HD09	149	1,7	57	HS09	90	3	39	BD09	62	2,7	99
HD10	172	1,7	39	HS10	149	5	71	BD10	82	5,9	100
HD11	170	1,8	55	HS11	120	2,4	41	BD11	87	3	98
HD12	179	1,4	17	HS12	148	1,6	30	BD12	185	3,4	99
HD13	176	2,8	69	HS13	150	1,5	43	BD13	183	1,8	88
HD14	141	1,4	29	HS14	187	1,3	26	BD14	193	1,7	81
HD15	119	5,1	86	HS15	142	1,3	43	BD15	134	9,2	98
				HS16	130	1,1	31	BD16	151	5,3	88
				HS17	135	1,1	41	BD17	174	1,1	69
				HS18	152	1,3	25	BD18	202	1,1	92
				HS19	128	1,4	36	BD19	142	13,9	100
				HS20	130	1,6	61	BD20	209	1,2	90
				HS21	157	1,3	29	BD21	153	1,9	75
								BD22	164	6	91
								BD23	124	7,6	91

HD: Dynamic vessel in harbour area moving)
HS: Static vessel in harbour area (mooring)
BD: Dynamic vessel at Baltic Sea

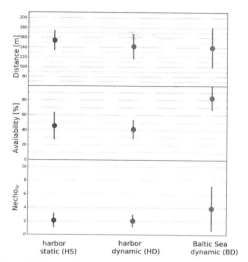

Figure 9. Mean value and standard deviation of minimum distance, Necho, and availability provided for vessel groups with identical status

AVAILABILITY OF ARPA ECHOES

The middle panel in Figure 9 shows the average availability of ARPA during our measurement campaign separated into the three regions. It can be seen that there is a significant improvement of availability in the Baltic Sea area in comparison to both subsamples in the harbor area. The calculated statistical values are 45% ± 18% (mean value ± standard deviation) for static vessels in the harbor, 41% ± 12% for dynamic vessels in the harbor, and 83% ± 15% for vessels en route at Baltic Sea. The better availability in the Baltic Sea might be explained by a better automatic acquisition of ARPA targets in an environment with a smaller number of unintended reflections.

NUMBER OF ARPA ECHOES PER AIS TARGET

As schematically illustrated in Figure 6 the possibility exists that one AIS target has more than one ARPA representation. The lower panel in Figure 9 shows the average number of extracted ARPA multiplicity again separated into the three subsamples harbor static and dynamic as well as Baltic Sea. The estimated statistical values are 2.0 ± 1.0 targets (mean value ± standard deviation) in the static harbor area, 2.3 ± 0.9 targets in the dynamic harbor area, and 4.7 ± 3.2 targets in the Baltic Sea.

In the case of perfect data we would expect 3 to max. 4 ARPA messages per time slot of 10 seconds, depending on the alignment of the ARPA extraction time and AIS position time. The expected Number of ARPA multiplicity is exceeded in the Baltic Sea area which might be explained by additional reflections from the wake behind the vessels. This effect depends on the velocity of the ship and will be

stronger with the higher speed in the open sea producing larger wakes. In the harbor area the observed multiplicity is smaller than expected, which was probably caused by the lower availability for this subsample.

8 CONCLUSION

In this paper we analyzed 70.000 AIS and 200.000 ARPA messages from a 12 hour measurement campaign in the Port of Rostock and the Baltic Sea. Since this analysis is based on a single vessel on a single voyage it is important to keep in mind that the data is one sample and specific equipment. During this campaign we received AIS signals from 59 vessels with at least one associated ARPA counterpart.

The following list summarizes first conclusions of this paper:

- There are AIS targets with no detected ARPA counterpart.
- There are ARPA targets with no AIS counterpart.
- The average ARPA availability for a given AIS target is 56% in overall (Harbor: 41 - 46%; Baltic Sea 82%).
- The average separation (distance) between ARPA and AIS targets is independent of the area and in the order of 150 ± 30 m.
- The existence of multiple (on average three) ARPA targets per AIS target can be observed.

Additionally we conclude that the fusion of automatic ARPA extracted echoes with AIS targets from the test vessel can be used to gain integrity information with a position accuracy of 150 ± 30m, which is two to three times the AIS position accuracy for vessels with an update rate of 6s and a speed of 20 knots (60m), assuming perfect positioning with the GNSS sensor.

Furthermore we are able to show that ARPA detects most of the AIS objects, but struggles with the task of finding all available AIS targets. The analysis of the availability of the ARPA extractions during our measurement campaign has shown that it was area dependent because the ARPA performance was decreasing in areas where unwanted disturbances from other reflecting sources were present. These results leads to the answer of the title question in the following way: "Within the setup described in the paper ARPA is suitable for automatic assessment of AIS targets with an accuracy of around 150m in minimum distance and an availability of an ARPA signal around 40% in the harbor area and 80% in the Baltic Sea.". On a more general note we would like to emphasize that the present design of the ARPA TTM messages does include distance and bearing information of the detected targets, but does not include any kind of error estimate which will be important especially when it comes to faint radar targets. In this case where the radar reflections are weak and the interpretation is difficult part of the image will be useful in order to use the additional information which is provided by shape of the target and/or its surroundings.

The presented analysis showed that the main source of error seems to be the time synchronization. In order to improve on this fact in further studies it is necessary to analyze the time behavior of all involved sensors. To improve the AIS position accuracy it is necessary to increase the time resolution of the included timestamp, which is 1 second at the moment of writing this paper. The time behavior of the ARPA TTM message on the other hand depends on the currently unstudied and manufacturer dependent performance of the extraction algorithm and signal runtime, as well as the fixed update rate caused by the antenna rotation. The signal runtime error of the fusion process can be removed by adding additional information of the extraction time from the TTM message.

The next planned step is the improvement of the time synchronization in order to improve the fusion accuracy by the development of a complete error analysis model for the radar and AIS fusion process. This will help to gain integrity information from an AIS and radar fusion process with the final aim of the introduction of data and system integrity into the maritime traffic situation aboard a vessel as well as ashore.

REFERENCES

Banyś, Paweł, Thoralf Noack, and Stefan Gewies 2012. *The assessment of AIS vessel position report under the aspect of data reliability.* Annual of Navigation. Akademia Marynarki Wojennej, Gdynia. ENC 2012, 25.-27

IMO RESOLUTION A.823(19). 1995. *Performance standards for automatic radar plotting aids (ARPAs).*

Krol A., Stupak T., Wawruch R., Kwiatkowski M., Paprocki P., Popik J.: *Fusion of Data Received from AIS and FMCW and Pulse Radar - Results of Performance Tests Conducted Using Hydrographical Vessels "Tukana" and "Zodiak".* TransNav - International Journal on Marine Navigation and Safety of Sea Transportation, Vol. 5, No. 4, pp. 463-469, 2011

Kwiatkowski M., Popik J., Buszka W., Wawruch R.: *Integrated Vessel Traffic Control System.* TransNav - International Journal on Marine Navigation and Safety of Sea Transportation, Vol. 6, No. 3, pp. 323-327, 2012

Lisowski J.: *The Sensitivity of Safe Ship Control in Restricted Visibility at Sea.* TransNav - International Journal on Marine Navigation and Safety of Sea Transportation, Vol. 6, No. 1, pp. 35-45, 2012

Neumann T.: *Multisensor Data Fusion in the Decision Process on the Bridge of the Vessel.* TransNav - International Journal on Marine Navigation and Safety of Sea Transportation, Vol. 2, No. 1, pp. 85-89, 2008

NMEA 0183. *Standard for Interfacing Marine Electronic Devices.* National Marine Electronics Association.

Rec. ITU-R M.1371-3. *Technical Characteristics for an Automatic Identification System Using Time Division Multiple Access in the VHF Maritime Mobile Band.*

SOLAS. 1974. *Safety of Life at Sea Conventions.* SOLAS.

Identification of Object Difficult to Detect by Synchronous Radar Observation Method

M. Wąż, S. Świerczyński & K. Naus
Polish Naval Academy, Gdynia, Poland

ABSTRACT: This article presents a method of identifying object difficult to detect by radar facilities. Synchronous radar observation allows creating a complete radar image of a particular area. This image is created by an aggregation of multiple images received from different radar stations. This radar image not only provides full information showing a complete picture of the area under observation, but it also enables the detection of low signal echoes, featuring the application of digital image processing and filtering methods.

1 INTRODUCTION

Radar navigation is an essential navigational aid on vessels. Equipping ships with radio-location gear is defined by regulations and convention requirements, such as the SOLAS Convention, which lays down the minimum requirements regarding equipment, depending on the type of vessel and its tonnage.

An on-board radar is used for conducting safe navigation which is linked to the process of observation, tracing echoes and anticollision maneuvers. Radar is also used to determine position. To ensure safe navigation, a radar image should provide information about all the moving vessels and fixed objects. The analysis of accidents reveals an insufficient degree of modern radars' detecting capabilities with regard to vessels with low radar cross-section (especially when the signal returns in bad weather). Often weak echoes detected by radar are not recognized by an inexperienced operator and are treated as noise and interference. Using TVG (Time Variable Gain), the operator attenuates noise, which may ultimately lead to a complete loss of information about the relative position of an object.

After processing the original radar signal within the image, one is left solely with echoes from objects characterized by a strong signal whose voltage is greater than the threshold voltage (cut-off).

The paper (Wąż & Naus 2012) describes a method of strengthening weak echoes.

In the case of fixed objects, weak echoes can be strengthened by successive summing of multiple radar images. As a result of summing, a weak signal is amplified above the threshold. Random noise and interference attain average values not exceeding the set threshold value (Wąż & Naus 2012). Restricting the use of this method applies to moving vessels.

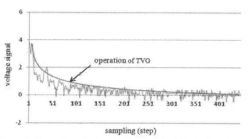

Figure 1. Operation of Time Variable Gain (Worked out by the Authors)

However, the method can be used upon simultaneous registration of radar signal from several radars observing an area. For this reason, we use the so-called Synchronous Radar Network described in the articles (Wąż 2011a,c). Presented below is a method of detecting difficult to recognize weak echoes which appear due to interference and noise in the radar image. The issue is important in terms of the safety of navigation. Early detection of a moving vessel which gives weak radar echoes can affect decisions made by a navigator and consequently, the whole process of navigation.

2 THE CONCEPT OF SYNCHRONOUS RADARS NETWORK (SRN)

The concept of the SRN is about the integration of technical systems of observation with on-board radars working in the same water area. It is assumed that as a result of fusing radar images from multiple radars located in the area, one can make a complete image, which provides almost full or full information about the area under observation.

SRN also sends this information to the "user" and displays it on the radar indicator (Wąż 2011a,c). Another problem is obtaining radar images from the radar included in the SRN and then processing them into a form which permits the detection of weak echoes. This issue is described in (Wąż 2011b).

The main concept behind the construction of the Synchronous Radar Network (SRN) is the need (Wąż 2011c):
- for creating a complete and real radar image, formed by summing real radar images from subordinate radar stations;
- for increasing the safety of navigation by providing the "user" with complete information about the situation in water areas and on the coast (eg. the movement of vessels, ice cover, hydrometeorological conditions, etc.);
- for supplying coastal technical observation systems with full and real radar information necessary for the tasks of radar traffic control and the protection of the coastal zone.

The basis of the concept is a multi-source data fusion which can provide the most effective solution to problems which have common characteristic features.

The fusion uses data from multiple sensors in order to achieve a result which would not be possible with the use of a single sensor. One intends to integrate not only the numeric data but also the image data. Coastal radars as a kind of radar sensors should supply information in order to yield a complete radar image of the whole water area.

The image can be delivered to other elements of the SRN, using contemporary methods of video signal transmission. The SRN can be classified into:
- fixed objects (permanent radar stations),
- mobile components (on-board radar),
- the central component (the CPU),
- component "user" (recipient of the data) (Wąż 2011a,c).

The central component is responsible for collecting, processing and fusing data. The network's most important task is to send and deliver to users complete radar information which will be displayed on radar indicators.

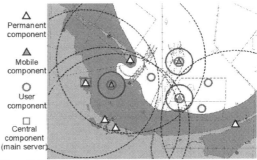

Figure 2. Components SRN (based on Wąż 2011c)

3 COLLECTION AND TRANSMISSION OF RADAR IMAGES IN SRN

Collecting radar images from multiple sources (radars) and sending them to the server in order to conduct a fusion of information are the main tasks of the SRN.

To that end, at the Institute of Navigation and Hydrography of the Polish Naval Academy a test stand was built to collect (in real-time) images from two independently operating radars and to send them to a computer which will enable their fusion or their display on one monitor for comparative purposes. The test stand was built on the basis of two video converters BlackBox (IR2 - BB) produced by Nobeltec. The basis of the test stand is a radar network, a type of server – user. Both video converters IRC2-BB are used to convert the analog signal from the radar into digital form, which is sent through a computer and server software to LAN network as UDP (User Datagram Protocol) communication protocol packets. The Nobeltec software installed on a computer allows a presentation of both radar images on the monitor of the central computer. As far as the distribution of video signals is concerned, the system has an unlimited range because it uses WAN/LAN/WiFi as a medium of the computer network. The system has been designed in such a way that in the future there will be a possibility of further development of the user components. Each of these user components must be equipped with a PC with a network card, the IRC2-BB converter and the Nobeltec software.

A block diagram of the radar network has been presented below.

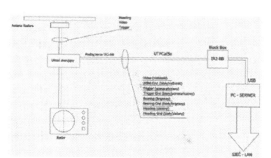

Figure 3. Diagram of the radar network (Radar – PC) (Worked out by the Authors)

A schematic diagram of the test stand shown in figure below.

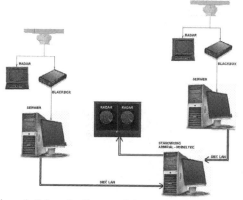

Figure 4. Schematic diagram of the test stand relay network. (Worked out by the Authors)

Figure 5. The implementation of the test stand in the laboratory (Institute of Navigation and Hydrography at the Polish Naval Academy) (based on Wąż 2011c)

To perform marine research it is necessary to have a "mobile radar kit" for recording radar images. The mobility of the test stand can be achieved using the "Maris PC Radar Kit" radar interface. The interface is a specialized computer card which is used to connect a radar to a computer. Together with the card the producer provides software for pre-processing and capturing individual radar images as well as video sequences. The PC Radar Kit Card includes functions of a network card. This gives great opportunities for operating interface: performing radar observations, registering radar images using a network (wired WAN / LAN or WiFi). A specialized card "Maris PC Radar Kit" enables the visualization of the radar signal on a computer monitor. The signal from the radar is connected to the input of the card which is one of the components of a computer system. It is shown below how to connect the card to a radar.

Figure 6. Connecting a PC Radar kit Card to a Radar (Worked out by the Authors)

Sending a radar image to a computer requires connection to the following signal cards:
- Video,
- Trigger,
- Bearing,
- Heading.

a)

b)
Figure 7. Mobile test stand for recording and processing of radar images
a) in the PNA laboratory , b) on the vessel (based on Wąż 2011c)

One should pay attention to the possibility of using a WiFi network for the transmission of radar images. This feature is very useful for measuring on vessels at sea.

In this configuration, the computer set with the specialized "Maris PC Radar Kit" card acts as a server and communicates with the user via the router. In order to verify the concept of the radar video image transmission through a wireless network, research into sea conditions was conducted.

On the vessel a test stand (radar, server and router) was installed with the specialized "Maris PC Radar Kit" card and receivers (laptops with the software) which were a kind of radar indicators - user component. A mobile test stand allowed conducting measurements.

First, radar images were registered in real conditions, and then they were sent through a router to component users through WiFi network. A mobile test stand working as a server provides an unlimited number of users with a radar image through a WiFi network.

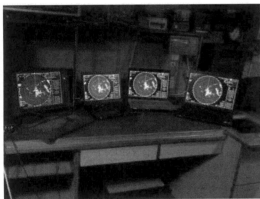

Figure 8. The "central component" of test stand with "Maris PC Radar Kit" , and three a user components on the vessel (based on Wąż 2011c)

4 COMPLETING PARTS OF RADAR IMAGE USING THE METHOD OF SUMMING IMAGES.

After delivering to the central component SRN radar images from different positions, one should sum the radar images in order to obtain a full picture. It should be noted that all the radar images should have the same type of display and stability presentation.

When radar images are recorded at different ranges of observation, transformation of images should be made in order to align the scale. Having the same image resolution will lead to the equalization of GSD (Ground Sampling stands for

Distance – the pixel size in a given area) for all the radar images. One should also pay attention to setting the same parameters of display on the indicators. This concerns in particular pulse length, pulse duration, pulse width and gain parameters (Gain, Rain, Sea). An attempt at summing test radar images from three radars in one picture is presented below.

Each radar was located in a different place in the area of the Gulf of Gdańsk and worked on a different range. The transformation of a common "denominator" was necessary. Standardized: the range, the type of display, the structure bitmaps and other parameters. The summed radar image was created without map background, based on GPS positions recorded at the moment of observation.

Figure 9. Images taken for summing (Worked out by the Authors)

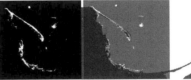

Figure 10. The radar image which is the result of summing images and radar image of the map background (Worked out by the Authors)

5 RECORDING THE INITIAL VIDEO SIGNAL

To deal with the above-mentioned problems of detecting objects characterized by a low level of reflected signal, the authors have developed a concept of creating imaging which will increase the probability of detecting radar echoes reflected from these objects. The concept assumes the possibility of recording radar images for the sake of their digital processing which allows the extraction of all their components (radar echoes) having an impact on navigation. Recording rastered images (bitmap) described above allows creating summed radar images for the SRN. These images are pre-processed by the primary and secondary processing module (it results from the fact that the images are obtained directly from the radar indicator block). In order to obtain information about the objects whose echo is on a par with noise level, a different method of recording radar images should be used. There is a need to obtain radar video signal just before digital signal processing where irreversible changes in the signal structure (filtering, noise reduction, etc.) are made.

There are several places in the receiving channel where one can connect a "measuring apparatus" and observe video changes. One of the instruments used for this purpose is an oscilloscope. Using modern digital oscilloscopes, one can record the observed signal on a computer drive in the form of measurement sequences which can be edited in a tabular and graphic form.

The graphic form of the measurement data is helpful in the completion of the measurement process, as well as in the subsequent phases of preparing database files. Its characteristic fragments may be additionally stored in the system and used, for example, for positioning or navigating in its different phases (Wąż 2010a,b).

For the registration of the original video signal, the test stand presented in the figure below was used. The test stand consists of: a navigation radar, an oscilloscope (DPO TEKTRONIX 3012) and a computer.

Figure 11. The test stand of initial video recording from navigation radar (based on Świerczyński 2011)

Using the test stand video signal was measured in real conditions. An example of one video signal "band" which was recorded is shown in the figure below.

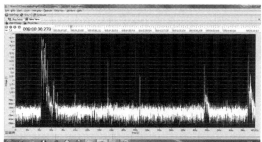

Figure 12. The original video signal recorded from the navigation radar (based on Świerczyński 2011)

Another research problem is processing the recorded video signal into a usable signal enabling the detection of weak echoes. The video signal coming from vessels should have constant position in relation to the position of radar antenna (assuming that the object does not move or moves slowly and the next signals were recorded once per one rotation of the antenna).

Noises and interference are random. Using the method of summing images from the same position and on the assumption that the observed objects are stationary or moving slowly, the quality of detecting objects against the background of random noises can be improved. One of the main techniques used to remove random noises is summing multiple images of the same object. Assuming that each image contains a random defect, summing them will yield an output image containing noises of all the components. However, the level of the unwanted signals in relation to the useful signal will be much lower than the original one. Increasing the number of added images will further increase the difference between the useful signal and the noise. The presented process of summing images is about adding the values of all video signals from one position, followed by performing normalization, the purpose being to scale the signal level to the range limits. The result of the summing is shown below.

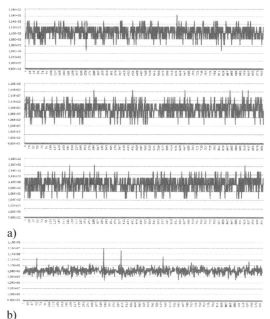

a)

b)

Figure 13. Summing video signals (observation of vessels which have weak reflected echoes)(based on 2011b)

In the figure above in part a) there are three examples of waveforms of recorded video signal. Part b) shows the result of summing and normalizing the value of video signals for 150 recordings. This is equivalent to approximately 7.5 minutes of recording time. In figure 13 b), we can clearly see the extracted video signal voltage values which are hypothetical echoes reflected from objects which have not been detected before.

6 CONCLUSION

The described concept of the Synchronous Radar Network (SRN) should include elements of existing systems. The SRN can be built and developed on the basis of currently operating technical observation points of the Navy, Border Guard and the Maritime Administration (VTS).

Radar information from all the stations would be managed by the central component of the SRN. It would sum and process all radar images incoming simultaneously from all the coastal radar stations and vessel radars, and create a radar image showing the situation in the area covered by all the radar stations and send the information to the vessels moving in the area.

The work of the Synchronous Radar Network would improve the safety of navigation in coastal areas by providing watch officers with necessary information to ensure safe navigation. This method of preparing and modeling information about moving vessels can create a new qualitative assessment of the safety of navigation, thanks to receiving complete and more detailed information about the movement of objects in the area covered by the system of radar stations.

Weak echoes reflected from the objects whose signal level is comparable to the noise level are often "ignored" by the radar. Modern methods of recording and digital processing of primary video signals make it possible to detect weak echoes whose signal level is comparable to the noise level of the radar station receiver.

REFERENCES

Świerczyński, S. 2011. *Rejestracja sygnału wizyjnego z radaru nawigacyjnego – opis sesji pomiarowej.* Opracowanie do projektu KBN: Automatyzacja procesu wyznaczania pozycji jednostki pływającej w nawigacji radarowej 2108/B/T02/2007/33, AMW Gdynia 2011.

Czaplewski K. & Świerczyński S. 2012a. *Determination of the precise observed ship's position using the traffic control systems and applying the geodesic estimation methods,* Zeszyty Naukowe Akademia Morska w Szczecinie, nr 32(104), 2012

Czaplewski K. & Świerczyński S. 2012b *Wykorzystanie geodezyjnych metod wyrównania obserwacji nawigacyjnych do zwiększenia dokładności określania pozycji statków w systemach VTS,* TTS Technika Transportu Szynowego, nr 9/2012 numer ISSN 1232-3829, 2012

Wąż, M. 2010a. *Navigation based on characteristic points from radar image.* Scientific Journals Maritime University of Szczecin, 20 (92), 2010, s. 140-145.

Wąż, M. 2010b. *Problems with Precise Matching Radar Image to the Nautical Chart.* Annual of Navigation 16/2010, Gdynia 2010

Wąż, M. 2011a. *Multi-Source Radar Information service,* 2nd European Conference of Systems (ECS'11); Puerto De La Cruz, Tenerife, Spain 2011, pp.80-84

Wąż, M. 2011b. *Rejestracja i przetwarzanie zobrazowania radarowego w celu zwiększenia możliwości detekcji ech o niskim poziomie sygnału,* Logistyka nr 6/2011, pp. 3887-3894

Wąż, M. 2011c. *Synchroniczna Sieć Radarowa,* Zeszyty Naukowe AMW, Gdynia 2011; ZN nr 186A

Wąż, M. & Naus, K. 2012. *Wykrywanie ech radarowych o małej amplitudzie sygnału.* Logistyka nr 3/2012 str. 2397-2400

Wąż, M. & Stupak, T. 2007. *Amplitude detection of weak radar signal.* Barcelona 5th The International Congress on Maritime Technological Innovations and Research, Barcelona, November 23, 2007;

Selected Methods of Ultra-Wide Radar Signal Processing

M. Džunda & Z. Cséfalvay
Technical University of Košice, Košice, Slovakia

ABSTRACT: The contribution is dealing with the selected stages of Ultra-Wide Band (UWB) radar signal processing. To this purpose, it provides the definition of UWB and describes some of the phases of the UWB radar signal processing. Also dealt with are the procedures of preliminary data processing and ways of static background subtraction.

1 INTRODUCTION

The abbreviation of UWB (Ultra-Wide Band) [6] is applied to the technology using signals covering an extremely wide band of the frequency spectrum. In [1, 2, 3], UWB technology is defined as follows: Absolute width of band 0 B $\geq$ 500 MHz, at attenuation – 10dBm, or relative width of band $\eta \geq$ 20 %. Relative width of band represents the proportion of absolute width of band to medium frequency.

Selected procedures of signal processing will be illustrated on M- sequence radar. Stimulation of the signal transmitted by this type of radar results in a pseudo-random binary sequence called M-sequence.

The advantages in using M sequence: (Maximal–length-binary-sequence)

- use of periodic signals avoids bias errors, and allows for linear averaging of noise suppression,
- M-sequence has a low crest factor which enables use of the limited dynamics of real systems
- signal acquisition may be carried out by under-sampling
- the signals may be sampled by using low cost, commercial Analog to Digital Converters (ADC) in combination with sampling gates.

The selected stages of radar signal processing will be illustrated on a radar with pseudo-random sequence – M (maximal–length-binary-sequence). Fig. 1). UWB radar signal processing is a complex process usually involving several stages [3, 4].

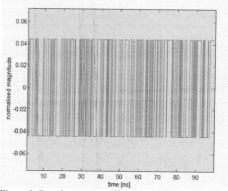

Figure 1. Pseudo-random sequence – M [7]

2 PRELIMINARY DATA PROCESSING

Preliminary data processing is aimed to eliminate or sufficiently reduce the influence of the equipment (radar) on the received primary (raw) data. The notion of primary data is to be understood as a set of responses to a stimulating signal obtained from the surrounding environment the signal is propagating. At a radar with pseudo-random sequence (Maximal length binary sequence), the received signal is correlated with the originally generated M-sequence. The resulting two-dimensional (2D) picture arranged into a plane in which the vertical axis of propagation of the stimulating signal (t) and the horizontal axis of observation (s) is called a radargram.

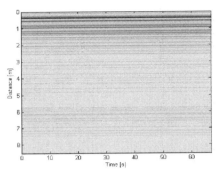

Figure 2. Radargram before data processing

2.1 *Set-up time < 0 >*

Time < 0 > is defined as the moment when sending of the first elementary pulse u (chip) of the pseudo-random sequence – M is started by the transmitting antenna. The given parameter is dependent on the length of the antenna cable and the position of the pseudo-random (PN) sequence chip, from which the transmission of the pseudo-random sequence is started. This is a random position obtained following each switch-on. Determining time < 0 > is meant to rotate the file of received responses so that the first pulse of the received PN sequence corresponds to the position of the antenna in the area.

Precision of determining the time <0> is important for the precision of the target localization. One of the methods of determining time <0> is making use of the direct signal of the transmitting-receiving antenna. The radargram following the preliminary data processing is illustrated in Fig. 3.

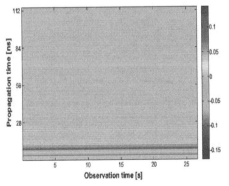

Figure 3. Radargram following preliminary data processing

2.2 *De-convolution*

The method of de-convolution can be of advantage at target trace estimation [2, 3] complex scenarios. The path of the signal in the time region can be described by the scenario (Fig. 4) mathematically in the following way [2, 3]:

$$M(t) * \left[h_{ET}(t) * h_{AT}(t) * h_T(t) * h_{AR}(t) * h_{ER}(t) \right] = {}^{\wedge}r_T(t) \quad (1)$$

when the pulse responses (except for the target responses) are marked as:

$$h_R(t) = h_{ET}(t) * h_{AT}(t) * h_{AR}(t) * h_{ER}(t) \quad (2)$$

and simultaneously the radar output is marked $h'_T(t)$, thus the convolutional relation is obtained

$$h'_T(t) = h_R(t) * h_T(t) \quad (3)$$

where $h_T(t)$ represents the real primary output of the radar. Deconvolution can be expressed as the process of obtaining eand estimate to the real priamry output of the radar $^{\wedge}h_T(t)$ while knowing $h'_T(t)$. Response $h_R(t)$ is obtained through measurement with stationary components. are stationary. Mathematically, the given process can be expressed as follows:

$$dekonvol\left[h_R(t) * h_T(t) \right] = {}^{\wedge}h_T(t) \approx h_T(t) \quad (4)$$

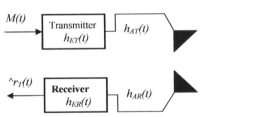

Figure 4. Deconvolution

3 ELIMINATING THE BACKGROUND

The purpose of eliminating the background is increasing the signal-to-noise ratio at the received data following preliminary processing. Background subtraction can help remove the stationary clutter, such as the antenna feed-back, clutters caused by impedance inadaptability an the *static* clutter of the surrounding environment. It enables detection of the responses from the moving objects.

3.1 *The basic principle of eliminating background*

Techniques of subtracting background help remove the time invariant background of the received radar data. *Background* is understood as signal, which is present in all responses to the stimulating signal. Background signal is a direct signal between the transmitting and receiving antenna (cross – talk) , which, by its performance, substantially exceeds the performance level of the electro-magnetic waves reflected from targets and of the static objects. Illustration of the radargram, in which the vertical axis represents the time of signal propagation stimulation t [ns] and the horizontal axis the time of observation τ [s], as in Fig. 5. Response to a single stimulating pulse $h(t,\tau)$ (pulse response) for $\tau = const.$ is represented by one vertical line (Fig. 5a). Direct signal between the transmitting and the receiving antenna is illustrated by a unipolar mono-pulse (blue) signal reflected from static objects by a bipolar dual pulse (black) and by a bipolar mono-pulse (red) signal reflected from moving targets. Ideal result of background subtraction is presented in Fig. 5b.

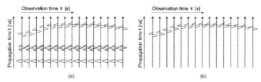

Figure 5. Illustration of signals (a) before background elimination (b) following background subtraction (by ref. [7]).

3.1.1 *Mathematical expression :*

Signal reflected from target: $s(t,\tau)$, Noise: $n_o(t,\tau)$ Clutter: $c(t,\tau)$ Jamming: $j(t,\tau)$.

Model of primary data:

$$h(t,\tau) = f\left[s(t,\tau), n_o(t,\tau), c(t,\tau), j(t,\tau)\right] \quad (5)$$

Assuming no presence of jamming and linkage of undesired components, i.e.

$$n(t,\tau) = n_o(t,\tau) + c(t,\tau) : j(t,\tau) \equiv 0 \quad (6)$$

a simplified linear model is obtained:

$$h(t,\tau) = s(t,\tau) + n(t,\tau) \quad (7)$$

The basic principle in subtracting the background is in estimating the stationary component of the primary data from the set of responses to the stimulating signal and its subtraction (separation) from the primary data [4, 5, 6]. Due to the fact that the primary signal between the transmitting and receiving antenna as well as the signals reflected from static objects ($n(t,\tau)$) in real case cannot be separated from the desired signal ($s(t,\tau)$), the result always contains a certain error in subtracting (estimating) the background. The status can be mathematically expressed as:

$$h_b(t,\tau) = h(t,\tau) - {}^\wedge n(t,\tau) = s(t,\tau) + \left[n(t,\tau) - {}^\wedge n(t,\tau)\right] \ (8)$$

where

$$^\wedge n(t,\tau) = g\left[h(t,\tau)\right]_{\tau_1,t_1}^{\tau_2,t_2} \quad (9)$$

represents the estimation of the stationary component of the set of responses to the stimulation signal

$$if : t_1 = t_2 = const. \Rightarrow {}^\wedge n(t,\tau) = g\left[h(t,\tau)\right]_{\tau_1}^{\tau_2}$$

The magnitude of the error in subtracting the background (level of distortion) depends on the method of background subtraction.

3.2 *Basic methods of background subtraction*

The fundamental methods of background subtraction can involve calculating the **mean value**, **median** or **modus**. Background estimation makes use of n number of previous pulse responses. [4]:

$$^\wedge h(t,\tau) = E\left[h(t,\tau)\right]_{\tau_1}^{\tau_1+n}, \quad (10)$$

$$^\wedge b(t,\tau) = median\left[h(t,\tau)\right]_{\tau_1}^{\tau_1+n},$$

The essence of the exponential method of calculating the mean value is in making use of the value of the previous estimate of the background $^\wedge b(t,\tau-1)$, to calculate the actual value of the background estimate $^\wedge b(t,\tau)$ on the basis of the new pulse response $h(t,\tau)$ based on the following relations:

$$^\wedge b(t,\tau) = \alpha \, {}^\wedge b(t,\tau-1) + (1-\alpha)h(t,\tau) =$$
$$= {}^\wedge b(t,\tau-1) + (1-\alpha)\left[h(t,\tau) - {}^\wedge b(t,\tau-1)\right] = \quad (11)$$
$$= {}^\wedge b(t,\tau-1) + (1-\alpha)h_B(t,\tau),$$

where α is the scalar weight factor acquiring the values within $(0 \div 1)$ and τ denoting the actual time of observation. New estimation therefore is representing a certain part of the previous and a certain part of the actual estimate. It helps emphasize the previous events, or attenuate (smooth out) the frequently occurring fluctuations (variations) thereby revealing long-term trends of background estimates. Vector $h_B(t,\tau)$ for $\tau = cont.$ represents the signal with the subtracted background estimate i.e. reflection from the moving targets. In view of the simplicity in and low requirements in terms of memory capacity, the given method is suitable for real time signal processing.

3.3 *Adaptive methods of background subtraction*

Compared to the previous methods, there is an important difference in using the rules of decision-making, which enable selection of the analyzed pulse responses. The rules will decide on whether the given response $h(t,\tau)$ will be evaluated as background, or as a potential target (foreground).

Selective methods also include the **adaptive Gaussian model** [4, 5] of background subtraction. It assumes a background with Gaussian distribution of probability $N(\mu,\sigma^2)$. The rules of decision – making are based on knowing the standard deviation of the background $\sigma(t,\tau)$. In case when $|h(t,\tau) - \mu(t,\tau)| \le k\sigma(t,\tau)$ where k is

241

a positive real number, model parameters are updated by way of:

$$\mu(t,\tau+1) = \alpha h(t,\tau) + (1-\alpha)\mu(t,\tau)$$
$$\sigma^2(t,\tau+1) = \alpha\left[h(t,\tau)-\mu(t,\tau-1)\right]^2 + (1-\alpha)\sigma^2(t,\tau) \quad (12)$$

In the opposite case, i.e. when $\left|h(t,\tau)-\mu(t,\tau)\right| > k\sigma(t,\tau)$, the actual pulse response $h(t,\tau)$ is considered as the potential target and all the parameters remain at original values:

$$\mu(t,\tau+1) = \mu(t,\tau)$$
$$\sigma^2(t,\tau+1) = \sigma^2(t,\tau) \quad (13)$$

In view of the fact that all variables are explicitly given by the relations used and memory requirements are acceptable, the method is suitable for real time application. A disadvantage to this method is in its ability to operate only with a uni-modal background with limited precision of modeling.

Radargram with subtracted background for $R_x 1$

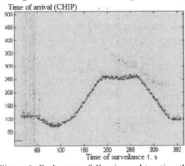

Figure 6. Radargram following subtracting the background by way of exponential method

Under ideal conditions, the following quality of preprocessed data picture can be accomplished. Look at fig. 7.

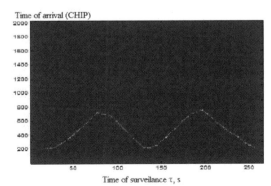

Figure 7. Idealized radargram for $R_x 1$

4 CONCLUSION

The references provided offer a relatively large amount of methods of subtracting static background, based on non-parametrical or combined multi-model models, and further methods based on Kalman filters etc., which meeting the application requirements of special cases. Usually, they are typical for high complexity of calculations. Use of some of the methods above is dependent on special requirements set for signal processing. Among them can be real-time operation and precision of location. Selection of the suitable methods is also dependent on the possibility of detecting static or moving targets. In case of detecting static targets, it is appropriate to use the method of background modeling. The above mentioned methods will be used in our further research when investigating the use of UWB radars for locating fixed or moving targets behind obstacles. In practice, it can be applied for example for determining the movement of persons behind the wall.

REFERENCES

[1] *Second report and order and second memorandum opinion and order (2004) Federal Communication Commission,* FCC 04 - 285, ET Docket No. 98 - 153, Washington D.C., USA, 55. Available online at : http://hraunfoss.fcc.gov/edocs_public/attachmatch/FCC-04-285A1.pdf.

[2] KOCUR, D., ROVŇÁKOVÁ, J., ŠVECOVÁ, M., AFTANAS, M., DRUTAROVSKÝ, M., GAMEC, J., M. GAMCOVÁ , M.: UWB Radar Signal Processing for Trought Wall Moving Target Tracing. Multimedia Communications, Technical University of Košice, Presentation in Univ. of Ilmenau, Dec. 2008.

[3] PICCARDI, M.: *Background subtraction techniques: a review,* in Proc. IEEE International Conference on Systems, Man and Cybernetics, October 2004, pp. 3099–3104.

[4] ROVŇÁKOVÁ, J.:*UWB Signal Processing for Moving Target Detection,* Thesis to the dissertation examination, Technical University of Košice, Department of Electronics and Multimedia Communications, Slovak Republic, December 2007.

[5] FONTANA, R.J, FOSTER, L.A., FAIR, B., WU,D.: *Recent Advances in Ultra Wideband Radar and Ranging Systems,* IEEE International Conference on Ultra-Wideband , Singapore, September 2007.

[6] ZETIK,R., CRABBE,S., KRAJNAK, J., PEYERL, P. J. SACHS, J., THOMA, R.: *Detection and localization of persons behind obstacles using M-sequence through the-wall radar,* Proceedings of Sensors and Command, Control, Communications and Intelligence (C3I), Technologies for Homeland Security and Homeland Defense, Vol. 6201, May 2006.

Radar, ARPA and Anti-Collision
Advances in Marine Navigation – Marine Navigation and Safety of Sea Transportation – Weintrit (ed.)

A Discussion on the Duty of an Anchored Vessel to Avoid Collision

P. Zhang
Shanghai Maritime University

ABSTRACT: It is very common to see a collision at sea between an anchored vessel and another vessel underway. When determining liability between an anchored vessel and the other vessel in a collision, the anchored vessel may usually be held not liable, especially in the judgment of the Maritime Authority or Maritime Court. The author discusses some issues in respect of a collision between an anchored vessel and other vessels under the International Regulations for Preventing Collisions at Sea, adopted by the International Maritime Organization in 1972 (COLREG) , and then considers a number of opinions regarding duties of anchored vessels to avoid collision at sea under COLREG.

1 INTRODUCTION

Collisions between anchored vessels and vessels underway are bound to occur from time to time in the navigation and operation of vessels. However, there are no direct or clear rules in COLREG regarding the duties of anchored vessels to avoid collision. Whether or not an anchored vessel has the obligation to act to avoid a collision will be directly related to liability identification issues after a collision. In many cases, when people determine the collision liability ratio between an anchored vessel and the other vessels in a collision, they always believe that the vessel underway shall accept all or most of the responsibility before conducting a detailed analysis of the reason for and process of the collision. To some extent, this mistaken thinking inertia results in the anchored vessel taking no action or negative action when the risk of a collision occurs.

2 THE CONCEPT OF THE ANCHORED VESSEL

Anchoring is a method of vessel manoeuvring. There are many reasons for vessel to anchor, such as anchoring when waiting for a berth, anchoring during bad weather conditions, anchoring when receiving fuel oil, anchoring when loading or discharging cargo. Also, vessels conduct operations by using the anchoring method: large vessels use an anchor to assist berthing and turn around, or drag the anchor during sailing, etc.

The vessel's vulnerability to risk at sea and ability to avoid a collision with other vessels will differ according to the method of operating the vessel. All of these operations are included within the broad category of anchoring. On the aspect of vessel operation, using the anchor is a way to control a vessel, i.e. controlling the motion of vessel with the anchor's holding power. Generally speaking, the restricted definition of anchoring refers to the anchor fluke holding the sea bed firmly, with no change of anchor position, with the vessel only moving around the anchor position under wind force within the length limit of anchor chain.

The COLREG regulations do not set the rules for what kind of specific status is referred to as anchoring. It only sets four categories of vessel status in the definition of "underway" , i.e. underway status, anchoring status, berthing status and aground status[2]. Most of the items in the COLREG are guidance for taking related actions, and normally apply to underway vessels. Therefore, how to determine the dividing line between underway vessela and anchored vessel is very important[3].

According to COLREG, if a vessel is not defined as anchoring, berthing or aground, then the vessel

---

[2] Article 3, term 9 of COLREG: the word 'underway' means that a vessel is not at anchor, or made fast to the shore, or aground.
[3] This article will not discuss the differences among underway, made fast to the shore and aground in details.

should be defined as underway. The definition of 'anchoring' under COLREG generally refers to a motion status where the anchor controls the vessel firmly under its holding force[4]. The anchoring status is ended due to anchor aweigh or anchor dragging. Weighing anchor is an active way to end the anchoring status, making the anchor break away from the sea bed by operating the windlass; however, dragging the anchor is a negative way to end the anchoring status, resulting from the force of an external influence, such as wind and wave. In the case of negatively ending anchoring status, the holding force of the anchor in the seabed becomes loose; the holding power of anchor decreases. Therefore, it is impossible to control the vessel in a position firmly. Both of the two circumstances will finally change the vessel status from anchoring to underway. In addition, an anchor is also frequently used to assist in vessel operation, such as sailing with anchor dragging, dropping the anchor for a turn around or berthing/departing the berth, etc. Under these circumstances, anchor usages are only a method for assisting vessel operation. At that moment, the vessel still has the status of underway, not anchoring[5].

The beginning of the anchoring status is when the anchor catches the seabed firmly. When the anchor breaks away from the seabed during weighing anchor, or when the ship drags the anchor in the sea, it is regarded as the end of anchoring status. There must be a process during the beginning or end of anchoring. Only when the anchor holds the seabed firmly and is able to control the vessel by anchor and anchor chain, is the vessel regarded as in the anchoring status[6]. The following diagram describes identification of, and changes in the different motion statuses:

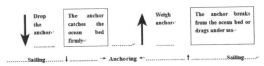

Figure 1.

3 DUTIES OF AN ANCHORED VESSEL TO AVOID COLLISION UNDER COLREG

The correct identification of underway status and anchoring status is relates to the application of rules and regulations. For example, the vessel should still be regarded as in underway status even when she is dropping anchor. Under these circumstances, she should comply with the related rules applicable to underway ship in COLREG. The vessel will be in anchoring status when the anchor catches the seabed[7]. After that time, the vessel should hoist an anchor ball or turn on the anchor light immediately. Similarly, when the anchor breaks away from the seabed during weighing anchor, the vessel should lower the anchor ball or turn on the navigation light[8]. If an anchored vessel does not properly follow the requirements of COLREG to show its anchoring status and finally a collision occurred due to her fault, then she should be liable for the collision, or even the main liability. In the collision case between the two vessels Presto and Llanover, Presto was staying in the anchoring status and whistled, but she did not turn on the anchor light and this failure caused a collision with a sailing vessel Llanover. In the court case which followed, the Judge (PilcherJ) thought that although Presto was in the anchoring status, it meant that she was responsible for the main liability in the collision[9].

In the COLREG regulations, there are no direct stipulations instructing anchored vessels how to take action to avoid collision with other vessels. Moreover, an anchored vessel is excluded by the Article 18 of COLREG, in which the responsibilities of various kinds of vessels are stipulated. Therefore, it is very difficult to identify an anchored vessel's responsibility when she encounters an underway vessel at sea. Judging from the general principle of COLREG, the vessel which has higher maneuvering ability for avoiding a collision should give way to the vessel which has relatively lower maneuvering ability for avoiding a collision. An anchored vessel anchors at sea by the holding force of the anchor and anchor chain. The vessel will make a turning motion around the anchor position under wind and wave force. Is this condition of the vessel bound to leave it in the position of lower maneuvering ability of avoiding collision with another vessel ? The answer is definitely no.

3.1 *Collision risk between anchored vessels*

When anchoring, the vessel must make sure that there is ample water room for her safe turning

[4] Please refer to *Shallow Discussion on the Concept and Correct Operation of 'Anchoring'*. CHEN Yu-peng. *Tianjin Navigation* (1997.02).

[5] Please refer to *Discussion on the status of vessel motion under "COLREG"*. WANG Zhi-ming. *Word Shipping* (1994.05)

[6] The Esk and Gitana (1869) L.R.2A. &E.350.See the Nortonian (1935)51 LI.L.Rep.317 at 318. It seems that she is 'at anchor' while being towed up to her anchor, so long as the anchor has not broken out of the seabed.

[7] After the anchor chain is on the brake, observe the anchor chain status for judging whether the anchor is holding firmly or not. If the anchor chain pulls tightly forward and moves stably and rhythmically on the sea, then works little loose, this means the anchor has held firmly.

[8] Please refer to The Romance [1901] P.15. The Judge thought that the vessel is in the anchoring status if the vessel winds the anchor but it has not left the soil. In 1967, Judge Carl Minsky of the English Court pointed out that once the anchor leaves the soil, the vessel shall lower the anchor ball, when he dealt with the case of "Forest Lake".

[9] (1994) 77 LI.L.REP.198 at 202.

movement under the effect of wind and water. If a vessel chooses an anchorage position near another ship which has arrived there earlier, the newly-arriving vessel should take full consideration of the surrounding water area situation and weather condition, and leave enough water room to make sure she will not collide with another one under normal circumstances. "If one vessel anchors there, and another here, there should be that space left for swinging to the anchor that in ordinary circumstances the two vessels cannot come together. If that space is not left, I apprehend it is a foul berth."[10] If the distance between two anchored vessels becomes too close due to bad weather, and pose a risk of collision, then the vessel which dropped anchor later should take extraordinary action to make enough space for the former anchored vessel. If the vessel anchored earlier weighs anchor due to anchor dragging, and then drops the anchor again, it should be regarded as the late anchored vessel.

3.2 Collision risk between the an anchored vessel and a vessel not under command

There are many reasons for causing a vessel to be not under command, such as the propeller not working normally due to main engine failure, and vessel not changing course due to steering engine failure or rudder blade missing, etc. No matter what kind of reasons make the vessel not under command, they will reduce or deprive the vessel of the ability to avoid collision with another vessel. That is the reason why not under command vessels are listed at the end of Article 18 of COLREG, as those which cannot give way to another kind of vessels. If a collision risk occurs between an anchored vessel and a vessel not under command, it is clear that the anchored vessel is in a position which has higher ability to avoid collision. The anchored vessel can weigh her anchor immediately or change the length of anchor chain or use her main engine and rudder, so as to change her position, in this way reducing the possibility or consequence of collision.

3.3 Collision risk between anchored vessel and vessel underway

In this section, vessels underway refers to those vessels which are sailing in normal condition, excluded those which are not under command, dragging anchor, or restraint in their ability to maneuver.

Compared with an anchored vessel, vessels underway are in an advantageous position to avoid collision. In the case of the anchored vessel which complies with the related anchoring regulation of the local port and the COLREG requirements and fully shows its anchoring status[11], the underway vessel should take measures to keep clear of the anchored vessel. However, under circumstances where it is impossible to avoid immediate collision by the action of the under way vessel alone[12], then the anchored vessel has an obligation to take action as will best assist in avoiding the collision or reducing the consequence of collision[13]. If the anchored vessel does not take any positive action or behaves negatively, letting the collision happen where there is an immediate risk of collision, it is regarded as 'negligence' and the anchored vessel has to accept the subsequent collision liability[14].

4 CONCLUSION

Among the many processes involved in operating a vessel of any size, an especially important one is anchoring, using the vessel's anchor as a control mechanism in a variety of different situations. Relatively little attention has been given to this procedure in the academic literature. This article attempts to provide a comprehensive overview, showing that the concept of an 'anchored vessel' is more complicated than it might appear at first sight. Anchored vessels are not always completely stationary. In respect of the obligation of an anchored vessel to avoid collision at sea, it has been noted that there are no direct and clear stipulations in the COLREG regulations. However, based on the principles and good seamanship requirements of COLREG, the anchored vessel is still required to take active and positive steps, depending upon the specific circumstances, to avoid collision.

---

[11] For example, making good preparations for the main engine; having enough staff on duty; showing the anchor light or anchor ball; using the method of sound signal, etc. to warn other vessels in foggy weather.

[12] Please refer to COLREG Rule 17 (b): When, from any cause, the vessel required to keep her course and speed finds herself so close that collision cannot be avoided by the action of the give-way vessel alone, she shall take such action as will best aid to avoid collision.

[13] Please refer to Marsden on collision at sea. Simon Gault, Steven J. Hazelwood, A.M. Tettenborn Page.125⊡ A ship at anchor which obstinately refuses to move from her anchorage where she necessarily endangers other craft may be held in fault for a collision that follows. An anchored vessel may reasonably be required to do what she can to assist the other to clear her, either by sheering with her helm, using her engines, paying out chain or in any way possible and failure to do so may be held to be negligence. She must, however, beware of acting too soon.

[14] See the sellina (1920)5 Ll.L.Rep.216, The offin (1921)6 Ll.L.Rep.444.The Defender (1921) 6 Ll.L.Rep.392.See Paragraph.6-46.

---

[10] Please refer to Dr Lushington in The Northampthon (1853)1Sp.152,160 see The Diomed (1946) 79 Ll.L.Rep.526, Affd.(1947)80 Ll.L.Rep.164,CA. Distinguishing a bad berth, which may be described as a berth where a ship taking the ground is liable to be damaged owing to the unevenness of the bottom.

Data Integration in the Integrated Navigation System (INS) in Function of the Digital Processing Algorithms Used in the Frequency Modulated Continuous Wave (FM-CW) Radar

R. Wawruch

Gdynia Maritime University, Gdynia, Poland

ABSTRACT: Paper presents reasons of introducing of the Integrated Navigation System (INS) on sea going vessels, basic functions of this system and possibility assessment of the use of Frequency Modulated Continuous Waves (FM-CW) radar as its sensor. Detection possibility of the FM-CW radar in function of the type and parameters of the video signal digital processing algorithms utilised in the radar was subject particular analysis. Algorithms efficiency was assessed by comparison of detection possibilities of two radars installed in Gdynia Maritime University: FM-CW type CRM-203 and pulse Raytheon NSC34.

1 INTRODUCTION

Pursuit of the International Maritime Organization (IMO) to raise the level of safety of navigation and protection of the environment causes, inter alia, increasing number of required navigational equipment installed onboard seagoing vessels. Ships' owners pursuit to reduce the number of members of ships' crew at the same time. These are two main causes of the need for integration of the ship's navigational equipment. Technical progress, especially the development of electronics, automation and computer science makes this integration possible. On the bridges began to install Integrated Navigation Systems (INS), which increasingly replaces traditional mono-functional navigational equipment with dedicated monitors.

INS is introduced in order to relieve the officer of watch of the part of duties assigned to him, by accomplished automatically:

– Providing the data needed for voyage planning and control ship's movement;
– Evaluation of data derived from all connected sensors and linking these data to provide information warning in due time about the potential danger;
– Preventing the degradation of the correctness of the presented information; and
– Facilitation of comprehensive analysis and assessment of the current situation.

According to the recommendation of the IMO Resolution MSC 86(70) Annex 3 "Performance standards for an Integrated Navigation System (INS)" the purpose of this system is to enhance the safety of navigation by providing integrated and augmented functions to avoid geographic, traffic and environmental hazard and so called "added value" for the operator to plan, monitor and/or control safety of navigation and progress of the ship. The INS shall timely present correct and unambiguous information to the users and shall provide subsystems and subsequent functions within the INS and other connected equipment with this information. It is defined as such if work stations provide multifunctional displays integrating at least the following navigational tasks/functions:

– Route monitoring; and
– Collision avoidance.

Additionally it may provide manual and/or automatic navigation control functions.

An INS should combine process and evaluate data received from connected sensors and sources checking its availability, validity and integrity. To do that, it shall receive each data from at least two mutually independent sensors, acting, if possible in different ways and on different principles. Quality of information presented on the INS display unit depends on the correctness of the algorithms integrating data received from connected sensors with different degrees of sensitivity and accuracy.

Radar equipment still remains a sole fully independent technical source of data on the navigational situation around the ship. Safety of navigation of vessels with navigational bridge fitted with integrated systems as shown on Figure 1 depends largely on onboard radar detection capability and accuracy of presentation of the radar video signal on integrated navigation monitors. This issue is particularly important in the case of the FM-CW radars currently entered in operation as maritime radar. The question is the quality of the digital signal processing algorithms utilized in this type of radar, having effect on its sensitivity, and so on its detection range of surface objects, resistance on hydro-meteorological disturbances and possibility of integration of data received from FM-CW and traditional pulse radars.

Figure 1. Onboard INS display units.

In order to assess the quality of currently used algorithms of digital processing of the radar video signal experimental research was conducted using radar FM-CW type CRM-203 installed in the radar laboratory of the Gdynia Maritime University with antenna located at the height of 21 meters above sea level on the roof of the university building nearby the south entrance to the Gdynia Harbour.

2 RADAR AND EXPERIMENT DESCRIPTION

The prime function of the FM-CW radar type CRM-203 is detection of sea surface objects, estimation of their planar co-ordinates and automated tracking the selected ones to perform the surveillance tasks. Its basic parameters are presented in Table 1.

Table 1. Basic parameters of the FM-CW radar type CRM-203

Parameter	Value
Output power	1mW, 10 mW, 100 mW, 2W
Carrier frequency	9.3 – 9.5 GHz
Frequency deviation switched according to the required scale range:	54 MHz at 6 NM 27 MHz at 12 NM 13.5 MHz at 24 NM
Range scales	0.25 NM – 48 NM
Modulation	Direct Digital Synthetizer (DDS) based linear FM-CW
Sweep repetition period	1 ms
IF bandwidth	4 MHz
Frequency curve slope of IF amplifier	6 dB/oct; 12 dB/oct; 18 dB/oct
Beam width (horizontal/ vertical)	0.70°/22°
Polarisation	Horizontal
Gain	32 dB
Analog-to-digital converter	14 bits
FFT signal processing	8192-points FFT
Sampling frequency	8 MHz
Display resolution	1280 × 1024 pixels
Acquisition	Automatic up to 100 targets
Tracking	Automatic of all acquired targets
Range accuracy	1% of selected range or 50 m (whichever is greater)
Angle resolution	0.1°
Bearing accuracy	0.7°

Utilised radar is a prototype and user can define bigger number of its parameters than in serial one. He has to define following parameters of the scanner, transceiver and processing algorithms:
1 Scanner rotation speed between 12 and 30 rpm/min with increment of 1 rpm/min.
2 Anti clutter sea – three levels: 6 dB/oct; 12 dB/oct; 18 dB/oct.
3 Receiver gain level before signal digitalisation (automatic/manual) between 0 dB and 63 dB with increment of 1 dB.
4 Carrier frequency – 10 different values in the band 9300-9500 MHz defined by manufacturer and marked as F0 - F9.
5 Constant False Alarm Rate (CFAR):
   – Type of algorithm:
      – FIX (detection without CFAR, voltage threshold given manually by operator);
      – CMAP (clutter map);
      – CAGO (normalized ambient average left, greatest of);
      – CASO (normalized ambient average right, smallest of);
      – OS (ordered statistic); and
      – CA (duplex normalized ambient average)
   – Level of the voltage threshold between 0 and 255 with increment of 1;
   – Length of the cell: 4, 8, 16, 32, 64 and 128;
   – Attenuation level between 0 and 255 with increment of 1.
6 Type of window algorithm:
   – Bartlett;
   – Blackman;

- Gaussian;
- Hamming;
- Hanning;
- Harris;
- Kaiser;
- Rectangular;
- Triangle; or
- Von Hann.

7 Differentiation - switched on/off.
8 Pulse clutters correlation - switched on/off.
9 Fix clutters correlation - switched on/off.
10 Clutter chart - switched on/off.
11 Binary integrator with the "M-of-N" rule - switched on/off.
12 Binary integrator parameters - m, n - any value of integer between 0 and 30.

Radar structure, principle of work and tracking accuracy was presented in more detailed manner during previous TransNav Conferences (Król 2011, Kwiatkowski 2011, Plata 2009).

Detection possibility of tested radar were compared with detection distances of the pulse radar Raytheon NSC 34 installed in the same laboratory with scanner located a few meters from the FM-CW radar antenna position.

Table 2 presents pulse radar Raytheon NSC 34 basic technical parameters elaborated on the base of data found in the user manual delivered together with the equipment by its producer.

Table 2. Basic parameters of the radar Raytheon NSC 34

Parameter	Value
Output power	25 kW
Carrier frequency	9410±30 MHz
Range	0.25; 0.5; 0.75; 1.5; 3; 6; 12; 24; 48; 96 NM
Pulse length	0.06 µs, 0.25 µs, 0.5 µs, 1.0 µs,
Pulse repetition frequency	3600 Hz, 1800 Hz, 900 Hz
IF bandwidth	20 MHz, 6 MHz
Antenna length	2.1336 m
Beam width horizontal/vertical	1.0°/23°
Polarisation	Horizontal
Gain	29 dB
Rotation speed min/max.	22/26 rpm
Display size	28 inch
Resolution	1600 × 1200 pixels
Acquisition	Automatic up to 40 targets
Tracking	Automatic of all acquired target
Range accuracy	0.3% of selected range or 6.4 m (whichever is greater)
Angle resolution	0.3°
Bearing accuracy	1.0°

Figures 2 and 3 present positions of the radar antennas installed on the roof of the university building (Figure 2) and the field of view of these antennas (Figure 3).

Figure 2. Positions of the radar antennas on the roof of the university building.

Figure 3. The field of view of radar antennas.

Experimental study of the algorithms of digital signal processing in radar FM-CW CRM-203 was carried out in the spring and summer 2012. The purpose of the measurements was to examine in real conditions the efficiency of window and CFAR algorithms and other parameters of the digital processing by assessing their impact on the radar detection capability and quality of radar video signal presented on the radar display monitor and ECDIS Transas 4000. All tests were conducted in good hydro-meteorological conditions without precipitation and for wind from coast directions causing the state of the sea up to 2 degrees. Both radars: FM-CW type CRM-203 and pulse radar Raytheon NSC 34 were working on the range of 12 nautical miles (Nm) equal to 22224 m enabling to observe on the display monitors coastline of the Gulf of Gdańsk and ships and floating aids to navigation in the area of this gulf. Detection possibility of the FM-CW radar and quality of its video signal were assessed by comparison with detection distances for the same objects and quality of radar image received on radar Raytheon NSC 34 display unit.

Figures 4-9 present measuring positions and samples of radar FM-CW video signal presented on the display monitor of this radar and ECDIS.

Sample of radar image received during conducted test on the Raytheon NSC 34 display monitor shows Figure 10.

Figure 4. Console of the FM-CW radar CRM-203 (radar display monitor and kayboard on the left and console for entering radar parameters and chosing digital processing algorithms on the right of the stand).

3 CONCLUSIONS

On the basis of the measurements carried out in the optimal hydro-meteorological conditions, it was stated that the best FM-CW radar image and detection possibility were obtained after switching differentiation function on and turning functions of correlation and integration off, for frequency F3 (approximately 9360 MHz) and CASO CFAR algorithm.

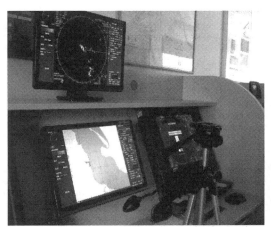

Figure 5. ECDIS Transas 4000 monitors and radar Raytheon NSC 34 display unit.

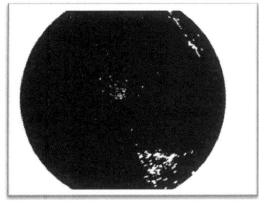

Figure 6. Sample of radar image on FM-CW radar display monitor on range 12 Nm (22224 m), carrier frequency F3, CFAR algorithm CASO.

Figure 7. Sample of radar image on FM-CW radar display monitor on range 12 Nm (22224 m), carrier frequency F3, CFAR algorithm CAGO.

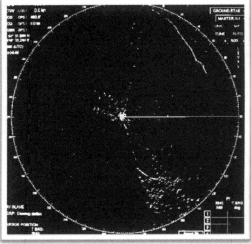

Figure 8. Sample of the FM-CW radar video signal presented on ECDIS Transas 4000 monitor (range 12 Nm (22224 m), carrier frequency F3, CFAR algorithm CAGO.

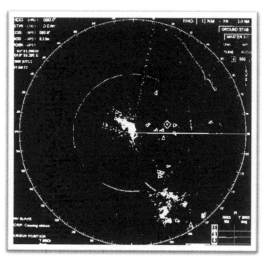

Figure 9. Sample of the FM-CW radar video signal presented together with AIS symbols on the ECDIS Transas 4000 monitor (range 12 Nm (22224 m), carrier frequency F0, CFAR algorithm CAGO.

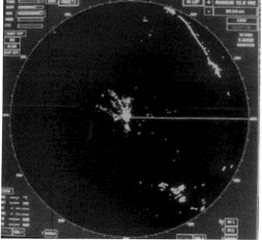

Figure 10. Sample of radar image received during conducted test on the Raytheon NSC 34 display monitor.

This conclusion corresponds to the expectation. In the absence of precipitation and sea clutters, smaller requirements for integration and correlation assure higher sensitivity of the device. Test of the integration and correlation rules conducted in adverse hydro-meteorological conditions may lead to other conclusion. At higher sea state, intense precipitation, etc., the values of the coefficients of correlation and integration functions should be increased to eliminate disturbances less correlated than echoes received from rear sea surface objects.

Research of the effect of different window algorithms on the radar detection capability demonstrated that the quality of radar image depends on the type of used algorithm in favourable weather conditions very slightly. The particular algorithms differ in efficiency of reduction of the interferences from the second radar set only.

It should be point out that during the research were obtained the same detection distances of coastline, floating aids to navigation and ships by tested FM-CW radar type CRM-203 and the modern pulse radar Raytheon NSC34. It means that data obtained from these devices can be considered as received from two independent and equivalent sensors and merged in order to present on monitor of Integrated Navigation System.

REFERENCES

Król A., Stupak T., Wawruch R., Kwiatkowski M., Paprocki W., Popik J. 2011. Fusion of Data Received from AIS and FMCW and Pulse Radar - Results of Performance Tests Conducted Using Hydrographical Vessels "Tukana" and "Zodiak". In Adam Weintrit (ed.), Marine Navigation and Safety of Sea Transportation. Navigational systems and simulators: 159-165. Leiden: Balkema.
Kwiatkowski M., Popik J., Buszka W., Wawruch R. 2011. Integrated Vessel Traffic Control System. In Adam Weintrit (ed.), Marine Navigation and Safety of Sea Transportation. Transport System and Process: 183-187. Leiden: Balkema.
Plata S., Wawruch R. 2009. CRM-203 Type Frequency Modulated Continuous Wave (FMCW) Radar. In Adam Weintrit (ed.), Marine Navigation and Safety of Sea Transportation: 207-210. Leiden: Taylor & Francis Group.

Chapter 8

Watchkeeping and Safety at Sea

See More – Analysis of Possibilities of Implementation AR Solutions During Bridge Watchkeeping

D. Filipkowski

Gdynia Maritime University, Gdynia, Poland

ABSTRACT: One of the main tasks of the e-navigation, in accordance with the definition proposed by the International Maritime Organization (IMO, International Maritime Organization), is to increase the safety of the ship while sailing from point A to B. The analysis of marine accidents indicate that the main cause of accidents is the so-called human element. The same analysis shows that navigator keeping watch on the bridge makes decisions that eventually lead to tragedy. In this article, the author tries to show how the solutions specified as Augmented Reality (AR) through a better presentation of information can help improve the current situation.

1 INTRODUCTION

One of the main tasks of the e-navigation, in accordance with the definition proposed by the International Maritime Organization (IMO), is to increase the safety of the ship while sailing from point A to B. The experience of previous years, in particular the spectacular sea accidents of the twentieth century, indicate that very often first must come to the tragedy, before the changes in the rules, procedures and equipment were introduced. This time, in line with e-Navigation system, group of people working on the creation of system, wants to be one step ahead of a possible tragedy. Based on experience, statistics and reports on incidents from previous years, they are trying to design a system that will actually prevent accidents. In this paper, author describes, how the introduction of e-Navigation and the use of Augmented Reality solutions during a navigational watch on the bridge can increase the level of safety of ships at sea. Additional information, which navigator can get, it may affect his decisions and reduce the number of collisions and groundings, which are among the most serious incident at sea. Not only threaten to destroy the ship and cargo but very often directly threaten the lives of the crew and the environment.

2 COLISSIONS AND GROUNDINGS

Analyzing statistical data from end of the century, you can get the impression that safety at sea is improving. Data published by the International Union of Maritime Insurance (IUMI) for vessels that due to various circumstances have been lost, indicate that in 2006, complete loss of ships dropped from 0.5% to 0.1% (% of all vessels) compared to 1990. The same report also includes comparative data on the loss of ships in 2006 and 1994. They refer to a specific number of lost ships. According to them, in 2006, there were less lost than compared to 1994 (respectively 2006 – 80 lost vessels, while in 1994 - 180 lost vessels). The data appear to be satisfactory, but more detailed analysis of the causes of accidents are more worrisome. Total losses caused by the groundings and collisions in 2002-2006, compared to the same period in 1997-2001 indicates that there is an increase that number by about 4-5%. Statistical data seems to be covered by reports, studies and analyzes carried out by the Marine Accident Investigation Branch (MAIB). The conclusions of these reports seem to confirm the hypothesis captain Nick Beer (member of the UK MAIB) that the vast majority of these accidents were caused by human error. He says that in almost half of cases (43%) when it comes to collisions, officer of the watch was completely unaware of the impending danger until the collision, or discovered when the threat was no longer in any way to avoid a collision. The e-Navigation, through tools such as Augmented

Reality, will help reduce the number of situations in which the navigator alone, decide on the basis of scanty, unverified information. This paper presents examples of applications of e-Navigation and the use of Augmented Reality in the pursuit of a navigational watch.(Patraiko, 2010)

2.1 *Colisions*

The data presented in Figure 1 as the main causes of collisions indicate an incorrect assessment of the situation by the navigator (24%) and lack of or inadequate visual observation (23%). In the next row will be unconscious of the presence of another vessel until the crash (13%), poor communication on VHF (9%), fatigue (8%), failure to comply with COLREG (8%) and alone watchkeeping (7%). Analysis of the results of these studies raises the hypothesis that the main cause of conflict is the wrong assessment of the situation by the navigator, which in turn results from the isolation of the ship's OOW and deprive it of the ability to consult and review their decision. Not without significance is the lack of information needed to make the right decision due to the amount of the obligations imposed on the navigator acting alone watch. None of this information makes it responsible for the safety of the entire ship officer is often unaware of the risks involved. Regulation 7 of COLREG says "Assumptions shall not be made on the basis of scanty information, especially scanty radar information." (Gale, 2007)

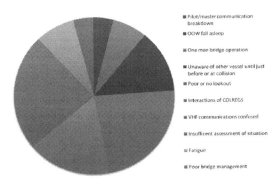

Figure 1. Causes of collisions (Nautical Institute)

Nowadays, radar and ARPA have become the main tools for assessing the risk of collision and collision avoidance. Navigator cannot forget that every electronic device has some limitations. The user being aware of these limitations is unable to trust the unconfirmed information obtained from it. Officer of the watch would not make decision only on basis on this uncertain information. In accordance with good practice and with COLREG, visual observation is still the primary mean of risk assessment and avoidance of collisions at sea. (Amato, 2011)

2.2 *Groundings*

Statistics presented in Figure 2 indicate that the main reasons of groundings. The most common cause are bad management and lack of cooperation on the bridge (18%) and lack of or improperly preparing voyage planning (17%). For the next group it can qualified: fatigue (14%), insufficient control of ship's position (14%), problems with the exchange of information with pilot (14%) and ineffective or lack of visual observation(12%.). Because officer of the watch falling asleep, according to statistics, only 8% of vessels will go aground, and because the lack of adequate procedures only 3%. (Patraiko, 2007)

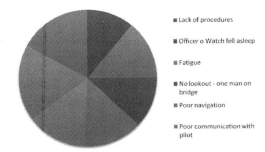

Figure 2. Causes of groundings (Nautical Institute)

3 E-NAVIGATION

The representatives of maritime administrations of Japan, Norway, Marshall Islands, Netherlands, Singapore, United Kingdom and United States, at a meeting of the Maritime Safety Committee (MSC) of the International Maritime Organization, which was held in December 2005 jointly undertook a discussion on the need introduction of the new system, which later became known as e-Navigation.

3.1 *Definition*

Discussion on the implementation of e-Navigation was taken in 2005, but it is believed that the origins should be sought much earlier, in the events that took place 40 years ago. In January 1971 two tankers collided in conditions of poor visibility in the area of Golden Gate in San Francisco. The result was a spill of about 800 000 barrels of oil in the bay of San Francisco. During the case, the US Coast Guard observing the whole event due to an experimental system HARP (Harbour Advisory Radar Project) and they were completely helpless. Subsequent attempts to establish a connection with vessels running into danger have failed and there was a collision. Events described above led to the creation of the Vessel Traffic Service (VTS) in this area and is considered by some as a milestone on the way to create a system that would treat all the issues of navigation and vessel traffic management in a holistic way. Creation of e-Navigation system is the answer to the need for a strategic vision for the use of new tools, in particular electronic tools to enhance maritime safety and protection of environment. In 2009, the MSC has approved a plan to implement the e-Navigation according to which the introduction of new technologies should be in correlation with the changes in the rules and take into account user requirements for both shore-based and those working at sea. International Association of Marine Aids to Navigation and Lighthouse Authorities (IALA), created a coherent and at the moment the official definition of the e-Navigation.(IALA, 2010)

"The harmonized collection, integration, exchange, presentation and analysis of marine information onboard and ashore by electronic means to enhance berth to berth navigation and related services for safety and security at sea and protection of the marine environment."

According to the above definition, e-Navigation system would help ships sail in a safe and efficient manner friendly to the environment. User and his needs would be one of the main drivers of the system, and new technologies that enable effective communication and safe navigation would help him send, receive and analyze information and make good decisions. (Lee, 2009)

3.2 *Examples of the application of e-Navigation*

If e-Navigation would be the strategic vision aimed at enhancing the safety of navigation, system developers should consider how the e-Navigation can lead to improvements in the areas listed in the first chapter, According to research of Nautical Institute they are the main causes of collisions and groundings. All of the analyzed events were caused by so-called human factor, which is responsible for about 75% of all accidents at sea. Here are some concrete examples of how e-Navigation can provide a significant improvement in these areas, focusing on eliminating errors of a single person:

- e-Navigation will include elements of decision support system, the decision must be verified either by another person, or by the system itself, depending on the technical capabilities,
- e-Navigation on the bridge will allow easy and continuous access to relevant information presented in a clear and unambiguous manner, using standardized symbols and coding systems for controls and displays,

- e-Navigation will indicate the operational status of automated functions of integrated components, systems and / or subsystems,
- e-Navigation will enable fast, continuous and effective information processing and decision-making,
- e-Navigation will help to prevent or minimize the amount of excessive and unnecessary work and the conditions that lead to distraction on the bridge, these factors can lead to fatigue or weaken vigilance,
- e-Navigation will help to reduce the risk of human errors as well, but if they occur, detect them through monitoring and alarm systems, in order to officer on the bridge and the pilot could in time take appropriate action,
- e-Navigation will include a system to detect hazards and alert the user, which will consist of both technical (network of sensors, devices analyzing the data and alarming threat exists) as well as with the relevant procedures that you must implement when there is a risk, or the possibility of existence risk is high,
- e-Navigation will be in a clear and transparent way to present the information hierarchy by which the information will be presented must minimize the risk of distraction or misinterpretation, at the same time paying attention to the really important information,
- e-Navigation will provide effective exchange of information between ship and shore to help navigator on board, figuratively speaking will be an additional member of the crew, it would eliminate situations where the watch officer makes decisions in isolation and is not verified in any way.

Statistics prove that working within the VTS station, where the actions of the vessels are monitored by land users and their vessels are informed of the potential hazards in these areas, the amount of incidents and accidents is much smaller. It is all the more significant factors that VTS stations are usually located in areas with difficult navigation characteristics like Malacca Strait or areas with heavy traffic as the Strait of Dover. (Weintrit, 2007)

4 AUGMENTED REALITY

Augmented Reality (AR) is a system that combines the presentation of information about the real world with computer-generated objects. Augmented Reality is a tool, or rather a set of tools that enhance the user's perception and his interaction with the real world. Virtual objects that are displayed, provide the information that the user can not directly get to the observation using only his senses (hearing and sight). What seems to be crucial not only for navigation but for the industry in general, as the information provided by the tool in the form of augmented reality virtual objects help the user to perform real tasks.

4.1 *Definition*

Augmented Reality is not an entirely new concept. It seems that the need for its development was to disseminate such technologies as the Internet, mobile phones, etc. Miniaturization and personalization of PCs and other electronic devices (eg, laptops, i-Pads, and digital cameras, etc.) have enabled more users to read of AR. Devices that are an integral part of Augmented Reality to some extent defines areas where Augmented Reality has already achieved success.

Some researchers define Augmented Reality as the system associated with special displays worn on the user's head. Dr. Ronald Azuma of the University of North Carolina, one of the authorities in this area, to avoid restrictions to specific technologies, defines AR as the system:

– combines real world and virtual reality;
– interactive in real-time;
– with three-dimensional elements (3D).

Figuratively speaking, Augmented Reality is a system that expands users view of the world and complements the reality with virtual (computer-generated) objects that appear to coexist in the same space and time. Dr. Azuma definition makes AR the open system with new technology and is not limited to specific technical solutions. There are several approaches to Augmented Reality and there is no clarity on technical solutions. According to one of the system concept includes the processing and presentation of data, as well as other data collection and transmission. (Azuma, 1997)

MIXED REALITY

REAL ENVIRONMENT   AUGMENTED REALITY   AUGMENTED VIRTUALITY   VIRTUAL ENVIRONMENT

Figure 3. Reality/Virtuality Continuum (Miligram, 1994)

In 1994, Paul Milgram (University of Toronto) and Fumio Kishino (Osaka University) tempted to find the definition of Mixed Reality (MR). They identified it as a reality wherever located between the borders of Virtuality Continuum, extending from the object completely real, to occur solely and entirely in a virtual environment. The consequence of this is the location of Augmented Reality and Augmented Virtuality between these boundaries.

These continuum is examined in Figure 4 in two dimensions. Virtuality level is shown in the data presented on the horizontal axis. In the literature the vertical axis often means Mediality. Mediality is a concept quite enigmatic and vague because the author treats Mediality as affordability and usefulness of the information provided. We can change the level by using different ways of presenting information. Defined as a measure of media attention is the user's subjective assessment of suitability and affordability of information presented. If we assume that the point R is unmodified to the axis of the reality of virtuality, we can put both Augmented Reality, but also Augmented Virtuality. With the increase in the value of the abscissa (M axis) increases the level of usability and accessibility to information. Above AR and VR (Virtual Reality), we can find a Mediated Reality or Mediated Virtuality or any combination of this two. According to this approach further up and to the right we have a virtual environment in which the information is presented in a very mediated way. That makes it accessible to the user and makes it more useful. Figure 4, which is a two-dimensional version of the continuum of virtuality/reality shows not only the concept of Mixed Reality, but also, in addition to the impact of additions (extensions), also takes into account the effects of build-up over time deliberately modified (reduced) reality. What's more, this diagram shows that the reality can be modified in different ways and to different degrees. It also takes into account that not only the reality may be extended by the addition of virtual elements, but also the virtual world can be augmented by adding elements of reality. The presentation of information is not limited to the selection of equipment and techniques used but also refers to the more fundamental issues, such as the number of available sources of information that he received. The ability to determine suitability of the information presented in a way will help determine the suitability of the presentation. With appropriate assumptions will also determine the suitability of equipment used in the processing and presentation of information. (Mann, 2002)

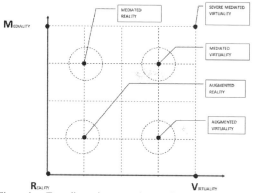

Figure 4. Two-dimension version of Reality/Virtuality Continuum (Miligram, 2007)

4.2 *Application of AR during watchkeeping*

Information about the navigational dangers and presentation of data about other ships facilitate for navigator decision making in excessive close-ups. Additional use of decision support systems would give the effect of visible safe sailing sectors and proper course alteration to avoid collision (with appropriate assumptions about speed and motion parameters of other objects). Information about the seabed in the form of three-dimensional bathymetric maps and profiles facilitate anchoring. Information on the current weather and traffic in specific regions optimize the planning of the voyage. Virtual buoys are solutions already used for example in the port of Antwerp. Going a step further, navigator would be able to generate virtual temporary and preliminary corrections, virtual port approaches, virtual areas such as anchorages or fairways. As in the ECDIS these virtual corrections would appear in the form of layers applied to the actual chart, AR objects would appear on a computer or on a special screen installed on the bridge.

4.2.1 *Safety*

Applications installed on mobile phone or other device belonging to a passenger on the ship would show him the shortest escape route during an emergency. On the other hand, rescue team would coordinate the action from the outside, with access to the camera image captured (camera installed for example on the helmet of one of the lifeguards). At the same time they could compare it with a virtual layout of the ship and could have a much easier way to coordinate action (shorter time to reach the victims and casualties). Coordinator with information about the location of victims could be using virtual directions or voice lead Search and Rescue team to the target, even in total darkness or very low visibility (very common condition of the shares and search and rescue, especially when we have to dealing with a fire).

4.2.2 *Protection of Sea Environment*

Information about special zones would facilitate proper application of the International Convention for the Prevention of Pollution from Ships (MARPOL), in particular the regulation relating to pollution by oil (Annex I) and the regulations concerning the storage and disposal of waste (Annex V). Navigator could also have information on the regions where the endangered species of marine animals, special areas (nature reserves, etc.) and other related to the protection of marine environment.

4.2.3 *Cargo*

Depending on the type of cargo and the ship type you could for example see through the walls of a container correctly identifying located in it cargo or to see a specific place in (accordance with stowage plan) where container should be loaded. Information collected by the sensors of temperature, pressure, etc. would facilitate better use of loading equipment and increase control over the cargo operations. Information about the ballast, and recent data from the loading computer would give the officer supervising the loading progress a complete review of the current situation and increase the effectiveness and safety of cargo operations.

4.2.4 *Training*

In some Universities in the U.S., where much emphasis in the education process is placed on enrichment experiences for students, teachers uses Augmented Reality to activate cognitive processes. The components of many courses are educational trips suplemented with mobile technology what makes lesson even more attractive. An example is the Massachusetts Institute of Technology, in which the participants of the "Environmental Detectives," learn about the ecosystem searching for clues and gradually revealing the secret. Students in addition to AR use e.g. GPS. In the process of training crews and land-based personnel of maritime transport there are very often used simulators, which are nothing but a virtual environment. Training of seamen with AR simulators could take training literally into new dimension.

5 CONCLUSIONS

Efthimios Mitropoulos (Secretary-General of IMO), said that the e-Navigation should not reduce the role of navigator only to a passive viewer monitoring system. It should provide support in the form of information that will allow him to make faster decisions in accordance with the principles of good seamanship. Reducing inputs grounding and collisions caused by human error with the support of the technology, e-Navigation and Augmented Reality can be a key enabler for improving safety at sea. However as says COLREG the officer of the watch must comply with regulations and avoid collisions. No matter how advanced technology he will have a corresponding observation must lead "by all possible means ... in order to avoid a collision."

REFERENCES

Amato F., Fiorini M., Gallone S., Golino G., 2011, *e-Navigation and Future Trend in Navigation.* TransNav - International Journal on Marine Navigation and Safety of Sea Transportation, Vol. 5, No. 1, pp. 11-14,

Azuma R., 1997, *A Survey of Augmented Reality,* Teleoperators and Virtual Environments, str. 355–385,

Azuma R. T., Baillot, Y., Behringer, R., Feiner, S., Julier, S., & MacIntyre, B.: *Recent Advances in Augmented Reality,* IEEE Computer Graphics and Applications, 21(6), 34-47.

European Commission, Directorate-General for Mobility and Transport, 2010, *The EU e-Maritime initiative,*

Filipkowski D., Wawruch R., 2010, "Concept of "One Window" Data Exchange System Fulfilling the Recommendation for E-Navigation System", Transport Systems Telematics, Springer

Filipkowski D., 2011, "Informatyczne elementy systemu e-Nawigacji", Logistyka Nr 6

Gale H., Patraiko D., 2007, *Improving navigational safety. The role of e-navigation.*, Seaways, str.4-8,

Hagen E.K., 2012 *Why eNavigation ?*, Seaways, str.14-16,

IALA e-Navigation Comitee, 2010, *e-Navigation Frequently Asked Questions (Version 1.5),*

Lee A., 2009, *e-Navigation and Electronic Charting: Implications for Hydrographic Community,* US Hydrographic Conference 2009, Norfolk, VA,

Mann S., 2002, *Mediated Reality with implementations for everyday life.,* Teleoperators and Virtual Environments,

Milgram P., 1994, *Augmented reality: a class of displays on the reality-virtuality continuum.,* Warszawa, Telemanipulator and Telepresence Technologies,

Milgram P., Takemura H., Utsumi A., Kishino F., 2007, *Augmented Reality: A class of displays on the reality-virtuality continuum,* Proceedings of Telemanipulator and Telepresence Technologies,

Pardel P., 2009, *Przegląd ważniejszych zagadnień rozszerzonej rzeczywistości,* Zeszyty Naukowe Politechniki Śląskiej, seria INFORMATYKA, Tom 30, Numer 1(82),

Patraiko D.,2007, *The Development of e-Navigation.* TransNav - International Journal on Marine Navigation and Safety of Sea Transportation, Vol. 1, No. 3, pp. 257-260,

Patraiko D., Wake P., Weintrit A., 2010, *e-Navigation and the Human Element.* TransNav - International Journal on Marine Navigation and Safety of Sea Transportation, Vol. 4, No. 1, pp. 11-16,

Sub-Committee on Safety of Navigation, Session 85, 2009, *Strategy for the development and implementation of e-Navigation,* IMO, London,

Sub-Committee on Safety of Navigation, 2007-2009, Sessions 53-55: *Development of e-Navigation strategy,* IMO, London,

Weintrit A., Wawruch R., Specht C., Gucma L., Pietrzykowski Z., 2007, *An approach to e Navigation,* "Coordinates", Delhi, Vol.III, Issue 6, pp. 15-22.

Investigation of Officers' Navigation and Port Watches Exposed to Excessive Working Hours

H. Yılmaz, E. Başar & E. Yüksekyıldız
Karadeniz Technical University, Maritime Transportation & Management Engineering, Trabzon, Turkey

ABSTRACT: Today, seafarers working time is also increased with the increasing speed of the operation of vessels in parallel with the developments in the world maritime trade. This increase has brought with it insufficient resting hours, especially in short sea transport. It is known that seafarers are exposed to fatigue due to hard and stressful work environments that human errors have been increased. In this study, the officers' working hours are simulated in full-mission simulator at short sea transport to investigate errors made during the navigation and port watches. For this purpose, the data were obtained from 7 volunteers watchkeeping officers with the help of video monitoring and check lists in the full-bridge simulator system and also, system records in cargo handling simulator. With this study, introduced what types of errors made during navigation and port watches by watchkeeping officers under the intense pace of work. And the errors are evaluated under the legislation of international maritime. As a result, to comply with the limitations of working hours, it is necessary to increase the number of officers responsible for operations on short sea transport.

1 INTRODUCTION

With the ever increasing emphasis placed on the human factor in marine accidents in the recent years, the studies of International Maritime Organization (IMO) and other researchers with human focus have demonstrated a rise. Fatigue and human errors naturally are within the scope of such studies. It is known that about 70-80 % of the marine accidents originate from human errors (Arslan and Er, 2007). The maritime accidents, indeed, may be defined as catastrophic incidents when considered the number of people in the ships, environmental damages and the values of the ships and carried cargos (Chauvin, 2011).

According to the Marine Accident Investigation Branch (MAIB) data, 2031 vessels, 100 GT and greater commercial vessels registered in UK sea registry, have participated to the marine accidents between 1998-2011 years and 20 vessels among such accidents have been lost (MAIB, 2012). French Marine Accident Investigation Office (BEAmer) have reported a total of 418 marine incidents for the period 2004-2011 (BEAmer, 2004-2011). According to the Federal Bureau of Maritime Casualty Investigation (BSU) data, 60 marine incidents have occurred including commercial vessels in German waters in the year 2010 and 89 incidents in 2011 (BSU, 2011). Marine Casualty Investigation Board (MCIB) has reported that the number of ships participating in marine incidents through 2002-2010 had been 123 (MCIB, 2011). The Standing Commission for Maritime Accidents and Incident Investigations (CIAIM) has stated 41 marine incidents in 2009 and 91 marine incidents in 2010 (CIAIM, 2010; 2011). The Swedish Maritime Administration (SMA) data show us that 1294 accidents had occurred in Swedish registered ships between 1997 and 2006 and also 1555 near miss incidents (SMA, 2007). According to Latvia Division for Investigation of Marine Accidents (DIMA) data, the total number of accidents/incidents is 140 between the years 2004-2009. 92 of these incidents have been realized in Latvian waters and 48 out of Latvian waters. Again DIMA reports that 55 incidents have been faced in relation with fishing vessels between the years 1993 and 2009 (DIMA, 2009). Danish Maritime Authority (DMA) has stated 435 marine incidents in Danish waters between 1998 and 2007 (DMA, 2008). 466 incidents have been experienced by the commercial vessels, and 325 marine incidents by fishing vessels, with Denmark and Greenland registry between the years 1999 and 2008 (DMA, 2009). The Accident

Investigation Board of Finland (AIBF) has reported 59 accidents for the period between 2002 and 2006 (AIBF, 2007). The more we deepen the statistics and data sources related to marine accidents the more we may understand the real numbers of accidents are not limited with the aforementioned figures.

The losses of live and commodity have been faced in such marine incidents as well as caused environmental pollution and economic damages. The subjects being focused on happened to include the originating environmental problems and examination of marine accidents, in time. The decisions taken and arrangements performed in maritime world have always, from ever history, based on the previous incidents, accidents and losses. The importance of human factor in the marine accidents, which cause great damages on nature, economy and maritime companies, have been officially considered with the acceptance of the subjects *"the requirement of focusing on human activities for the safe operations of the vessels and the need for providing a high standard of safety, security and environmental protection aiming to obtain an eminent reduction of maritime accidents"* and *"high priority of human factor issues in the working program of the Organization since plays an important role in the prevention of sea accidents"* by IMO (IMO, 2004).

The human based errors cause more marine accidents when compared to equipment based errors (Shea and Grady, 1998). The human error has been started to be accepted as the main reason of collision, running aground and petroleum leakage etc. marine incidents. At the same time, according to many research and analyses, nearly 70-80% of the marine incidents have been caused by the human errors (Arslan and Er, 2007). The statistics of IMO related to the global figures of marine accidents have established the share of human error as between 80% and 85% (Ece, 2008).

The concept "Error" is being defined as "the fault performed without intention, purpose or desire"; as well as the definition "the person or the institution not able to demonstrate the behaviour in compliance with the requirements of the situation and time" (Çakmakçı, 2001). According to Human Factors Analysis and Classification System (HFACS), the human error coverage also includes the unsafe/dangerous actions of the operators, the preparatory conditions of unsafe/dangerous actions, incompetent audit and organizational effects. The human error may be classified as errors based on skills, decision errors and perception errors. The reasons of the errors are physical and technological surrounding factors, psychological and physical situation of the operator, human resources management and personal underlying/preparing factors such as rest and alcohol usage and organizational factors such as insufficient allocation and maintenance of resources and incompetency of audits. If we consider the watchkeeping officer as the performer of any error or breach during the voyage of a vessel, then we may also take the individual factors such as stress, physical and psychological fatigue, motivation, work load and working hours etc. and the surrounding factors such as angle of view and length of the bridge, lighting and visual obstructions etc. as the factors affecting the error (Reason, 1990; Shappell and Wiegmann, 2000).

1.1 *Watchkeeping Officer and Working Hours*

Certain functions shall continue uninterruptedly 24 hours in the vessel environment and the works to be performed by the humans shall be arranged as to provide resting periods. Thus, working in shifts in accepted for such kind or works. The watchkeeping officer is the deck personnel with complete information about the locations and functions of all the safety and navigational aids on the vessel. The watchkeeping officer is responsible for the safe navigation of the vessel, under the general instructions of the master, as for preventing the collisions and running aground. These officers are also responsible for performance of the operations in a manner not jeopardizing the persons, vessel, cargo and the port in their port watches as well protecting the marine and surrounding environment (IMO, 2011).

The working hours and resting periods of the watchkeeping officers are arranged under the International Convention on Standards of Training, Certification and Watchkeeping for Seafarers (STCW) with 2010 Manila amendments (IMO, 2011). Seafarers' Hours of Work and the Manning of Ships Convention (C180) also provides arrangements on working hours in compliance with STCW. The working hours of the seamen have also been arranged under European Union Directives and national regulations of the countries. Table 1 summarizes the information related to the working hours and resting periods of the seamen under STCW and C180 conventions, European Union Directives Nr.1999/63/EC (European Union, 1999) and Nr. 2003/88/EC (European Union, 2003) and applicable Turkish Seamen Regulations (Official Journal, 2002).

It may be the case as noncompliance to the working hours aforementioned, over working and, accordingly, fatigue. It is known that over working is a factor that triggers the fatigue and, in relation, the human error (IMO, 2001).

Table 1. Regulatory arrangements related to the working hours of the seamen

Legislation	Daily resting hours	Daily resting periods and intervals	Weekly resting hours	Daily working hours	Weekly working hours
STCW	Min. 10 hours	– Max. two periods, one of which shall be lest 6 hours – Intervals between periods shall not exceed 14 hours	Min. 77 hours		
C180	Min. 10 hours	– Max. two periods, one of which shall be lest 6 hours – Intervals between periods shall not exceed 14 hours	Min. 77 hours	Max. 14 hours	Max. 72 hours
1999/63/EC	Min. 10 Hours	– Max. two periods, one of which shall be lest 6 hours – Intervals between periods shall not exceed 14 hours	Min. 77 hours		
2003/88/EC	Weekly working hours shall be max. 48 hours and comply with STCW and C180.				
Seamen Regulations	Min. 10 hours	– Max. two periods, one of which shall be lest 6 hours – Intervals between periods shall not exceed 14 hours	Min. 70 hours (Including emergency situations)		

2 METHODS

It is a known fact that the increased working hours and difficulties of the working environment is an important factor towards the tiredness of the seamen. This study has been conducted to determine the errors of the watchkeeping officers subject to excessive working hours during their watches; and, camera monitoring system and check lists have been used in the full-mission bridge simulator environment whereas the system records have been used for liquid cargo handling simulator. This study has been performed with healthy 7 male volunteers, between 22-24 age interval (average age 22,3), who have the competency of unlimited watchkeeping officer. During the data collection stage, the tanker simulator has been used for total 245 hours and the bridge simulator for total 154 hours. Each volunteer is worked with total 8 days of which 1 day to be preparation day. The volunteers have been observed during a one week program carried out.

2.1 Weekly Working and Resting Program

The studies related to the topic of fatigue in maritime profession have shown that the fatigue is more eminently observed in short sea shipping (Smith et al., 2006; Uğurlu et al., 2012). This study also has considered, for the formation of the working hours program, the working hours of certain tankers working in short sea shipping in Sea of Marmara. While determining the working hours data, the results of an interview performed with 8 masters and watchkeeping officers, working in the tankers navigating between the İzmir Tütünçiftlik region platforms and İstanbul Ambarlı region platforms in the Sea of Marmara, have been also based on (Uğurlu et al., 2009). The periods required by the third mate to perform its duties, of whom the working hours are to be simulated according to the obtained data, have been shown in Table 2.

According to such hours, if the scenario starts with the watchkeeping by third mate, the minimum working hours indication shall happen as shown under Figure 1. Each division represents the 10 minutes part of a day. The meanings of the color codes have been explained below.

Table 2. The periods required for the third mate to perform its duties (Uğurlu et al., 2009)

Duties	Average time
Receive documents from agent before departure	0 h 30 m
Preparing the bridge for unberthing	0 h 30 m
Unberthing maneuver	0 h 40 m
Navigation watch	4 h 10 m
Preparing the documents for agent in arrival port	1 h 00 m
Preparing the bridge for berthing	0 h 30 m
Berthing maneuver	1 h 30 m
Delivery the documents to agent after berthing	0 h 30 m
Passage planning and preparing ISM (International Safety Management) documents (at discharging port)	3 h 00 m
Preparing the ISM documents (at loading port)	1 h 00 m
Port watch (if the cargo operation of the vessel up to 24 hours, two watches in day)	4 h 10 m
Preparing the bridge before anchorage	0 h 30 m
Anchor watch	4 h 10 m

(m: minute, h: hour)

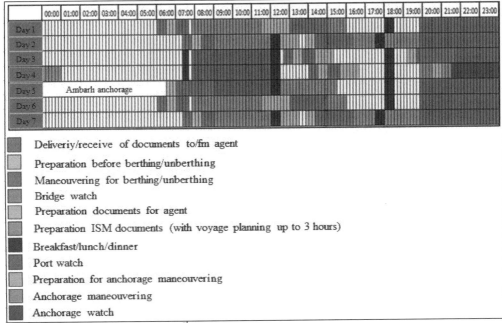

Figure 1. The minimum working hours of 3rd mate

The 3rd officers working on such vessel whom has duties in Figure 1, has approximately 30 minutes for breakfast, lunch or dinner in a day. White areas means the officer's resting times, but the officers use only long resting periods, not short period such as 1 or 2 hours in real life due to next duties. The color codes present the minimum times required for duties. In practical working life, any person may need more time. Therefore, the rest periods can be shorter than expected.

2.2 Establishment of Navigation Scenarios on the Bridge and Observation Criteria

The navigation scenarios have been realized in SINDEL MISTRAL® 4000-Full-mission bridge simulator centre. The voyage area in the bridge simulator, according to the program, has been Sea of Marmara and the voyages are round trips from İzmit Bay's Tütünçiftlik refinery area to İstanbul Ambarlı platform area. Considering the vessels in the scenario are tankers, it has been planned the loading in the refinery in İzmit Tütünçiftlik and unloading in a platform in İstanbul Ambarlı. The route distance between Tütünçiftlik and Ambarlı is 56 nm whereas takes about 3 hours 45 minutes with an average speed of 15 knt. However, the duration of voyages reaches up to 4 hours due to speed reduction points on the route. The general view of route is given under Figure 2.

İzmit Tütünçiftlik platform has been selected as the loading point and the time lapsing in the loading port has been established as about 20 hours

including berthing and departure manoeuvres. A platform in Ambarlı area has been taken as the unloading point and the time lapsing in the unloading point has been established as about 24 hours including berthing and departure manoeuvres. Despite 24 hours pilotage services in the İzmit Tütünçiftlik area, Ambarlı area permit manoeuvring only in daylight and the pilotage services are estimated to be obtained only between the hours 07.00-18.00.

While implementing the scenarios, the operations required to be performed by the watchkeeping officers on the deckhouse have been followed by using control forms. The operations performed by the volunteer, in the role of watchkeeping officer, in certain scenario parts are more significant when compared to the generality of all the operations. These are preparing the vessel to manoeuvre before the departure of the vessel and the period in which the command is delivered to the watchkeeping officer during the navigation.

Under the coverage of 1 week program applied to the volunteers, voyage is performed from Ambarlı to Tütünçiftlik terminal on the 1st day and from Tütünçiftlik terminal to Ambarlı on the 2nd day. During the following course of the program, 6th day scenario, in which the officer is expected to be tired, is made parallel to 1st day and the 7th day scenario in parallel to 2nd day.

The periods in which the watchkeeping officer is monitored/observed during the Ambarlı-Tütünçiftlik route of the vessel are as follows:

- Preparing the bridge before the departure of the vessel
- The internal and external communications during and after the preparations
- The operations required to be performed after the boarding and leaving of Ambarlı pilot
- The passing movements with the vessels during the separation pass
- The passing movements with the vessels from separation exit till Darıca pilot point
- The operations required to be performed after the boarding of Darıca pilot
- The deckhouse routines to be performed all through the navigation
- The communications to be performed all through the navigation

The periods in which the watchkeeping officer is monitored/observed during the Tütünçiftlik-Ambarlı route of the vessel are as follows:
- Preparing the bridge before the departure of the vessel
- The internal and external communications during and after the preparations

- The operations required to be performed after the boarding and leaving of Darıca pilot
- The passing movements with the vessels faced from Yelkenkaya Cape till the separation entrance
- The passing movements with the vessels during the separation pass
- The operations required to be performed after the boarding of Ambarlı pilot
- The deckhouse routines to be performed all through the navigation
- The communications to be performed all through the navigation

The navigation assessment lists are formed to ease the monitoring of the bridge operations and recording the data. Also vessel passing lists are formed for assessing the passes by the target vessels. The Table 3 demonstrates the decided control points and related regulations. Abbreviations associated with the legislation and categories meanings given under the table.

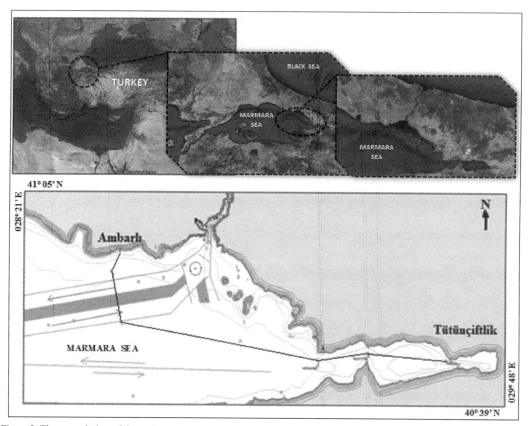

Figure 2. The general view of the route

Table 3. Control points and related regulations

Nr.	Control points	Legislation	Categori
1	Did he put the maps in order?	ISM	P.E.
2	Is the map to be used ready?	ISM	P.E.
3	Is the steering gear adjusted to "control console"?	SOLAS	P.E.
4	Are the steering gear pumps ready?	SOLAS	P.E.
5	Is the steering gear tested?	SOLAS	P.E.
6	Does he manually control the steering gear?	SOLAS	P.E.
7	Are the navigation lights open?	COLREG	P.E.
8	Is the Bravo Flag hung?	COLREG	E.C.
9	Is the whistle tested?	COLREG	P.E.
10	Are the wind panel adjustments performed?	ISM	P.E.
11	Is the "meter" adjustment is performed in Echosounder panel?	ISM	P.E.
12	Is the radar on?	COLREG	P.E.
13	Are the radar "day/night" adjustments performed?	COLREG	P.E.
14	Are the VHFs open?	SOLAS	P.E.
15	Is the MF/HF open?	SOLAS	P.E.
16	Did he turn INM-C to "power" ?	SOLAS	P.E.
17	Did he turn INM-C to "login"?	SOLAS	P.E.
18	Did he prepare pilot card?	ISM	D.P./R.
19	Did he complete the form of navigation with pilot?	ISM	D.P./R.
20	Did he complete the control form before departure?	ISM	D.P./R.
21	Did he ask to the engine room the time required for being ready?	ISM	I.C.
22	Did he call the master and asked for confirmation to call the pilot?	ISM	I.C.
23	Did he call the pilot from VHF Ch 12?	N.L./P.I.	E.C.
24	Did he inform the master after contacting with pilot?	ISM	I.C.
25	Did he inform the engine room?	ISM	I.C.
26	Did he inform the deck?	ISM	I.C.
27	Did he hang the hotel flag?	COLREG	E.C.
28	Did he record the time of "Pilot on Board"?	ISM	D.P./R.
29	Did he present the pilot card to the pilot?	ISM	M./N.
30	Did he record the time of "Disembark of Pilot"?	ISM	D.P./R.
31	Did he pulled down the hotel flag?	COLREG	E.C.
32	Did he inform Sector Marmara from Ch 10 while entering the sector?	N.L./VTS	E.C.
33	Did he inform the pilot before arrival from VHF Ch 12? (1 hour)	N.L./P.I.	E.C.
34	Did he inform Sector Marmara from Ch 10 while leaving the sector?	N.L./VTS	E.C.
35	Did he inform the master after contacting the pilot?	ISM	I.C.
36	Did he inform the master before arriving to the pilot station? (30 m)	ISM	I.C.
37	Did he inform the engine room before arriving the pilot station/separation? (30 m)	ISM	I.C.
38	Did he inform Sector Marmara while entering the separation?	N.L./VTS	E.C.
39	Vessel passes	COLREG	M./N.

ISM	: International Safety Management Code	N.L.	: National Legislation
SOLAS	: International Convention of Safety of Life at Sea	P.I.	: Port Instruction
COLREG	: The International Regulations for Preventing	P.E.	: Preparing Equipment
	Collisions at Sea	E.C.	: External Communication
GMDSS	: Global Maritime Distress and Safety System	D.P./R.	: Document Preparation/Records
VTS	: Vessel Traffic Service	I.C.	: Internal Communications
		M./N.	: Maneuvering/Navigation

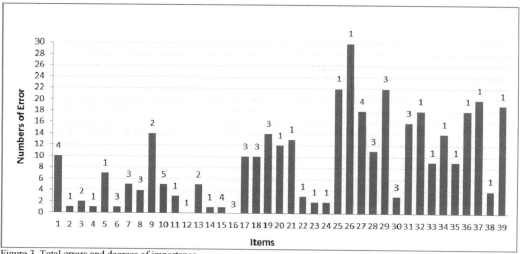

Figure 3. Total errors and degrees of importance

2.3 Create of Liquid Cargo Handling Scenarios

The watchkeeping officers, besides their the bridge navigation watches, are observed under their real-time loading and unloading watches. These watches are performed in tanker simulation laboratory and have been named as port watches. The detailed instructions as formed and presented to the watchkeeping officers for the port watches to explain the operations to be performed by them and to prevent the errors that may originate from lack of experience. These watchkeeping officers are informed in line with the STCW convention requirements, tanker compliance and tanker operations certification. The monitoring of the officer during the port operations is performed and recorded automatically by the TRANSAS® LCHS 4000 OIL (V.1.0) LCC TANKER liquid cargo handling program being used in tanker simulator. Thus, the detection of the errors happens to be possible.

3 FINDINGS AND DISCUSSIONS

3.1 Navigation Scenarios

The weekly total working and resting hours of the volunteers, during the conducted study, have been in Table 4. The total resting periods of the volunteers, after performing their minimum duties in the Table 2, have varied between 57 hours 30 minutes to 63 hours 30 minutes. The working hours in turn have changed about between 95 hours and 102 hours. When considered the legal regulations given under Table 1, it may be seen that the legal 72 hours working limit is exceeded and the resting hours are left under the 77 hours which is the legal minimum limit. Thus, under the light of aforementioned facts,

it may be said that a watchkeeping officer of tanker at short sea shipping is subject to over activity and accordingly under the effect of the fatigue.

Table 4. Weekly total resting and working periods

Volunteers	Total Resting Hours	Total Working Hours
1	61 h 00 m	98 h 50 m
2	61 h 25 m	98 h 10 m
3	61 h 10 m	95 h 30 m
4	63 h 30 m	94 h 40 m
5	63 h 10 m	96 h 30 m
6	57 h 30 m	101 h 50 m
7	57 h 50 m	100 h 00 m

The errors performed by the volunteers subject to the working hours and resting periods in Table 4 have been provided under Figure 3 by means of control points and significance priorities. The degree of significance for the navigation safety has been shown on top of the graphic columns that include the error figures of each point. The significance degrees are listed from 1 to 5 (1 as the most important and 5 as the least important).

According to the Figure 3; the volunteers have made errors most frequently, during their navigation duties under the program, under the titles internal and external communications and passing manoeuvres with the other vessels. The distribution of the total errors by means of their significance degrees has been provided under Table 5.

Table 5. Error figures over the significance degrees

Significance Degrees	Total Errors	Percentage (%)
1	208	58,0
2	21	5,8
3	96	26,7
4	29	8,1
5	5	1,4
TOTAL	359	100

267

The errors seen in Table 5 have occurred, under the one week working program, through the navigation watches of the watchkeeping officers subject to over activity (4 navigation watches for each volunteer); it is a serious situation that 58% of the total error points has been of 1st degree importance group. When considered that the errors under 1st degree importance are navigational anoeuvres and vessel passages, it is understood that a watchkeeping officer subject to over activity, and tired accordingly, may make errors in very important operations that may endanger the safety of the vessel.

When the control points are assessed against the legal regulations, it can be seen that, among the total 359 errors made by the volunteer watchkeeping officers, 58% have been under ISM coverage, 23% under COLREG coverage, 13% under local regulations and 6% under SOLAS coverage (Figure 4).

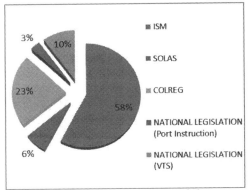

Figure 4. Distribution of the errors over regulations

When the control points are assessed against the classifications under Table 3, it can be seen that, among the total 359 errors, 33% relates to internal communication, 24% relates to external communication, 18% relates to equipment preparation, 14% relates to document preparation and records, and 11% relates to manoeuvre and navigation (Figure 5).

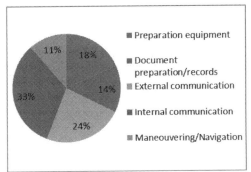

Figure 5. Distribution of Error Classes

3.2 Liquid Cargo Handling Scenarios

When the system records related to the load operations performed by the volunteer watchkeeping officers under port watches, as in tanker simulation laboratory, are examined, it may be seen that the volunteers have reduced the valves, stopped the pump before decreasing the speed of the cargo pump, opened mistaken valves, forgotten some valves open after the operation, closed the load entry valves while the terminal continues to function, closed the entry valve of the tank while the ballast pump is in operation during the ballast intake, opened mistaken manifolds, functioned the mistaken pumps, performed unnecessary valve opening/closing transactions, performed unloading with double pumps contrary to the operational instructions, performed ballast operations without a command, tried to close the already closed valves, re-instructed without being aware of the already functioning terminal, started the cargo pump before the discharging line is ready and not complied with the criteria related to the order of operations given under operational instructions.

Valve reduction causes the increase of pressure in the line and damage on the same. It is a known error also general in practice. It may also be named as a breach since performed against the rules. Stopping the pump without decreasing the pump speed may damage the respective pump. The functioning of the discharging pumps of the vessel before the lines are ready will mean damage on the pump. A valve not being opened in the line means the line is not ready. Forgetting the valves open after the operation then shall result with uncertainty about the position (ready or not) of the line in a subsequent operation. Opening the mistaken manifold valve or establishing a mistaken line connection is a serious situation which may result with discharging the liquid loads to an environment other than the line or to the sea. In the same manner, closing the respective loading valves while the terminal pumps continue to load into the vessel may cause explosion of the manifold or line as well as damage on the terminal pump and environmental pollution.

4 CONCLUSIONS

The fast transport of the products from the producer to the consumer have gained importance together with the technological advances. For the fast transport of commodities, fast loading, fast unloading and port durations have been reduced and the navigation speeds of the vessels have increased. Thus, the working periods and intensity of the crew have increased in parallel with the increased maritime operations. Many actions have been taken together with such increased working load as well as many accepted rules, however, this could not prevent the human errors to play an important role in the occurred accidents. Establishing the effects of the fatigue on the risk of errors by the seafarers will be important by means of prevention of marine accidents.

It is a known fact that the intensity of the operations and works in the tankers at short sea shipping cause the increase of the working loads of the watchkeeping officers assigned in such vessels. This study has established the errors realized by watchkeeping officers under fatigue by making use of bridge and tanker simulators. Considering the results of this study, it is thought a linear relationship exists between the fatigue and error making tendency. Compliance with the resting periods provided to the watchkeeping officers by the international and national regulations has an utmost importance. Not complying such periods or dividing such periods into parts shall mean not providing the required resting periods to the officers. Thus, it requires uninterrupted resting periods. It can be said that working in breach of rules increase the tendency of the officer to make error.

Working time regulations must be complied with to the maximum. To comply with the limitations of working hours, it is necessary to increase the number of officers responsible for operations such vessels.

REFERENCE

AIBF, 2007. Accident Investigation Board of Finland, Annual Report 2006, Helsinki, Finland (http://www.emsa.europa.eu/annual-reports.html)

Arslan, Ö., and Er, İ. D. (2007). Effects of Fatigue on Navigation Officers and SWOT Analyze for Reducing Fatigue Related Human Errors on Board. International Journal on Marine Navigation and Safety of Sea Transportation, 1(3), 345-349.

BEAmer, 2004. Rapport d'activité 2003 Second semestre & 2004 Année (in French) http://www.general-files.com/download/source/ gs58083370h32i0

BEAmer, 2005. Rapport d'activité 2005 (in French) http://www.general-files.com/download/source/ gs58083370h32i0

BEAmer, 2006. Rapport d'activité 2006 (in French) http://www.general-files.com/download/source/ gs58083370h32i0

BEAmer, 2007. Rapport d'activité 2007 (in French) http://www.general-files.com/download/source/ gs58083370h32i0

BEAmer, 2008. Rapport d'activité 2008 (in French) http://www.general-files.com/download/source/ gs58083370h32i0

BEAmer, 2009. Rapport d'activité 2009 (in French) http://www.general-files.com/download/source/ gs58083370h32i0

BEAmer, 2010. Rapport d'activité 2010 (in French) http://www.general-files.com/download/source/ gs58083370h32i0

BEAmer, 2011. Rapport d'activité 2011 (in French) http://www.general-files.com/download/source/ gs58083370h32i0

BP, 2010. Deepwater Horizon Accident Investigation Report http://www.bp.com/liveassets/bp_internet/globalbp/globalb p_uk_english/incident_response/STAGING/local_assets/do wnloads_pdfs/Deepwater_Horizon_Accident_Investigation _Report.pdf

BSU, 2011. 2011 Annual Report http://www.bsu-bund.de/SharedDocs/pdf/EN/ Annual_Statistics/annual_ statistics_2011.pdf?__blob=publicationFile

Chauvin, C. (2011). Human Factors and Maritime Safety. Journal of Navigation, 64(04), 625-632

CIAIM, 2010. Annual Report 2009 http://www.emsa.europa.eu/annual-reports.html

CIAIM, 2011. Annual Report 2010 http://www.emsa.europa.eu/annual-reports.html

Çakmakçı, M., 2001. Medical Errors: Definitions and Importance of the Topic, ANKEM Journal, 15, 3, 247-249 (in Turkish)

DIMA, 2009. Summary of Marine Accidents and Incidents in 2009 http://www.emsa.europa.eu/annual-reports.html

DMA, 2008. Marine Accidents 2007, Copenhagen, Denmark http://www.dma.dk/SiteCollectionDocuments/Publikationer /Maritime-accidents/Accidents-at-Sea-2007.pdf

DMA, 2009. Marine Accidents 2009, Copenhagen, Denmark http://www.dma.dk/SiteCollectionDocuments/Publikationer /Maritime-accidents/Accidents%20at% 20Sea%202009.pdf

Ece, J.N., 2008. Maritime Accidents in History and Countermeasures (in Turkish) http://www.denizhaber.com/ index.php?sayfa=yazar&id=11&yazi_id=100278

European Union, 1999. Council Directive 1999/63/EC of 21 June 1999 concerning the Agreement on the organisation of working time of seafarers concluded by the European Community Shipowners' Association (ECSA) and the Federation of Transport Workers' Unions in the European Union (FST), Official Journal of the European Union, L 167/33-37

European Union, 2003. Directive 2003/88/EC of The European Parliament and of The Council of 4 November 2003 concerning certain aspects of the organisation of working time, Official Journal of the European Union, L 299/9-18

Greenpeace, 2010. BP Deepwater Oil Disaster (in Turkish) http://www.greenpeace.org/turkey/Global/turkey/report/201 0/5/bp-deepwater-petrol-felaketi.pdf

Hoch, M., 2010. New Estimate Puts Gulf Oil Leak at 205 Million Gallons http://www.pbs.org/newshour/rundown/ 2010/08/new-estimate-puts-oil-leak-at-49-million-barrels.html

IMO, 2001. Guidance on Fatigue Mitigation and Management MSC/Circ.1014, London, http://www.imo.org/OurWork/HumanElement/VisionPrinci plesGoals/Documents/1014.pdf

IMO, 2004. Resolution A.947(23), Human Element Vision, Principles and Goals for The Organization http://www.imo.org/blast/blastData.asp?doc_id=12252&fil ename=A%20947(23).pdf

IMO, 2011. International Convention on Standards of Training, Certification and Watchkeeping for Seafarers STCW Including 2010 Manila Amendments, IMO Publications, London.

MAIB, 2012. Annual Report 2011, Southampton, UK http://www.maib.gov.uk/cms_resources.cfm?file=/MAIB_ Annual_Report_2011.pdf

MCIB, 2011. Annual Report 2010 http://www.mcib.ie/_domain/media/file//PDF/MCIB%2020 10%20Annual%20Report%20(English).pdf

Official Journal, 2002. Seafarers Regulation, Turkey Prime Minister Publications (No. 24832).

Reason, J., 1990. Human Error, Cambridge University Press, UK, 302 s

Rothblum, A.M., 2000. Human Error and Marine Safety, In Proceeding of the Maritime Human Factors Conference 1-10, College Park, Maryland. http://www.bowles-langley.com/wp-content/files_mf/humanerrorandmarinesafety26.pdf

Shappell, S.A. and Wiegmann, D.A., 2000. The Human Factors Analysis and Classification System-HFACS, 19 s, U.S. Department of Transportation, Office of Aviation Medicine, Report No DOT/FAA/AM-00/7, Washington D.C.

Shea I.P. and Grady N., (1998). Shipboard Organisational Culture in the Merchant Marine Industry, Proceedings of the Safe Navigation Beyond, Gdynia.

SMA, 2007. Notice of the Swedish Maritime Administration, No. 1 2007, Summary of Reported Marine Casualties, Near Accidents and Accidents to Persons – Swedish Merchant and Fishing Vessels 2006 http://www.emsa.europa.eu/annual-reports.html

Smith, A., Allen, P. and Wadsworth, E., 2006. Seafarers' Fatigue: the Cardiff Research Programme, Centre for Occupational and Health Psychology Cardiff University, Cardiff, 87 s

Uğurlu, Ö., Köse, E., Başar, E. and Yüksekyıldız, E., 2009. Interview Study on Working Hours of Deck Officers on a Product Tanker.

Uğurlu, Ö., Köse, E., Başar, E., Yüksekyıldız, E. and Yıldırım, U., 2012. Investigation of Working Hours of Watchkeeping Officers on Short Sea Shipping: A Case Study in an Oil Tanker. The 2012 International Association of Maritime Economists Conference, September 2012, Taipei

Enhancing of Carriers' Liabilities in the Rotterdam Rules – Too Expensive Costs for Navigational Safety?

P. Sooksripaisarnkit
School of Law, City University of Hong Kong

ABSTRACT: The United Nations Convention on Contracts for the International Carriage of Goods Wholly or Partly by Sea (the 'Rotterdam Rules') was adopted by the General Assembly of the United Nations on 11 December 2008. The Rotterdam Rules contain two oft-criticised changes from the existing regime governing international carriage of goods widely adopted among maritime nations, namely the International Convention for the Unification of Certain Rules Relating to Bills of Lading, Brussels, 25 August 1924 (the 'Hague Rules') and the subsequent Protocol in 1968 (the 'Visby Protocol' or the 'Hague-Visby Rules'). These changes are (1) an extension of the carrier's obligations to maintain seaworthy vessel through the voyage (Article 14) and a deletion of an exclusion of carrier's liabilities due to negligent navigation (Article 17). This paper addresses implications of these changes and assess whether ship-owners or ship-operators can comply with these without having to incur excessive additional expenses.

1 INTRODUCTION

The 'United Nations Convention on Contracts for the International Carriage of Goods Wholly or Partly by Sea' (the 'Rotterdam Rules') was adopted by the General Assembly of the United Nations on 11 December 2008. The Signing Ceremony was held on 23 September 2009, with immediate positive response from various nations who have become the Signatories. At the time of writing, Spain and Togo have ratified.[15] It will come into force within a year after ratification by the twentieth nation.[16] It marked an end to a long journey of hard work – started in 1990s as a joint project of the United Nations Commission on the International Trade Law (UNCITRAL) and the Comite Maritime International (CMI).[17] The world waits for the day it comes into force with excitement since throughout the entire history of international legal regimes governing international transportation of goods by sea, a uniformity of international rules has never been achieved. A million dollar question is whether such a day will *ever* come.

Among the current imperfect state of uniformity, the International Convention for the Unification of Certain Rules Relating to the Bills of Lading, Brussels, 25 August 1924 (the 'Hague Rules') along with its subsequent Protocol of 1968 (the 'Visby Protocol' or the 'Hague-Visby Rules') has received warmest welcome among maritime nations, including Norway, United Kingdom, Poland, Singapore, and Hong Kong SAR.[18] The Hague-Visby Rules were negotiated and signed, soon after the end of the Second World War. Several new nations had come into existence by then. They had not participated in the process leading up to the formulation of the Hague Rules. They strived for competitiveness and they viewed principles as enshrined in the Hague Rules to be in favour of already developed countries with large maritime trade – which they found to be unacceptable.[19] The Hague-Visby Rules have been perceived as 'ship-owing friendly' for they allow ship-owners to negate

---

[15] See United Nations Treaties Collection <http://treaties.un.org/pages/ViewDetails.aspx?src=TREATY&mtds gno=XI-D-8&chapter=11&lang=en> accessed 13 December 2012.
[16] Article 94.
[17] Kirval L. 2012 European Union's Stance on the Rotterdam Rules *TransNav International Journal on Marine Navigation and Safety at Sea Transportation* 6:555-562.

[18] See Comite Maritime International, 'CMI Yearbook 2010' <http://www.comitemaritime.org/Uploads/Yearbooks/Yearbook%20 2010.pdf> accessed 13 December 2012.
[19] Frederick David C. 1991 Political Participation and Legal Reform in the International Maritime Rulemaking Process: From the Hague Rules to the Hamburg Rules *Journal of Maritime Law and Commerce* 22: 81-117.

their liabilities due to negligence of their servants or their agents in navigation of ships (Article IV(2)(a)).[20] Such exception no longer exists in the Rotterdam Rules (Article 17). At the same time, in contrast to the ship-owners' obligations in the Hague-Visby Rules to maintain seaworthy ship 'at the beginning of the voyage (Article III(1)), the Rotterdam Rules have extended this obligation throughout the entire voyage (Article 14). The purpose of this article is to examine these two significant changes and evaluate whether these are feasible for ship-owning interests to comply and whether such changes should be welcomed by ship-owning interest nations. To achieve this aim, this article will be divided into three parts. In the first part, the author will explore the ambit of the shipowners' obligations to maintain seaworthy vessel under the Hague-Visby Rules and also the ambit of the exception of negligent navigation. Afterwards, in the second part, the author would trace the rationale for changes made in the Rotterdam Rules along with offering his analysis on practical implications of such changes. The final part will be concluded by evaluating the feasibility of such changes in light of modern maritime practices.

2 HAGUE-VISBY RULES: SHIPOWNERS' OBLIGATIONS TO MAINTAIN SEAWORTHY VESEL AND NEGLIGENT NAVIGATION EXCEPTION

As mentioned above, two significant aspects of the Hague-Visby Rules will be briefly outlined in this part: shipowners' obligations to maintain seaworthy vessel and negligent navigation exception available to shipowners.

2.1 Shipowners' obligations to maintain seaworthy vessel

To begin with, Article III(1) of the Hague-Visby Rules should be re-cited here in full:

The carrier shall be bound before and at the beginning of the voyage to exercise due diligence to:
1 Make the ship seaworthy.
2 Properly man, equip and supply the ship.
3 Make the holds, refrigerating and cool chambers, and all other parts of the ship in which goods are carried, fit and safe for their reception, carriage and preservation.

Two points should be emphasised here. First, the duration of the obligations is only 'before and at the beginning of the voyage'. Secondly, the standard of such obligations is rested upon 'due diligence'. The authority in point as to the duration of the duty is the decision of the Privy Council in *Maxine Footwear*

Co. Ltd v Canadian Government Merchant Marine Ltd[21] In this case, cargoes had already been loaded when the ship's pipes were found to be blocked by ice. The Master ordered the use of an acetylene torch to thaw out the ice. Due to negligence of the ship's officers, the ship caught fire causing the loss of cargoes.[22] One of the issues in this case was whether the duty to provide a seaworthy vessel (or to be more precise the duty under Article III(1)(c) of the Hague Rules applicable at the time) ended upon the goods loaded onto the ship. The Privy Council explained the phrase 'before and at the beginning of the voyage' to mean 'the period from at least the beginning of the loading until the vessel starts on her voyage'.[23] As for the standard of due diligence, this is explained as equal to the duty to take reasonable care in common law.[24] The modern authority which demonstrates this standard is in the case of *The Eurasian Dream*.[25] The case involved a car carrier which was destroyed by fire. The claimants, whose cargoes were destroyed, alleged unseaworthiness. Citing *The Amtelslot*,[26] Cresswell J. explained that '[t]he exercise of due diligence is equivalent to the exercise of reasonable care and skill: "Lack of due diligence is negligence; and what is in issue in this case is whether there was an error of judgment that amounted to professional negligence"'.[27] Such obligations to exercise due diligence to provide a seaworthy vessel are said to be 'non-delegable' obligations. This is a well-established legal position since the decision of the House of Lords in 1961 in *The Muncaster Castle*.[28] The case involved damages to cargoes due to the sea water entered the cargo hold. It was discovered that the inspection cover was not properly tightened due to the negligence of the fitter employed by the independent contractor.[29] The House of Lords, taking into account the history of the Hague Rules and the need to maintain uniformity in the interpretation of international conventions, unanimously held that the ship-owner in this case failed to exercise due diligence. In the passage of Lord Radcliffe, '[w]hat is stressed throughout is that the obligation of the carrier is "not limited to his personal diligence"…The carrier's responsibility for the diligence of those whom he employs to

---

[20] See Baughen, Simon, 2012. *Shipping Law*. Oxford: Routledge.

[21] *Maxine Footwear Co. Ltd v Canadian Government Merchant Marine Ltd.* [1999] A.C. 589.
[22] Ibid., 593.
[23] Ibid., 603.
[24] Girvin, Stephen, 2007. *Carriage of Goods by Sea*, Oxford: Oxford University Press, para 26.20.
[25] *Papera Traders Co Ltd and Others v Hyundai Merchant Marine Co. Ltd and Another (The "Eurasian Dream")* [2002] EWHC 118 (Comm); [2002] 1 Lloyd's Rep. 719.
[26] *Union of India v N.V. Reederij Amsterdam* [1963] 2 Lloyd's Rep. 223.
[27] *The Eurasian Dream*, para 131.
[28] *Riverstone Meat Company, Pty., Ltd v Lancashire Shipping Company Ltd* [1961] 1 Lloyd's Rep. 57.
[29] Ibid., 65-66.

discharge his own primary duty has been stated and recognized…'[30]

Such entire scheme of the shipowners' obligations to provide a seaworthy vessel has been well-understood and consistently applied for over eighty years. One would not doubt a prime significance of such obligations for the language of the Hague-Visby Rules makes it clear that shipowners can only rely on a list of exceptions provided to them in Article IV(2) if they fulfilled their duty under Article III(1). This is apparent from the language of Article IV(1): '…Where loss or damage has resulted from unseaworthiness the burden of proving the exercise of due diligence shall be on the carrier or other person claiming exemption under this article'.

Article IV(2) of the Hague-Visby Rules contains the list of seventeen grounds for shipowners to seek exemption from their liabilities. The widest ground is provided in Article IV(2)(q) which is dubbed by academic commentators as a 'catch-all exception', gives exemption for 'any other cause arising without the actual fault or privity of the carrier, or without the fault or neglect of the agents or servants of the carrier…' The Rotterdam Rules, whilst do not spell out this exception so explicitly, retain this 'catch-all exception' as in the language of Article 17(2): 'The carrier is relieved of all or part of its liability…if it proves that the cause or one of the causes of the loss, damage, or delay is not attributable to its fault or to the fault of any person referred to in Article 18'. Indeed, as can be seen from Article 17(3), the Rotterdam Rules retain list of exceptions in the Hague-Visby Rules, except one – exception as per Article IV(2)(a) – mentioned earlier. This particular exception reads: '(2) Neither the carrier nor the ship shall be responsible for loss or damage arising or resulting from – (a) Act, neglect, or default of the master, mariner, pilot, or the servants of the carrier in the navigation or in the management of the ship'. To what extent shipowners are negatively affected by such a change brought about by the Rotterdam Rules? It is this question which the paper now turns to address.

2.2 *Negligent navigation exception*

The negligent navigation exception has come to a sharp focus again following a fairly recent decision of the Supreme Court of New Zealand in *The Tasman Pioneer*.[31] An extreme situation in this case indeed leads to some doubts whether debates on the need for this negligent navigation exception which have persisted since the negotiation leading to the

drafting of a less successful piece of international convention on transportation of goods by sea, namely the United Nations Convention on the Carriage of Goods by Seas, 1978 (the "Hamburg Rules") should actually be revisited. As a re-collection, the Hamburg Rules took a cargo-owning orientated approach and there is no similar list of exceptions as provided in the Hague-Visby Rules. Instead, shipowners are presumed to be liable for loss or damage to cargoes.[32]

In *The Tasman Pioneer*, the Master chose to navigate the vessel via the shorter route in order to make the voyage on schedule. Unfortunately, the vessel hit the rocks causing damage to the hull of the vessel letting in seawater. Instead of reporting incidents to the shipowner and the relevant authorities for necessary measures to be arranged on time, including salvage, the Master proceeded with the voyage. He tried to change the route back to the original longer one the vessel should have taken. He also instructed the alteration of navigational chart and asked his colleagues to lie as to the cause of the damage.[33] Due to the delay in taking appropriate measures, the plaintiffs' cargoes were damaged by seawater. Nevertheless, the Supreme Court of New Zealand held in this case the shipowner is entitled to invoke a defence under Article IV(2)(a) of the Hague-Visby Rules.[34]

The only qualification to limit the applicability of the aforementioned Article IV(2)(a), as explained further in *The Tasman Pioneer*, is when there was a 'barratry',[35] drawing reference from Article IV(5)(e) depriving carriers' servant or agent from raising a defence or entitlement to limit liability again if the losses or damages were 'resulted from an act or omission of the servant or agent done with intent to cause damage…' In this respect, to prove barratry, it must be established that 'damage had resulted from an act or omission of the master or crew done with intent to cause damage, or recklessly and with knowledge that damage would probably result'.[36] This qualification, plausible as it is, may attract certain difficulties in practice. The burden of proof falls upon consignees who hold constructive possession to goods by way of a transfer of a bill of lading.[37] These persons did not involve in the

---

[30] Ibid., 84.

[31] *Tasman Orient Line CV v New Zealand China Clays and Others* (*The "Tasman Pioneer"*) [2010] NZSC 37; [2010] 2 Lloyd's Rep. 13.

[32] Article 5(1) of the Hamburg Rules provides: 'The carrier is liable for loss resulting from loss of or damage to the goods, as well as from delay in delivery, if the occurrence which caused the loss, damage or delay took place while the goods were in his charge as defined in article 4, unless the carrier proves that he, his servants or agents took all measures that could reasonably be required to avoid the occurrence and its consequences..

[33] *The Tasman Pioneer* (n 17) [1]-[4].

[34] Ibid., [31]-[32].

[35] Ibid., [10]-[12].

[36] Ibid., [13].

[37] Section 2(1) of the Carriage of Goods by Sea Act 1992: '(1) Subject to the following provisions of this section, a person who becomes – (a) the lawful holder of the bill of lading…shall (by

voyage and did not have any means of access to communications between shipowners and crewmen. Unless in rare cases of clear circumstances, it is submitted that consignees are likely to fail in their burden of proof and shipowners would be able to resort to their exception under Article IV(2)(a).

3 SHIPOWNERS' OBLIGATIONS TO MAINTAIN SEAWORTHY VESSEL AND NEGLIGENT NAVIGATION EXCEPTION: CHANGES MADE BY THE ROTTERDAM RULES

Unlike the Hague-Visby Rules, the Rotterdam Rules take into account reality of modern transportation of goods. This can be seen from a definition of a 'contract of carriage' in Article 1(1): '…a contract in which a carrier, against the payment of freight, undertakes to carry goods from one place to another. The contract shall provide for carriage by sea and may provide for carriage by other modes of transport in addition to sea carriage'. In a sense, the Rotterdam Rules acknowledge increasing use of multi-modal transportations. Specifically in relation to the carriage by sea, as mentioned earlier, the Rotterdam Rules extend the shipowners' obligations in this respect throughout the voyage. The opening language of Article 14 reads: 'The carrier is bound before, at the beginning of, and during the voyage by sea to exercise due diligence…' This change necessarily has a bearing on the allocation of risks in sea carriage between shipowners on the one hand and cargo-owners (shippers or consignees of the bill of lading) on the other hand. In other words, there is an alteration in the mode of determining 'who pays for damage done to cargo during the movement of the goods by [shipping] industry…'[38] This, however, may not have any impacts on shipowners' current practices. As Nikaki explains, the rationale of the Rotterdam Rules in this regard is nothing more than to align shipowners' liabilities with those responsibilities they currently have under the public law.[39] A reference to the public law here is a reference to the 'International Management Code for the Safe Operation of Ships and for Pollution Prevention' (The "ISM Code"),[40] in particular Articles 6 and 10. These articles read:

4 RESOURCES AND PERSONNEL

1 The Company should ensure that the master is:
   – properly qualified for command;
   – fully conversant with the Company's safety management system; and
   – given the necessary support so that the master's duties can be safely performed.

2 The Company shall ensure that each ship is manned with qualified, certificated, and medically fit seafarers in accordance with national and international requirements.

3 The Company should establish procedures to ensure that new personnel and personnel transferred to new assignments related to safety and protection of the environment are given proper familiarization with their duties. Instructions which are essential to be provided prior to sailing should be identified, documented and given.

4 The Company should ensure that all personnel involved in the Company's safety management system have an adequate understanding of relevant rules, regulations, codes and guidelines.

5 The Company should establish and maintain procedures for identifying any training which may be required in support of the safety management system and ensure that such training is provided for all personnel concerned.

6 The Company should establish procedures by which the ship's personnel receive relevant information on the safety management system in a working language or languages understood by them.

7 The Company should ensure that the ship's personnel are able to communicate effectively in the execution of their duties related to the safety management system.

5 MAINTENANCE OF THE SHIP AND EQUIPMENT

1 The Company should establish procedures to ensure that the ship is maintained in conformity with the provisions of the relevant rules and regulations and with any additional requirements which may be established by the Company.

2 In meeting these requirements, the Company should ensure that:
   – inspections are held at appropriate intervals;
   – any non-conformity is reported, with its possible cause, if known;
   – appropriate corrective action is taken; and
   – records of these activities are maintained.

---

virtue of becoming the holder of the bill or, as the case may be, the person to whom delivery is to be made) have transferred to and vested in him all rights of suit under the contract of carriage as if he had been a party to that contract'.,

[38] Sweeney, Joseph C. 1991 UNCITRAL and The Hamburg Rules – The Risk Allocation Problem in Maritime Transport of Goods *Journal of Maritime Law and Commerce* 22: 511-538.

[39] Nikaki Theodora. 2010 The Carrier's Duties under the Rotterdam Rules: Better the Devil You Know? *Tulane Maritime Law Journal* 35: 1-44.

[40] The ISM Code 'was adopted by the International Maritime Organisation (IMO) on November 4, 1993, and later incorporated

into the International Convention for the Safety of Life at Sea Convention (SOLAS) 1974'. Ibid.

3 The Company should identify equipment and technical systems the sudden operational failure of which may result in hazardous situations. The safety management system should provide for specific measures aimed at promoting the reliability of such equipment or systems. These measures should include the regular testing of stand-by arrangements and equipment or technical systems that are not in continuous use.

4 The inspections mentioned in 10.2 as well as the measures referred to in 10.3 should be integrated into the ship's operational maintenance routine.

It must be noted that the ISM Code provides a set of modern regulations concerning ship safety. It came out amidst modern shipping practices. This is in contrast with the Hague-Visby Rules which came out just after the Second World War. This is not to mention that the rules on seaworthiness of ship did not change from that of the Hague Rules which came out in 1924. There was a rationale back then to limit obligations to provide seaworthy vessel to the period at the beginning of the voyage considering difficulties in communications and controls between ships and shores. Such a scene has completely changed. In modern shipping environment, with developments of communications and satellite systems, ship staffs and shipowners along with their personnels on shores are in frequent communication. Shipowners have access to a network of shipping agents worldwide. This necessarily means that any defects on any parts of the vessel occurred during the voyage which may compromise the safety and seaworthiness of the vessel can be corrected at the nearest port. Repairs of either permanent or temporary nature can be expediently arranged. Specific training institutions are available to train up ship staffs. A decade of the existence of the ISM Code means that shipowners are familiar with the requirement under the ISM system and have implemented and embraced this into their firms' culture. Taking all these into account, the author is inclined to agree with Nikaki that the Rotterdam Rules did not impose any additional demands upon shipowners. Shipowners are unlikely to incur additional costs to make themselves in compliance with their obligations to provide seaworthy vessel under the Rotterdam Rules. Compliance with the ISM Code renders a favourable presumption on the part of shipowners that due diligence was exercised. In contrast, a failure to comply with the ISM Code would provide a pistol for cargo interests to point that there was a lack of due diligence.[41]

The actual increasing costs for shipowners, if the Rotterdam Rules are in force, come from a new liability of 'negligent navigation', which they would not have met under the Hague-Visby Rules. This causes difficulty for shipowners in practice as even the most competent crewman can act negligently.[42] Shipowners are not always on board to control decisions made by ship staffs. '[T]he removal of the nautical fault defense would result in less efficient and less cost-effective management of the risks and division of the financial consequences between P & I Clubs and cargo insurers'.[43] Shipowners may face large claims from cargo-owners in the event of serious damages to goods due to large incidents such as a collision caused by negligent navigation.[44] However, this event appears to disregard balancing machanisms for shipowners to be able to limit their liability under either the Rotterdam Rules (Articles 59-60) or the Convention on Limitation of Liability for Maritime Claims 1976. Plus, it is questionable why shipowners should benefit from the privilege granted under the 'negligent navigation' exception when same is not available to carriers in other modes of transportation.[45] It may be the case that, in order to balance and cope with expected increase in liabilities, shipowners will not hesitate to increase freight. However, shipping industry has changed its scene from the time before the coming into force of the Hague Rules. At the period prior to the drafting of the Hague Rules, shipowners had a relatively stronger bargaining power. They had a practice of inserting exclusion clauses into bills of lading exempting their own liability for negligence. Courts in different jurisdictions reacted differently to such a contractual provision. In the United Kingdom where ship-owning interests were influential, the validity of such clause was upheld.[46] The fact is that shipowners no longer hold strong bargaining power. This is especially the case since the global economic downturn in late 2008 which has caused shipowners to fight for trade. Shipowners who unscrupulously increase the freight will be sanctioned by industrial mechanism. Protection and Indemnity (P & I) clubs, representing ship-owning interests, also need to bear this in mind. Any unnecessary increase in calls will face an objection by ship-owning representatives sitting on the Board of Directors of the P & I club in question. At the same time, P & I clubs also have their own mechanism in calculating their calls, taking into account also claim records of relevant shipowners. In order to minimise claim records, it is necessary for shipowners to tighten up their safety and security measures. This can be done, for

[41] Girvin, para 23.17.

[42] Bachxevanis, Konstantinos. 2010 'Crew negligence' and 'crew incompetence': their distinction and its consequence *Journal of International Maritime Law* 16: 102-131.

[43] Weitz, Leslie Tomasello. 1997-1998 The Nautical Fault Debate (the Hamburg Rules, the U.S. COGSA 95, the STCW 95, and the ISM Code *Tulane Maritime Law Journal* 22: 581-594.

[44] Ibid.

[45] Tetley, William. 2008 *Marine Cargo Claims Volume 1*, Quebec: Les Editions Yvon Blais Inc, 954.

[46] Yiannopoulos, Athanassios N. 1957-1958 Conflicts Problems in International Bill of Lading: Validity of Negligence Clauses *Louisiana Law Review* 18: 608-627.

example, by appointing owners' representatives regularly joining the vessels to provide ship staffs with guidance, especially when significant decisions in relation to navigation have to be made. Placing many procedure manuals on board for ship staffs to comply would no longer be efficient. Regular drillings should be conducted on shore and on board for ship staffs and shore staffs so they are familiar with decision making and they can co-ordinate with each other in emergency situations Promotion of ship staffs into significant ranks such as the Master needs to be carefully considered, taking into account experiences of relevant ship staffs. Instead of raising unnecessary concerns on the unavailability of negligent navigation exception, it is submitted here that shipowners should turn their focus on risk management aspects of their business. They should focus their concerns on how to avoid or reduce liabilities for negligent navigation.

6 CONCLUSION

Superficially, the Rotterdam Rules appear to increase shipowners' liabilities beyond those in the Hague-Visby Rules and of course one concern would be whether the Rotterdam Rules will gain support from shipping interests or ship-owning countries. However, upon closer scrutiny, the Rotterdam Rules do not seem to increase shipowners' obligations beyond acceptable limits. Obligations to provide a seaworthy vessel throughout the voyage are merely reflections of shipowners' current practices under the ISM Code. The deletion of shipowners' exception for negligent navigation is just a mere adjustment for a right balance between ship-owning interests and cargo-owning interests and an attempt to put liabilities of shipowners in line with liabilities of operators of other modes of transportation. It will not be unreasonable if any countries will change their law relating to international carriage of goods along this line, abandoning the usual approach as in the Hague-Visby Rules. However, whether the Rotterdam Rules should be adopted in full, giving a complex structure of 92 provisions in total, should be left for future discussion, which is beyond the scope of this article.

REFERENCES

[1] Bachxevanis, Konstantinos. 2010 'Crew negligence' and 'crew incompetence': their distinction and its consequence *Journal of International Maritime Law* 16: 102-131.
[2] Baughen, Simon, 2012. *Shipping Law* Oxford: Routledge.
[3] Comite Maritime International, 'CMI Yearbook 2010' http://www.comitemaritime.org/Upload/Yearbook%2010.pdf
[4] Frederick David C. 1991 Political Participation and Legal Reform in the International Maritime Rulemaking Process: From the Hague Rules to the Hamburg Rules *Journal of Maritime Law and Commerce* 22:81-117.
[5] Girven, Stephen, 2007. *Carriage of Goods by Sea*, Oxfod: Oxford University Press.
[6] Kirval, L. 2012 European Union's stance on the Rotterdam Rules *TransNav International Journal on Marine Navigation and Safety at Sea Transportation* 6: 555-562
[7] Nikaki, Theodora. 2010 The Carrier's Duties Under the Rotterdam Rules: Better the Devil You Know? *Tulane Maritime Law Journal* 35: 1-44.
[8] Sweeney, Joseph C. 1991 UNCITRAL and The Hamburg Rules – The Risk Allocation Problem in Maritime Transport of Goods *Journal of Maritime Law and Commerce* 22:511-538.
[9] Tetley, William, 2008 *Marine Cargo Claims Volume 1*, Quebec: Les Edition Yvon Blais Inc.
[10] United Nations Treaties Collection http://treaties.un.org/pages/ViewDetails.aspx?src=TREATY&mtdsg_no=X1-D-8&chapter=11&lang=en
[11] Weitz, Leslie Tomasello. 1997-1998 The Nautical Fault Debate (the Hamburg Rules, the U.S. COGSA 95, the STCW 95, and the ISM Code) *Tulane Maritime Law Journal* 22: 581-594.
[12] Yiannopoulos, Athanassios N. 1957-1958 Conflicts Problems in International Bill of Lading: Validity of Negligence Clauses *Louisiana Law Review* 18: 609-627.

CASES

[1] Maxine Footwear Co. Ltd and Another v Canadian Government Merchant Marine Ltd. [1959] A.C. 589.
[2] Papera Traders Co. Ltd and Others v Hyundai Merchant Marine Co. Ltd and Another (The "Eurasian Dream") [2002] EWHC 118 (Comm); [2002] 1 Lloyd's Rep. 719.
[3] Riverstone Meat Company, Pty., Ltd v Lanchashire Shipping Company, Ltd [1961] 1 Lloyd's Rep. 57.
[4] Tasman Orient Line CV v New Zealand China Clays and Others (The "Tasman Pioneer") [2010] NZSC 37; [2010] 2 Lloyd's Rep. 13.
[5] Union of India v N.V. Reederij Amsterdam [1963] 2 Lloyd's Rep. 223.

Chapter 9

Historical Aspects of Navigation

The Baltic Light Vessels in the Nineteenth and Twentieth Centuries

I. Pietkiewicz & A.F. Komorowski
Polish Naval Academy, Gdynia, Poland

ABSTRACT: In the given paper some results of scientific research in the field of pharology from last 4 years are shown. Light vessels built and used in Denmark, Sweden, Germany, Finland and Russia were analyzed. During the analysis of documents found in many archives, libraries and other institutions not only was a number of light vessels in given time and area established, but also their constructions, problems connected with navigation safety as well as crews and the impact of collisions on their safety. Some parts of the research were shown in different publications – articles and conference reports. The information concerning light vessels of Zalew Szczeciński (Szczeciński Lagoon) was published in a monograph entitled "Marine Signposts of the Polish Coasts" in 2011.

The given paper is an attempt to present the synthesis of the research team's results covering many aspects of exploitation of the Baltic, Danish Straits and Kattegat light vessels in 19th and 20th centuries.

1 INTRODUCTION

The research team, whose research results on the Baltic light vessels we are presenting, carried out a project between 2008 and 2010, dedicated to lighthouses of Polish coast, their origins and modern history, as well as exploitation and lighthouse keepers' work. The results of the research were introduced in many publications, including books, such as Iwona Pietkiewicz's "Jarosławiec Lighthouse" ("Latarnia morska Jarosławiec"), published by OPTI, Gdynia in 2010, "Darłowo Lighthouse" ("Latarnia morska Darłowo", published by OPTI, Gdynia 2011, "Ustka lighthouse and coastal rescue station" ("Latarnia morska i stacja ratownicwa brzegowego Ustka"), co-author A. Komorowski, published by OPTI, Gdynia 2011 as well as the already-mentioned "Marine Signposts of the Polish Coasts". The experience gained from the research allowed to aim the team's effort at getting to know the history of Baltic light vessels, even though Poland did not use such vessels. Due to this fact, the work concerned Danish, Finnish, German, Russian and Swedish vessels. As long as financial sources allowed, research was carried out in English, German, Swedish and Russian archives. Literature base (books and other publications) from other countries, namely Denmark and Finland, were also analyzed. Due to the research, the project ended, as it was planned, in 2012 and publishing work is being finalized as we speak. The synthesis of the light vessel's research of the above-cited countries along with accurate graphs and charts is shown in the conclusion.

2 LIGHT VESSELS DEVELOPMENT

The term lightship (light vessel) – describes all anchored vessels equipped with a light, emitting fog signals, radio signals, and equipped with navigational signs. Thus light vessels fulfil the role of lighthouses of other navigational lights. They are classified as floating navigational aids. As well as a light with a defined characteristic (they cannot use sector lights because of the vessel's movement on the water), they are equipped with day markers and a clearly visible name. Generally such vessels lack propulsion and are permanently anchored in waters where constructing a lighthouse is impossible or too costly (A.F. Komorowski at al. 2011). The light range of light vessel is rather small due to the relatively low height of the light above the sea level. Light vessels are becoming increasingly rarer, and are being replaced by light buoys, which are safer and cheaper to maintain. Old manned light vessels finally have been replaced with automated unmanned vessels and buoys.

The earliest light vessels were being used in North-Eastern European waters, probably as early as during the Middle Ages. Unlike contemporary ones, they were "occasional" navigational aids used prior to the arrival of important vessels. The first permanent lightship, "The Nore", warned against a dangerous sandbank at the Thames estuary since 1731. It is worth noting that she was deployed as a result of a private initiative. In 1734 her running was taken over by Trinity House.

Along the German coast light vessels were initially used as mobile pilotage stations, displaying lights at night in order to be easily identified by other vessels. The first vessel of this type in German waters was Lotsengaliot, anchored at the Elbe estuary in 1774. However, pilotage involves changing the vessel's position, which contradicts the very principle of lightships. As such, Germany's first "proper" lightship was "Seestern", anchored in 1816 on the Elbe. In European waters light vessels have also been used to mark entrances into ports and estuaries. Locations of light vessels were included on maritime maps to aid navigation. The vessels were also used for meteorological observations and transmitting signals.

Hulls of contemporary light vessels are usually painted red. However, the original vessels, made of wood, were the colour of agents used to preserve wood. Red was accepted as the standard colour of light vessels in the Southern Baltic region in 1833, with an exception for lightships marking the positions of wrecks, which were painted green. There were also occasional vessels painted black.

The first light vessels had wooden hulls, similar to other vessels at the time. Hull lines were adapted from sailing ships because of their seakeeping ability and the need for propulsion with sails. Vessels used in the Southern Baltic were relatively small – up to 36m along the waterline. The largest wooden light vessel was "Adlergrund II", extended to 41m after renovation. However, the main issue in their construction was ensuring their unsinkability. The sinking of "Seestern" with her entire crew in 1816, and numerous collisions with merchants' ships, were an inspiration for finding new solutions to the problem. Heavy wood beams were used as ballast, with empty barrels or wooden crates mounted above them in the hold to provide additional buoyancy. Such solutions can be seen on design sketches of German light vessels dating back to 1825 (A.F. Komorowski at al. 2011).

Vessels with a greater displacement were generally used on open seas. Sheltered areas, internal waters, lagoons etc. did not need ships that large due to shallow waters and to high costs. Such light vessels were about 15-20m long, and in the 19th century they were used in the Szczecin Lagoon among others. The last light vessel built along the lines of a sailing ship was "Bürgermeister O`Swald", deployed in 1948.

Steel-hulled light vessels first appeared around the mid-19th century. They were initially sail-powered, with propulsion changing to steam engines around 1906 and diesel engines after 1912, with a power of 300-500hp, giving a speed of approx. 9 knots. This was sufficient power in the event of drifting, or returning to the designated position, or returning to port for emergency repairs or inspections. Earlier sail-powered vessels were usually adapted to having two sails: a gaff sail at the stern mast and a triangular jib at the bow mast. In addition to propulsion, sails were also used to increase the vessel's visibility.

Development of technology in the construction of steel hulls led to the increased size of light vessels. Larger hulls improved safety while at sea, especially during stormy weather, and provided better conditions for the crews.

The majority of light vessels have been decommissioned after 1988 or replaced by unmanned vessels or buoys, which are cheaper to operate and maintain. There are no manned light vessels in operation today, although some can be seen in ports as museum pieces or serving new, completely different functions.

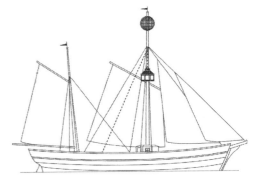

Figure 1. Wooden Swedish light vessel first generation with Argand's lamp

3 BALTIC LIGHT VESSELS AND THEIR LIGHT SYSTEMS

The structure of very first light vessels was not at all different from the goods vessels exploited at that time. They were rather small, wooden sailing ships lacking their own mechanic drive. The ships' hulls measurement was mostly 15 meter long, 5 meters wide and had 3 meter draught. There were also bigger vessels, able to hold their positions at open waters. The length of such ships was up to 40 meters, and some of them were as long as 50 meters. Not all Baltic countries exploiting light vessels had the same policy concerning those particular vessels.

Light vessels with steel hulls appeared in the middle of the 19th century. Their drive consisted usually of sails, later on, since 1906, it was a steam engine and since 1912 a diesel engine which had a power of 300-500 horsepower, which allowed the vessels to reach the speed of 9 knots. Such engine room was enough to come back to a desired position in case of drifting or to move to a harbor for maintenance or repair work. Earlier construction, based on a sails drive, was adapted to carry two sails: gaff rig on the back mast and a triangle foresail on the front mast. The sails, apart from drive, served as an element improving the vessel's visibility and, at the time of storm, they would lighten the anchors.

As the technology of building steel hulls developed , the measurements of the light vessels were becoming bigger. Bigger hulls ensured safer stay at sea, especially during a stormy weather, and allowed the crew to have better living conditions during work.

In Denmark, the very first light vessel was deployed near the island of Laeso in 1829. In 1888 there were already 12 vessels, in 1916 – 16 and in 1927 – 17. The characteristic thing about Danish lightships was their wooden, oak hulls, all of them had nearly exactly the same measures (about 30 meters). The hulls were painted red with a characteristic white cross on ships' sides, with the name of the vessel visible on them.

Danish drive-less light vessel's constructions were marked with numbers from I to XXI, and the engine vessels from I to IV. There was a total of 25 ships in the whole history of Danish lightships. The peak moment of Danish light vessels' activity was between 1945-1953 when there were 13 ships on their positions. The last one, with number XXI, was withdrawn from its position MON-SE in 1988. (L. Elsby, 2008).

Finland kept its light vessel fleet during the times of the Grand Duchy of Finland – about 9 vessels and at the time of independence 13-15 vessels. These were at first vessels with wooden hulls, later on steel ones. Their length was about 25 meters, but there were single light vessels longer than 30-40 meters. The peak of Finnish light vessels' activity was between 1918-1923, and the last of them, "Kemi" served until 1974. (L. Seppo, 2001).

Swedish light vessels (37 vessels located in 24 places) were kept in service on important maritime positions for over 130 years, from 1844. Among the vessels there were ones with wooden and ones with steel hulls. Some of the vessels were adapted to serve as light vessels, other were built specifically for that purpose. The ships had strong archive mechanisms and anti-ice buttress. The largest number of lightships on positions (22) appeared between 1933-1935, and the last of them, "Falsterborev Res." served until 1972. (W. Bjorn, 1999).

German light vessels were kept in service since 1816. They were usually sailing ships, adapted to serve as lighthouses, with 2 – 3 masts and initially, as in other countries, their hulls were wooden and their length was up to 40 meters. The vessels which were exploited on sheltered waters were obviously smaller; their length was up to 15-20 meters. Later on, special light vessels were built with steel hulls. (E. Wiese, 2000).

Most of the described light vessels were positioned on the Baltic Sea between 1915 and 1920. The last Baltic light vessel "Reserve Sonderburg" (Kiel, III), ended its service in 1986 and was rebuilt to become a sailing ship named "Alexander von Humboldt".

Russian light vessels were no different than the already-mentioned ones. In 1884 there was a total number of 5 and in 1914 twelve ships serving on their positions. The last light vessel ended its service in Irbe Strait in 1985.

The described vessels were marked with daytime marks, shaped like spheres or other geometric figures, placed on the top of the masts. An extra marking were flags showing the vessels' nationalities. Red hulls became a characteristic feature of light vessels distinguishing them from other vessels. At night, there were light signals placed on their masts.

Figure 2. Danish Motorized light vessel No I

A big problem concerning the exploitation of the Baltic light vessels was the Basin's freezing. Usually, during winter the vessels would remain in harbors and wait for better weather conditions. Another exploitation problem were anchor systems which, as a rule, ought to be very solid. Stormy weathers did not work in favor of lightships which would often drift away or change their positions due to the anchor chains break, despite being well-protected. According to the research, anchor chains had links as thick as 42 millimeters and were as long as 300-500 meters. The mushroom anchor's mass was between 800-1800 kilograms, and the whole set's mass was often about 15 tones. It is worth adding that the light vessels were usually equipped with three anchors, each of them having a different mass.

4 LIGHT SYSTEMS

The evolution of light signals exploited on light vessels was a consequence of the development of light systems used by lighthouses. Originally candles enclosed by glass. Such solution was rather impractical because candles' dim light was practically invisible at long winter nights. Another solution that was used on the lighthouses were oil lamps, powered by whale or rape oil (A.F. Komorowski & I. Pietkiewicz, 2011). They were put up on the masts with the help of lines attached to blocks; their layout created a so-called tackle, which allowed the light staff to pick them up more easily. Such simple lamps were used on the first German light vessels. The main problems caused by such solution were smoke caused by fire, constant need of cleaning the glass, as well as high price of good-quality vegetable oil.

A solution to these above-cited problems was introduced in 1755 by a Swiss physicist Amie Argand. A lamp constructed by him, due to the usage of a ring wick and a cylinder enclosing the flame, which allowed the air entering the flame to flow strongly, allowed any kind of oil to be burned in fair, non-smoldering flames. An improved version of Argand's Lamp with a multiple burner, placed in spherical or parabolic reflectors' focal, was used individually or in groups and were exploited on light vessels during the next hundred years. Such lamps were not, however, flawless. The lamps had to be lowered in order to regulate the wicks, which demanded a lot of work from the light vessel's crews.

Another problem that had to be dealt with was the one connected with the stability of light vessel's lights. In 1807, as Scottish building engineer, Robert Stevenson, solved the problem of a particular instability of light vessel's lamps by constructing a polygonal frame, inside of which he used Kardan's suspension. The construction was installed around a light vessel's mast, fixing nine oil lamps with parabolic reflectors to it. Such solution allowed the light to be emitted both vertically and horizontally. During a day when the light was not used and in order to perform maintenance, the lamps were lowered to a superstructure located on the lower part of the mast. The device was improved by a Swedish engineer Gustaf Dalén, who, after introducing the Fresnel lens system to light vessels, (Fresnel was a French physicist, a constructor of prismatic lenses), used earlier for light lighthouses' devices, constructed an independent optical suspension, in which the lens system is hung anglepoisely at the center of gravity, which did not allow any unwanted

light movement caused by the vessel's movement on water, to appear. (A.F. Komorowski & I. Pietkiewicz, 2011).

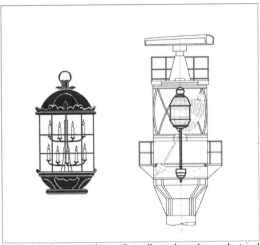

Figure 3. The comparison of candle and modern electrical light system

Unlike lighthouses, the Fresnel lens system was not used on light vessels until 1870's and 1880's. The reason of that was the prior lack of ways allowing to lift light systems with lenses onto the masts. The very first devices of such type were built with three arms, on which gimbals Argand lamps, located outside Fresnel cylindrical lens, were fixed. The whole device was pulled up on the mast. Due to the light sectors' cross, the light was seen around the mast. The modern lighthouses were much smaller and lighter than their cumbersome predecessors, built with Argand lamps and many parabolic reflectors.

In 1905, separate masts, used only for lanterns, were used on German light vessels, which allowed to use bigger light systems. In the middle of the ships a mast with an entrance possibility would be installed. On its upper part a platform would be situated, above which a light system would be put. Such solution allowed the necessity of lowering and lifting of light system to be eliminated. The crew would enter the platform to cover the lantern during the day and for maintenance, an action which made their job significantly easier. ((F.K. Zemke, 1995).

The next step for light systems used on such ships was initiated by the usage of gas as an energy fuel in lighthouses' lanterns. Transporting the gas on light vessels through pipes from containers placed on the land, similarly to the way it was done in the cases of lighthouses, was not possible, thus works had been taken up to use acetylene – a gas manufactured from carbide – to power up the light vessel's lanterns. The research concerning this matter was carried out in Sweden between 1849 and 1908, and allowed to obtain bright and very intense light with simultaneous massive fuel consumption. For the purpose of gas illumination, liquefied gas kept in cylinder was used. High exploitation costs of the lightning devices consuming economically wanton amount of the energetic factor, forced further research, which led to the construction of a device commonly known as "Dalén's lamp" in 1906. In 1916, Dalén constructed a device that would automatically change a burnt-out lamp cover into a new one. As a result of all carried out modernizations, Dalén's lamp, equipped with 24 exchangeable covers was able to work without any service for 12 months. (A.F. Komorowski & I. Pietkiewicz, 2011).

The introduction of electricity on light vessels demanded a significant financial input. The very first German light vessel on which electric light was used was „Bürgermeister O'Swald", destined to serve at position Elbe 1, built in 1912. An arc lamp, placed in a cylindrical lens with 250mm focal, served as a source of light. In order to gain proper characteristic of light, covers would move around the lens. A carbon-fiber light bulb was used as a back-up source of light. The device was gimbals hung on ball bearings which compensated the ship's heel up to 40 degrees. Batteries charged by two alternately working electricity generators were the sources of electricity. Such solution worked fine until the vessel sank in 1936.

Despite using generator-obtained direct current to power up electrical devices (such as radio, fog lights, anchor elevators and lights in rooms) on the Baltic light vessels as early as in 1927, the electrification of light vessel's light systems was not fully carried out until after World War II, when long-lasting light bulbs and DC generators were massively used. It is worth underscoring that the perfect solution of acetylene-powered Dalén's lamp was not abandoned and kept on being used as a back-up light system. (A.F. Komorowski & I. Pietkiewicz, 2011).

In order to gain intermittent light with a suitable characteristics assigned to a given light vessel when using electricity, electricity chopper was used.

5 SUMMARY

Since the introduction of light vessel's - marked navigational routes in the second half of the 18th century until the end of light vessel's exploitation era in the 20th century, the vessels underwent many changes and modernizations. They included both hull-construction matters (at first light vessels were nothing else but sail ships withdrawn from active service, and in the 20th century vessels specifically designed to serve only as light vessels were built) and light systems matters. Only in Germany in

the 20th century there were about 10 light vessels built, equipped in special lantern masts, and 4 others had such masts added during repairs and renovations. A light systems focal of 250mm or 300mm was used. Those were mostly cylindrical lenses, or circular lenses on a smaller scale, built from two or four walls. The characteristics of the light was obtained by turning the light system, spinning covers and, after the introduction of electricity, by using intermittent electricity.

The evolution of electrical light systems introduced on the lightships in the 20th century focused mostly on electrical devices' construction changes as well as the modernization of light sources. The dynamic development of electric energy sources, which resulted in the appearance of sun batteries and wind generators, led to an end of the intense exploitation of lightships and to their full automation. One of the main exploitation problems of the Baltic lightships were frequent breaking of anchor chains, the necessity of taking the ships back to the harbor during winter and its considerably high accident rate. It is suffice to say that one of Danish light vessels, serving on the Drogden position, suffered from 50 accidents and German Kiel III sank few times after collision with other ships.

To sum up, it is fair to say that during the whole time of their service, the Baltic light vessels did an excellent job. With over 110 vessels on duty, there were maximum 74 lightships serving on their 55 positions simultaneously (in 1920). During the later years of their activity their number decreased (see the diagram).

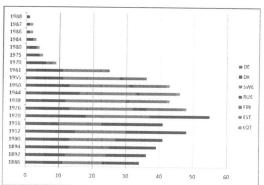

Figure 4. The list of the Baltic Sea light vessels' positions, based on individual volumes of lights lists from United Kingdom Hydrographic Office archives.

REFERENCES

Bjorn W. 1999. *Fyrskepp i Sverige*. Stockholm: Bo Nyman. Bundesarchiv Koblenz B 108/7364.
Elsby L. 2008. *Fyrskepp i nordiska vatten* [in]: Blänket .3 Arg. 12.

Komorowski A.F., Pietkiewicz I., Szulczewski A. 2011. *Marine Signposts of the Polish Coast.* Gdańsk: Fundacja Promocji Przemysłu Okrętowego i Gospodarki Morskiej, pp. 378-381.

Komorowski A.F., Pietkiewicz I. 2011. *Rozwój systemów świetlnych latarniowców.* Gdynia: Zeszyty Naukowe AMW 2011 nr 3 (186).

Latarnie morskie świata. Praca zbiorowa. 2000. Warszawa: Muza, pp. 65-70.

Laurell Seppo 2001. *Suomen majakat.* Helsinki: Gummesus Og.

Pietkiewicz I., Komorowski A.F. 2011. *Die Schifffahrtspolitik und die Markierung des Schifffahrtsweges Stettin-Swinemünde in den Jahren 1820-1945.* Szczecin: Studia Maritima. Volume XXIV.

Transpress Lexikon Seefahrt, VEB Verlag für Verkherswesen Berlin 1976 pp.142.

Wiese E. 2000. *Feuerschiffe.* Frankfurt: Heel.

Zemke F.K. 1995. *Feuerschiffe der Welt.* Hamburg: Koehler, pp. 60.

Chapter 10

Safety, Reliability and Risk Assessment

Safety, Reliability and Risk Assessment
Advances in Marine Navigation – Marine Navigation and Safety of Sea Transportation – Weintrit (ed.)

Maritime Risk Assessment: Modeling Collisions with Vessels Lying at an Anchorage

H.-C. Burmeister, C. Jahn & S. Töter
Fraunhofer Center for Maritime Logistics and Services CML, Hamburg, Germany

J. Froese
Hamburg University of Technology TUHH – Institute of Maritime Logistics, Hamburg, Germany

ABSTRACT: This paper proposes a collision model for ships underway and temporary objects as an extension to state-of-the-art maritime risk assessment. It gives a brief review of frequency modeling's theory and its applications, before it analogously derives a model to assess the risk of anchorage areas. Subsequently, its benefit is demonstrated by an example scenario.

1 INTRODUCTION

Maritime traffic volumes and ship dimensions are expected to increase further, requiring fairway and port designs being adapted to the new situation. In general, these design processes should be accompanied by an assessment of the risk of collision and grounding.

According to IMO (2007), "risk" is defined as the combination of number of occurrences per time unit and the severity of their consequences. The occurrence might be a collision or a grounding event. Its consequence is e.g. an oil leakage or a sinking ship which is mostly measured in monetary values. Thus, it implies the common risk definition as probability of a collision multiplied by its expected damage (Pedersen 2010).

To quantify the risk, the International Association of Marine Aids to Navigations and Lighthouse Authorities IALA recommends a probabilistic methodology based on frequency modeling (IALA 2009). Basically, the methodology distinguishes collisions between ships underway and grounding, which includes collisions between ships and fixed objects (in the following, "collision" includes grounding events, as they are methodologically similar to collisions with fixed objects).

However, in this specific case the risk of mooring dolphins in relation to an anchorage has to be assessed. While the former is clearly a fixed object, vessels lying at an anchorage are neither underway nor completely fixed objects. Thus, the anchorage's risk is difficult to determine with the proposed IALA methodology.

Based on a brief review of frequency modeling's theory in section 2, the collision type "ship-anchorage" is defined in section 3. In accordance with current maritime risk models, a frequency model for this kind of collision is derived in section 4. Section 5 applies the model on an example scenario and compares the results with alternative modeling based on current methodology. Section 6 closes with the drawn conclusions.

2 THEORY OF FREQUENCY MODELS

Collision probabilities are mostly determined by frequency models, which are based on the work of Macduff (1974), Fujii (1983) and Pedersen (1995). The methodology has been applied in several analyses e.g. in the Canary Islands (Otto et al. 2002), in the Øresund (Rambøll 2006) or in the Gulf of Finland (Kujala et al. 2009, Hänninen et al. 2012).

In these models, the frequency corresponds to the number of collision events N in a specific time. In principle, this number is calculated by multiplying the number of collision candidates N_a with the causation probability P_C:

$$N = P_C \cdot N_a \qquad (1)$$

2.1 Collision candidate

A "collision candidate" is a situation, which results in a collision, if no aversive maneuvers are made. For vessels underway, its number is calculated based on the geometric specification of the investigated sea area, the traffic volumes, vessel specifications and their lateral distribution under the assumption of "blind navigation". The latter implies, that initially a vessel choses its route independently of the current situation.

Depending on the type of meeting situation:
− Passing a fixed object,
− Head-on meeting,
− Overtaking,
− Crossing or
− Merging

different models are commonly accepted to determine the number of collision candidates (Pedersen 2010, IALA 2012). In case of object collisions, Pedersen (1995) proposes four categories to further classify the type of accident:
1 Ordinary, direct route at normal speed,
2 Fail to change course at given turning point,
3 Collision as result of evasive actions or
4 Other (e.g. drifting).

The situation of the first category is displayed on the bottom of figure 1. According to basic statistics, the number of collision candidates in a specific timeframe for this type can be estimated by:

$$N_a^I = \sum_i Q_i \cdot \int_{z_{min}-\frac{B_i}{2}}^{z_{max}+\frac{B_i}{2}} f_i(z)\,dz \qquad (2)$$

given that Q_i represents the number of passing vessels of type i in this time, B_i is the breadth of vessels of type i, $f_i(z)$ stands for their lateral distribution and z_{min} and z_{max} characterizes the dimensions of the fixed object.

The number of head-on collision candidates can be estimated in a similar way by:

$$N_a^{head} = L_W \cdot \sum_{i,j} \frac{v_i + v_j}{v_i \cdot v_j} \cdot Q_i \cdot Q_j \cdot P_{i,j}^{head} \qquad (3)$$

with j representing types of meeting vessels, v_i is the speed of vessels of type i and L_W is the length of the route or fairway, where head-on meetings are expected (see also figure 1). Finally, the probability $P_{i,j}^{head}$ for meeting vessels can be calculated by:

$$P_{i,j}^{head} = \int_{-\infty}^{\infty} \int_{z_i-(B_i+B_j)/2}^{z_i+(B_i+B_j)/2} f_i(z_i) \cdot f_j(z_j) \cdot dz_j dz_i \quad (4)$$

Analogously, estimation for the further types of collision can be derived, but as they are not further needed in this work, it is referred to Pedersen (1995).

2.2 Causation probability

The causation probability is defined as the fraction of collision candidates that results in a collision. In general, these factors differ depending on the situation types and are mainly derived from analytical methods or Bayesian networks. As this paper focuses on collision candidate determination and not on causation probability modeling, it is referred to e.g. IMO (2007), Hänninen & Kujala (2010), Pedersen (2010) and IALA (2012) for further information.

3 RISK FACTOR ANCHORAGE

Anchorages pose a risk to navigational safety, as there often anchor vessels, which then are an obstacle that others might collide with. Notwithstanding improving navigational aids and crew qualification, collisions still occur, like e.g. the collision between the "Katharina Siemer" and anchoring "Angon" on the Elbe River in November 2012 or between "Jinggangshan" and anchoring "Aeolos" near Gibraltar in May 2011. Thus, an appropriate consideration of anchorages during maritime risk assessments should be aspired.

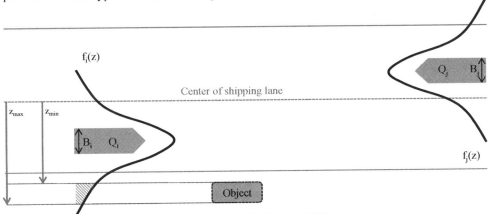

Figure 1. Head-on and fixed object category I situation (based on Pedersen 1995)

3.1 Anchorage characteristics

An anchorage is a limited area that is suitable for vessels to anchor. Those areas are highlighted in sea charts and might also be marked with buoys (BSH 2011). However, if the anchorage is not in use, it is not an obstacle for shipping as it normally does not necessitates any fixed infrastructure (except in the case of buoys).

The actual obstacles are the vessels lying at an anchorage, which one may collide with. In contrast to a berthed ship, those ones change their position by swinging at anchor depending on wind, waves and tide.

3.2 Collision types on an anchorage

Types of collision that may occur in relation to an anchorage are collisions between:
1. Anchored vessels,
2. Anchored vessel and vessel underway (e.g. with a transiting vessel),
3. Vessels underway (e.g. transiting vessel and a vessel leaving the anchorage) or
4. Other (e.g. drifting).

While the collision candidates for the third type can be determined with the help of a crossing or merging model, none is available for the upper categories, as the anchored vessels are neither fixed objects nor permanent obstacles. In the following, the first one is called "anchorage-" and the second one "ship-anchorage-collision".

4 PROBABILITY OF SHIP-ANCHORAGE-COLLISION

4.1 Mathematical model for collision candidates

To determine the ship-anchorage collision candidates similar to (2) and (3) first all meeting situations in a given timeframe must be calculated. If $\tau_{use,a}$ is the fraction of time that the anchorage is used by at least one vessel of type a and Q_a is the mean number of vessels of type a, which lay at the anchorage at the same time, then the total number of collision candidates is given by:

$$N_a^{S-A} = \sum_{i,a} Q_i \cdot \tau_{use,a} \cdot Q_a \cdot P_{i,a}^{S-A} \qquad (5)$$

where the first three elements determine the number of all meeting situations.

Afterwards, the probability $P_{i,j}^{S-A}$ of the underway vessel heading towards an anchored ship has to be calculated. This is done by $f_a(z)$, which is the probability density function of the anchoring ship's distance to the center of the shipping lane, and d_a representing its obstacle dimensions including paid out anchor chain perpendicular to the other vessel's moving direction (see also figure 2). If the vessels

underway are described as in (2), then the collision candidates can be estimated similar to (4) by:

$$P_{i,a}^{S-A} = \int_{-\infty}^{\infty} \int_{z_i-(B_i+d_a)/2}^{z_i+(B_i+d_a)/2} f_i(z_i) \cdot f_a(z_a) dz_a dz_i \qquad (6)$$

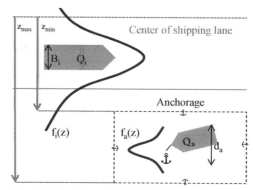

Figure 2. Ship-anchorage-collision

4.2 Causation probability

This work focuses on modeling of collision candidates. Thus, in the first instance the causation probabilities for fixed object collision could be used as an approximation for calculating the number of collision events.

4.3 Model variables

Of course, the different types of anchoring vessels allow modeling different sizes of ships. However, it furthermore allows incorporating swinging circle effects, as d_a and $f_a(z)$ depend on the actual weather and tidal constraints. Therefore the ship of type a is split into several types with different obstacle characteristics, while its likelihood is controlled by $\tau_{use,a}$, which is set according to the tidal conditions' fraction of time.

As the boundaries z_{min} and z_{max} of the anchorage are not part of the model, it has to be ensured by the chosen distribution function that the anchoring vessels are positioned within the anchorage area. However, the tails of $f_a(z)$ outside the anchorage could be used to model meeting situations between vessels underway and ships adrift because of a broken anchor or with ships swinging at anchor into the shipping lane.

5 RISK ASSESSMENT EXAMPLE

Due to confidentiality reasons the original case that inspired the extension can't be presented here. However, a simplified virtual decision situation shall demonstrate the utility of the ship-anchorage-collision model.

5.1 Decision alternatives

In an area of restricted tidal waters short-term berths are needed for ships waiting e.g. for a free berth at the pier or a locking. However, it is discussed to either display a narrow anchorage area next to the fairway or to construct several dolphins allowing for short-term moorings. As the second option is more costly its safety benefits should be analyzed.

5.2 Modeling as ship-anchorage-collision

The principal layout of the dolphin alternative corresponds to figure 1 and the one of the anchorage to figure 2. Table 1 gives an overview about the scenario variables. It is assumed that during 75% of the time the anchorage area or the dolphins are in use. Furthermore, the berths at the dolphins are on the fairway side, thus the obstacle dimensions increase if a vessel is moored. If there are several ships, they are moored in series; consequently the obstacle dimensions stay constant.

Table 1. Scenario variables.

Variable	Anchorage		Mooring dolphins	
Q_1/Q_2	20,000	Ships/a	20,000	Ships/a
B_1/B_2	20	m	20	m
$f_1(z)$	$N(50, 50^2)$		$N(50, 50^2)$	
$f_2(z)$	$N(-50, 50^2)$		$N(-50, 50^2)$	
z_{min}	150	m	300*	m
z_{max}	450	m	305	m
$\tau_{use,a}$	0.75	-	0.75	-
Q_a	2.5	Ships	2.5	Ships

* z_{min}=280m for dolphins if ships are moored

Due to tidal waters, the anchoring vessels swing at the anchor thus having different obstacle positions and dimensions over time. The latter strongly depends on the vessels angle to the fairway. If further weather effects are neglected and a tidal current parallel to the fairway is assumed, then the obstacle dimension of the anchoring vessel (perpendicular to the fairway) over time follows approximately the solid line in figure 3. It can be seen, that shortly after slack tide, when the dotted current line crosses the axis of abscissae, the changing current direction turns around the vessel until it lays again parallel to the fairway according to the new tide. During this time, the obstacle dimensions are of course much higher than in the dolphin case, where the vessels stay parallel independently of the tide.

Table 2. Obstacle dimensions in anchorage scenario

Relation	Orthogonal	In between	Parallel
Fraction of time	0.04	0.06	0.90
d_a	100 m	60 m	20 m
$f_a(z)$	$N(300, 40^2)$	$N(300, 20^2)$	$N(300, 10^2)$

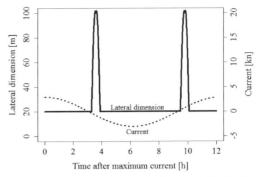

Figure 3. Lateral dimension of anchoring vessel (idealized)

Within this example, a turning rate of 5 degrees per minute is assumed and the turning process is divided into three different relations. Their characteristics are given in table 2 considering that the likelihood of anchored vessels outside the anchorage (e.g. due to drifting because of broken anchor) is below 1.0%. In reality, the position distribution as well as the lateral dimension should of course be derived by available information, as e.g. data from the Automatic Information System AIS.

5.3 Modeling as fixed-object-collision

Even though anchoring vessels do not fit the definition of fixed object collision models, two alternatives are presented based on the methodology described in Pedersen (1995) to allow for a comparison with the proposed model.

In the first alternative "Fixed Object: Anchorage" the whole anchorage is modeled as an object for 75% of the time, while the second one "Fixed Object: Anchored vessel" assumes that all obstacles lay in a row in the middle of the anchorage similar to the dimensions in table 2. Of course, the latter implies that not the whole anchorage area is used.

5.4 Comparison of results

Table 3 shows the estimated collision candidates for the decision alternatives. Using (2), the estimation of the collision candidates for the dolphin scenario can be performed directly according to accepted methodology and results in 0.084 collision candidates per annum.

Indeed, it is observable that the results for the anchorage scenario widely differ depending on the chosen model. If e.g. the whole anchorage area is modeled as a fixed object, then this results in 540 estimated collision candidates. This seems to be a very conservative approach as underway vessels traveling through this area do not necessarily be on collision course with an anchoring ship.

Table 3. Results of example

Model	Collision candidates	
Ship-Anchorage-Collision	2.472	p.a.
Fixed Object: Anchorage	540.040	p.a.
Fixed Object: Anchored vessel	0.111	p.a.
Fixed Object: Mooring dolphins	0.084	p.a.

In contrast to that assuming a fixed anchoring position might be too subjective due to the fact that the chosen position could strongly bias the results. As commonly used lateral probability distributions decline in the tails assuming a more distant anchoring position would strongly affect the calculated collision candidates and thus the risk assessment. If it is e.g. anticipated in this example, that all anchoring vessels lay next to the fairway-side border of the anchorage, the expected number of collision candidates would be close to the one in the "Fixed object: Anchorage" case.

As the result reacts very sensitive to the assumed anchoring position, it can be considered to be the most objective way to include the probability density function $f_a(z)$ of the anchoring position directly in the risk assessment by using the proposed model. Therefore, frequency distributions derived from recorded AIS-data provides an accurate base to determine the required functions $f_a(z)$ and $f_i(z)$.

Furthermore, the analysis of the different anchoring situations could also be of help by finding high risky situations. Table 4 shows the partial results of the ship-anchorage-collision-model in this example and it is observable, that nearly all collision candidates are expected during the orthogonal situation.

Table 4. Collision candidates' situation depending on tide

Relation	Orthogonal	In between	Parallel
Fraction of time	0.04	0.06	0.90
Collision cand.	2.257	0.108	0.107

6 CONCLUSION

Within this work additional collision types than those used in IALA (2012) have been defined, which are related to anchorage areas. A model for estimating collision candidates between vessels underway and vessels lying at an anchorage has been proposed, which is capable of taking into account information on the anchoring position's frequency distribution. Notwithstanding, the proposed model suffers similar drawbacks as frequency models in general, e.g. that vessel movements are not taken into account and that information about the exact collision situations is missing (Goerlandt & Kujala 2011).

Nevertheless, the proposed model goes in line with state-of-the-art frequency models for collisions between ships or ships and fixed objects to allow for comparison with other collision types. The method has been applied on an example case derived from a real problem to demonstrate the shortage of modeling anchorage areas as fixed obstacles. Additionally, the proposed model is capable to roughly consider swinging circle effects.

Indeed, further adjustments are necessary to establish a full risk model for anchorages. Next to the proposed geometric model, a deeper analysis of causation factors should be conducted for this type of accidents. Besides, an additional model for anchorage-collisions is needed to fully assess an anchorage's risk.

REFERENCES

BSH 2011. *INT 1 Symbols, Abbreviations, Terms used on Charts*. Hamburg, Rostock: Bundesamt für Seeschifffahrt und Hydrographie.

Fujii, Y. 1983. Integrated Study on Marine Traffic Accidents. *IABSE Reports* 42: 91–98.

Goerlandt, F. & Kujala, P. 2011. Traffic simulation based ship collision probability modeling. *Reliability Engineering and System Safety* 96: 91–107.

Hänninen, M. & Kujala, P. (2010): The Effects of Causation Probability on the Ship Collision Statistics in the Gulf of Finland. *TransNav - International Journal on Marine Navigation and Safety of Sea Transportation* 4 (1): 79–84.

Hänninen, M., Kujala, P., Ylitalo, J. & Kuronen, J. 2012: Estimating the Number of Tanker Collisions in the Gulf of Finland in 2015. *TransNav - International Journal on Marine Navigation and Safety of Sea Transportation* 6 (3): 367–373.

IALA 2009: *IALA Recommendation O-134 on the IALA Risk Management Tool for Ports and Restricted Waterways*. International Association of Marine Aids to Navigation and Lighthouse Authorities.

IALA 2012: *IWRAP Mk2 Wiki*. International Association of Marine Aids to Navigation and Lighthouse Authorities.

IMO 2007: *Formal Safety Assessment. Consolidated text of the Guidelines for Formal Safety Assessment (FSA) for use in the IMO rule-making process (MSC/Circ.1023−MEPC/Circ.392)*. International Maritime Organization.

Kujala, P., Hänninen, M., Arola, T. & Ylitalo, J. 2009: Analysis of the marine traffic safety in the Gulf of Finland. *Reliability Engineering and System Safety* 94: 1349–1357.

Macduff, T. (1974): The probability of vessel collisions. *Ocean Industry* 9: 144–148.

Otto, S., Pedersen, P.T., Samuelides, M. & Sames, P.C. 2002, Elements of risk analysis for collision and grounding of a RoRo passenger ferry. *Marine Structures* 15: 461–474.

Pedersen, P.T. 1995: Probability of Grounding and Collision Events. In Technical University of Denmark (ed.): *Accidental Loadings on Marine Structures: Risk and Response. 22nd WEGEMT Graduate School*. London.

Pedersen, P.T. 2010: Review and application of ship collision and grounding analysis procedures. *Marine Structures* 23: 241–262.

Rambøll 2006: *Navigational safety in the Sound between Denmark and Sweden (Øresund). Risk and cost-benefit analysis*. Online available at http://www.dma.dk/sitecollectiondocuments/publikationer/navigational_safety_oresund.pdf, last check 02.01.13.

Safety, Reliability and Risk Assessment
Advances in Marine Navigation – Marine Navigation and Safety of Sea Transportation – Weintrit (ed.)

On the Reliability of the Navigator – Navigation Complex System

V.G. Torskiy & V.P. Topalov
Odessa National Maritime Academy (ONMA), Ukraine

ABSTRACT: The provision of a sufficiently long reliably functioning "man-machine" system is the most important problem to be solved with the help of the "Reliability Theory". This is easily explained by the fact that a symbiosis of man and machine is a characteristic feature, that is, a man is an element of the system and performs a number of functions. Some fully automated systems are able to perform correctly their task without the participation of a man but his presence is considered to be necessary for assuring the system reliability.

The clauses of the Reliability Theory may be applied to the navigator keeping watch on the bridge of a ship. Indeed, during this period the watch officer is viewed as an element of the system "man-navigation complex", in which he performs the controlling functions. The reasonability of such approach is easily explained by the fact, that the principal causes of the disasters at sea are recognized to be the drawbacks and weaknesses characteristic both of the watch personnel and the technical devices of the ship, which becomes especially evident in the process of their interaction ("man-machine") in the complicated conditions of navigation at sea. In this connection it is evident that the increase of the reliability of the above complex is an important and pressing problem for the practical navigation, the solution of which will facilitate the decrease of the accident rate at sea.

1 INTRODUCTION

In the historical process of the technology development one can single out two major steps : the first one, started still in the Stone Age, consists in looking for the devices, allowing to increase the muscle strength of the man and the second one, which is still in progress nowadays, is characterized with designing the appliances enabling thousand fold increase of the human mental abilities. In the old times a man was valued according to the strength of his muscles; nowadays the experience and practical skills acquired by the people in a long process of labor of many generations are of a value.

The XIX – XX centuries were characterized by wide mechanization and automation of complicated production processes, which resulted in high growth of the productivity of labor. During only one century (1850 – 1950) the productivity of labor in the developed countries rose more than 15 times, while the productivity of the mental labor only doubled. The scientists made a conclusion that the so slow tempo of the mental labor growth was caused by a very low reliability of the instruments and devices used in this sphere of the human activity. And indeed, the first electronic numerical devices, consisting of a very large number of electronic valves, relays and resistors, often failed and were very unreliable assistants in performing very complicated calculations. Proceeding from the high industry demand for the above machines in the 60th of the last century the scientists proved that a highly reliable appliance may be assembled from a certain sufficient number of unreliable elements. A new theory, "Reliability Theory" (Longevity) was born in that period.

This Theory enables a man to predict the "fates" not only of technical devices, but also biological constructions – human beings. Thus, for example the longevity of a pine was found to be 300 years, of a turtle – 200 years, a crocodile – 250 years, a redwood (sequoia) lives some thousands years, a manta-fish – 100 years and a man - 70 years.

The longevity of technical appliances is comparatively short – that of a TV set – 8 years, a car – 8 years, a radio set – 12 years. The Egyptian pyramids "live" some thousands years.

The provision of a sufficiently longevity "man-machine" system may be applied to the watch officer on the bridge of a ship. Indeed, during the period of the watch he is viewed as an element of the system "man-navigation complex", which performs many diverse functions. The reasonability of a such notion is explained by the fact, that the main causes in most cases of disasters at sea are the deficiency both of the watch personnel and the navigational devices of the ship, which becomes perceptible in the process of their interaction especially in the complicated situations during the voyage.

In this connection, undoubtedly, that the increase of the reliability of the "Navigator—Navigation complex" system is an important problem for the practical of navigation, solution of which will facilitate the increase of safety at sea.

2 ANALYSIS OF THE LATEST RESEARCH AND PUBLICATIONS WHICH INITIATE THE SOLUTION OF THE PROBLEM

In connection with the considerable growth of the quantity and the complicity of the technical appliances (TA), the watch officer-operator has to interact with, on a modern ship, the necessity of solution of the two problems arose:
1 The distribution of functions between the operator and TA.
2 Man-technique symbiosis.

The first problem requires solving tasks of harmonious combination of human and technical factors. The problem of man and technique symbiosis on the bridge includes the research of man's qualities as a link of the system, in which the officer-operator in combination with TA provides for the efficient execution of the ship control functions. The reasonable distribution of functions between the watch officer and the technical appliances is a real source of the improvement of efficiency of the whole ship control system. But it would be erroneous to attempt to automate all the control functions leaving only observation functions for the watch officer, as those of an observer-operator or an inspector.

But the opposite extremity is also not desirable – claiming that all the "thinking" functions and decision making should never be entrusted to the automation equipment. That is, the optimum system, including a man and automation appliances should provide for the maximum possible efficiency of its operation by means of the best employment of the both components' opportunities and advantages. Nowadays the manufacturers of the navigation appliances are deeply engaged with this problem as well as professionals in ergonomics safety, ship designers and builders, classification societies, IMO and other maritime organizations. Hard work is

being affected in creation of the perfect designs of the bridge and its equipment, comfortable placing of the instruments, windows, doors and other objects, as this is the principal working place from which the ship movement under the various conditions is controlled. The general trend is considered to be the creation of the integrated systems of the watch personnel information provision, and the control of the processes.

The general purpose of the above system consists in the information support of the watch officer decisions by providing him with timely, sufficient and conveniently represented data, necessary for decision making. The results of the above aspirations are reported in the publications of the classification societies – the American Bureau of Shipping(ABS), Norwegian Register (DNV), Russian Register (RS), IMO publications: SOLAS-74 Convention, resolutions A.529 (13), A.815 (19), A. 869 (20) etc. It is evident that TA should fulfill those functions which the man either is not able to fulfill or performs inefficiently.

After the functions distribution between the watch officer and the technical appliances, in the ship control system has been done, the problem of their efficient interaction in the process of functioning arises. The efficient functioning of the ship's "man-machine" control system depends on too many various factors and it is often practically impossible to take into account in full degree their influence. With the purpose of the increase of the efficiency of the "Watch Officer -Navigation Complex" system functioning the works are nowadays being performed aimed at provision of high reliability of this complex elements, the provision of the most important automation means with the self-diagnostics devices and the means of warning the watch officer of the malfunctions and the failures occurred, and the additional training of navigators in usage and servicing TA. In particular an especial attention is paid to those questions when training watch officers in bridge resource management, in the process of simulation training and technical studies on shipboard. All the above is well proved by the publications in the Nautical Institute journal Seaways [1,2,3] and other publications [4,5]. It is worthy to note that the scientists (researchers), designers, manufacturers of the systems of navigation and ship control and their users will always face this problem. The characteristic for our (modern) time persistent process of improvement of the automation means of the working operations on board and widening their possibilities makes us to face the problem anew: division of functions between man and machine and the securing their efficient interaction. The least developed ones to our mind, are the problems of provision of the system of navigation and ship control functioning reliability, the watch officer, being one of the elements of the above system. The characteristic features of the man and the machine are of quite different nature, dynamics and the intensity of changes in time.

The purpose of the article: One of possible approaches to the determination of the reliability of the system "man-machine" as applied to the activity of the watch officer on the bridge of a ship.

3 THE MATERIAL OF THE INVESTIGATION

Let's observe some general points concerning the above problem. Thus, under the reliability of the system "man-machine" we understand in general the whole complex of characteristics of technical means and the men servicing them, ensuring the functioning of the system within the limits of the necessary requirements, faultlessness and reparability of the technical means, accuracy of operator's actions, his medico-biological reliability, satisfying his requirements in rest, the possibility of restoring his workable state, readiness of the technical means and the operators to further operation. As applied to the system "Navigator-Navigation Complex", the reliability of its functioning may be defined as probability of the fact, that the system would preserve workability, at least during the previously set period of time, when it is being used in certain conditions. The reliability of the system depends on the reliability of its elements of the way they are joined into one system and the function of each one.

The systems may be reserved and no reserved (i.e. the absence of doubled elements), reserved ("hot" reserve), the case when additional reserve elements are introduced with the purpose of the improvement of the system reliability. In case of the failure of the principal element, the system switches automatically on the reserved element. And at last there are reserved systems with light ("cold") reservation. In this case every reserved element is in the so called "cold" reserve, and a certain time is required for its putting into operation.

There are two types of failures in the Theory of Reliability: abrupt and gradual ones. The first one is instantaneous failure, which makes the operation of the system impossible. The gradual failure is connected with the gradual deterioration of the system characteristics. Further is this paper we are going to consider only abrupt failures, more common in the shipboard conditions, and we are going to present the system "Navigator-Navigation Complex" as a simple system (without reservation) in such system the failure of any element causes the failure of the whole system. The faultless operation of a simple system (fig.1) requires the faultless action of its every element. It may be assumed that

the elements N and NC may fail independently of each other.

-------------- N --------------- NC ----------
Figure 1.

According to the "Reliability Theory", the reliability of a simple system, consisting of independent elements equals to the reliability of its elements

P=Pn * Pnc

For the quantitative determination of the reliability of the given system we introduce the notion failure intensity λ (t), this one denotes the mean quantity of failures in a unit of time for one working element. We consider that failure of the operator is his inability to perform the required functions in the system "Navigator-Navigation Complex"; the failure of the system is the faults causing the loss of the ability to function in the required mode of operation.

Theoretically [6] the reliability is expressed in the intensity of the failures; in this particular case when λ (t) = const, the reliability (the probability of the faultless operation of a certain element in the given condition for length of time t and defined by the formula

$$P(t) = e^{-\lambda t}$$

This is the so called Exponential Law of Reliability. When the elements of the system are coupled serially the intensities of failures are added.

Originally analyzing the reliability of the system consisting of the elements N and NC, we believed that the failures of the elements occur independently of each other: Nevertheless this assumption is not always true: under the conditions of sea navigation, both elements may be influenced by one or a whole group of random factors (roll, vibration, wave blows, temperature), simultaneously influencing their reliability. These factors determine the mode of the system operation. Let us analyze the system consisting of two serially joined elements N and NC which is able to operate in one of the three modes R_1, R_2, R_3, the probabilities of which are $P(R_1)$ = 0.5; $P(R_2)$ = 0.3; $P(R_3)$ = 0.2. As a result of the action on the system differently characterized factors, the intensities of the failure streams of the elements N and NC in different modes of functioning of the system will be different. With a very high degree of conventionality we may accept the next intensities of the failure streams of elements N and NC : at regime R_1 they equal to 0.05 and 0.1 (failures an hour) respectively; at regime R_2 – 0.1 and 0.2; at regime R_3 – 0.2 and 0.3.

Now we will determine the reliability of the system "N – NC" for the period of a navigational watch: t = 4hrs

The conditional reliabilities of the system at the 3 modes are found in the following way

$$P(t/R_1) = e^{-(0,05+0,1)t} = e^{-0,15t};$$

$$P(t/R_2) = e^{-(0,1+0,2)t} = e^{-0,3t};$$

$$P(t/R_3) = e^{-(0,2+0,3)t} = e^{-0,5t};$$

The full (unconditional) reliability of the system P(t) is found:

$$P(t) = 0.5e^{-0,15t} + 0.3e^{-0,3t} + 0.2e^{-0,5t}$$

Assuming t = 4 we acquire:

$$P(4) = 0.5e^{-0,6} + 0.3e^{-1,2} + 0.2e^{-2,0}$$

Thus the reliability of functioning of the system in the period of a watch:

$$P(4) = 0.5* 0.549 + 0.3* 0.301 + 0.2* 0.135 \approx 0.4$$

If we assume that system "N − NC" will operate (function) in two modes: R_1 and R_2, with the probabilities 0.8 and 0.2 and the intensities of the failure streams of the elements equal: at mode R_1 − 0.05 and 0.1 at R_2 − 0.1 and 0.2 (failures an hour) using the calculation analogous to the above we will a acquire the value of the system operation reliability $P(4) \approx 0.5$.

4 CONCLUSION AND PROSPECTS OF THE FURTHER RESEARCH IN THIS DIRECTION

1 In spite of the accepted conditional values of the operation modes failure intensities for the system ("Navigator - Navigational Complex") and the failure intensities of each element (necessary statistical data are absent) the acquired results allow to come to the conclusion that the reliability of a system will be higher under two conditions:
   − high reliability of every element
   − minimization in intensity and the length of the action on the system of various nature factors, causing the failure of the operator, or/and technical means.
2 In case of the navigation complex the first condition may be assured by using highly reliable elements and the reservation ("hot reserve") of some parts as well as the systematic supervision and reliable maintenance. As to the operator-navigator the problem is much more difficult: the statistic data of the world fleet disasters evidence that the man is the weakest link of the above system: about 80% (per cent) of all the wrecks at sea happen due to human mistake, mainly those of navigators. That is why a watch officer should be sufficiently competent (knowledge and skills in operating and servicing the navigation complex, familiarity with the opportunities and limitations of its instruments and devices, which is always the result of the proper training (simulator course etc.) as well physically and psychologically ready for the performance of the required functions and efficient use of the technical means in the conditions of destabilizing factors.
3 The second condition is connected with the necessity of designing and manufacturing the navigation instruments and devices highly protected from the negative factors of nature and technical origin.
4 The compliance with these conditions, requiring solution of number of particular problems, will provide for the possibility of the efficient interaction of the man and the techniques in the system of the ship control, which will inevitably result in the decrease of disaster rate at sea .

REFERENCES

[1] M. Lutzhoft. The Technology's Great When it Works, Maritime Technology and Human Integration on the Ship's Bridge./ M/ Lutzhoft // Seaways, NI – June 2005, p. 21-23
[2] N. JoyKody, Liu Zhengiang. Modern Technology. Mariners awareness, competence and confidence/ N. JoyKody, Liu Zhengiang. // Seaways, NI – September 2009, p. 27-30.
[3] H. Mehrkens. Improving the Life at Sea / H. Mehrkens. // Seaways, NI – June 2003 p. 17-18.
[4] S. Ahvenjarvi. Overreliance on the user of Integrated Navigation and Control System can cause Accidents./ S. Ahvenjarvi // IAMU Journal – 2007. – Vol.5, #1, p. 273-276.
[5] Ch. Kuo. Safety Management and its Maritime Application./ NI, London, 2007, p. 289.
[6] Wentzel E. Operation`s research. Moscow, Soviet radio, 1972, 552 p.

Fuzzy Risk of Ship Grounding in Restricted Waters

P. Zalewski
Maritime University of Szczecin, Poland

ABSTRACT: The failure data available for power, navigation, propulsion and steering systems onboard ships are often accompanied with a high degree of uncertainty. For this reason the use of conventional probabilistic risk assessment methods may not be well suited. The approach described uses fault tree analysis to calculate the fuzzy probability of the systems failure. The risks associated with failure events are determined by combining their fuzzy probability and possible - fuzzy consequences to produce a fuzzy risk ranking. The parameters of fuzzy consequences model and the total risk ranking comprising grounding risk of seaworthy vessel and defective vessel are based on simulation trials run in the Full Mission Ship Simulator in Maritime University of Szczecin.

1 INTRODUCTION

Quantification of any risk generally considers two parameters:
– Probability of hazardous event occurrence,
– Consequence severity.

$$R = P_{Ac} C \qquad (1)$$

These two parameters are evaluated in many risk assessments utilised in contemporary marine FSA (formal safety assessment) (IMO 2002, 2007).

In case of ship's grounding event the data behind the resultant probability of accident is usually analyzed independently for initiating error (human or equipment failure) and resulting hazardous event. Then it is combined as (Krystek 2009, Ślączka 2011):

$$P_{Ac} = P_{Herr} P_T P_G \qquad (2)$$

where:
– P_{Ac} – resultant probability of accident,
– P_{Herr} – component probability of human error,
– P_T – component probability of technical failure,
– P_G – component probability of hazardous event – grounding for instance.

The equation (2) is usually simplified to:

$$P_{Ac} = P_T P_G \qquad (3)$$

while conducting full mission simulation studies with real navigators prone to human errors. In such case P_{Herr} is already included inside P_G. However, being strict, the cumulative probability of grounding accident is sum of grounding probability with and without technical failure occurrence, so eq. (2) takes form:

$$P_{Ac} = P_G(G \mid T) \cdot P_T + P_G(G \mid 0) \qquad (4)$$

where:
– $P_G(G|0)$ – component probability of grounding without any technical failure
– $P_G(G|T)$ – component probability of grounding in case of technical failure.

$G|T$ and T are assumed to be independent events and $G|0$ is treated exclusively of $G|T$.

In the case of data relating to equipment or material failure, the attributes of the equipment are often not recorded and insufficient data is given in the context of its use. Almost invariably, failures from historical data are assumed to be random in time, that is, the observed number of failures is divided by an exposure period to give a failure rate. In reality, some modes of failure are more common in the earlier or later years of the life of a component or a system, and some modes of failure are rectified in newer versions of a system though other can be generated. Even where data is of high quality, sample sizes are often small and statistical uncertainties are correspondingly large. As such, a fuzzy logic modelling approach may be appropriate to model the probability P_T (Pilay 2003).

The quantification of consequences severity can be accomplished in several ways, subjective reasoning and expert judgement is one of the common methods. As many accidents in the marine industry, especially on LNG carriers, can lead to catastrophic consequences, it may be difficult to quantify the severity of an accident strictly or "crisply". Once again the use of a fuzzy logic modelling approach integrating expert knowledge may be well suited for this purpose (Pilay 2003).

2 PROBABILITY OF FAILURE EVENT

After failure data collection, the fault tree construction is commenced – data must be grouped and sorted by its equipment/sub-system and finally the system to which the component belongs. The top event of the fault tree will be the failure of the equipment (e.g. gyrocompass), while the initiating events and basic events will be the component failures (e.g. too high internal DC power supply, exceeding limits of gyro rate of turn). It is best to construct a fault tree for equipment within a system separately as it enables data handling and analysis to be conducted. The individual fault trees can later be combined to analyse the total system failure.

In the structure selection phase, the linguistic variable is determined with respect to the aim of the modelling of an undesired critical event. Informally, a linguistic variable is a variable whose values are words or sentences rather than numbers. In case of technical failure event the linguistic terms to describe its occurrence likelihood can be, for example: Very High, High, Moderate, Low and Remote.

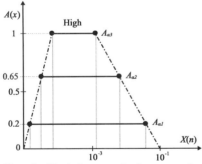

Figure 1. Vertical approach for a membership function determination based on α-cuts.

Six types of experimental methods help to determine membership functions: horizontal approach, vertical approach, pair-wise comparison, inference based on problem specification, parametric estimation and fuzzy clustering (Ying 2000). The membership function chosen must be able to represent the available data in the most suitable and accurate manner. As no mathematically rigorous formulas or procedures exist to accomplish the design of input fuzzy sets – the proper determination of design parameters is strictly dependent on the experience with system behaviour. For simplification of the arithmetic involved, the shape of the membership function suited would either be triangular or trapezoidal, therefore the horizontal or vertical approach for function determination can be applied.

Given the critical event or undesired condition (top event), a fault tree can be developed backwardly to create a network of intermediate events linked by logic operators (usually AND and OR operators) down to the basic events, and to enable fuzzy calculation. The fault tree itself is the logic structure relating the top event to the primary events. These primary/basic events may be related to human error (operators, design or maintenance), hardware or software failures, environmental conditions or operational conditions. The probability of the fuzzy set A as the expected value of the membership function for A can be defined as:

$$\widetilde{P}_A = \int_X \mu_A(x)dP \qquad (5)$$

on a discrete space $X = \{x_1, x_2, x_3 \dots x_n\}$ this gives:

$$\widetilde{P}_A = \sum_{i=1}^{n} \mu_A(x_i)P(x_i) \qquad (6)$$

Equations (5) and (6) define the probability of a fuzzy event as the summation over all elements, or as the probability the event occurs weighted by the degree to which the element is a member of the event. Alternatively, it can be viewed as the probability of the possibility of the fuzzy event.

In a parallel system of two components A and B with fuzzy probabilities the system probability failure will be the product of the individual component failure probabilities similarly as is traditionally assumed:

$$\widetilde{P}_{sys} = \widetilde{P}_A \widetilde{P}_B \qquad (7)$$

where:
- $\widetilde{P}_{sys}$ is fuzzy system probability of failure,
- $\widetilde{P}_A$ and $\widetilde{P}_B$ are fuzzy probabilities of failure events A and B.

In a series system, all constituent components must be operational in order for the system to work. Series systems are analysed in terms of their component reliabilities. The analysis of a series system using reliabilities is identical to that of a parallel system using failure probabilities. In terms of failure probabilities for a series system of two components A and B with fuzzy probabilities the system probability failure will be:

$$\tilde{P}_{sys} = [1 - (1 - \tilde{P}_A)(1 - \tilde{P}_B)] \qquad (8)$$

Fuzzy arithmetic operations are equivalent to basic operations on real numbers extended to those on fuzzy intervals. A fuzzy interval A is a normal fuzzy set on R (set of real numbers) whose α-cuts for all $\alpha \in (0,1]$ are closed intervals of real numbers and whose support is bounded by A. When α-cut representation is employed, arithmetic operations on fuzzy intervals are defined in terms of arithmetic operations on closed intervals. The exemplary addition operation on the α-cut: $\mu(x) = \alpha = 0.5$ is presented at Figure 2.

Using symbols $[a_1^\alpha, a_2^\alpha]$ and $[b_1^\alpha, b_2^\alpha]$ to denote for each α-cuts of fuzzy intervals A and B (where $\alpha \in (0,1]$), the exemplary addition operation on these intervals can be defined by the following formula:

$$A_\alpha + B_\alpha = [a_1^\alpha + b_1^\alpha, a_2^\alpha + b_2^\alpha] \qquad (9)$$

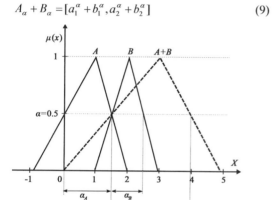

Figure 2. Addition operation on the fuzzy set α-cut

3 CONSEQUENCES EVALUATION

The consequences compiled for each event/failure have to be categorised or grouped for further analysis. In the general LNG carrier operations, four groups experiencing their consequences can be identified: personnel, equipment, operation and environment. For each event or failure, a rating describing the consequences of an event occurring in linguistic terms such as "Negligible", "Marginal", "Critical" and "Catastrophic" can be utilised. The significance of each of the ratings can be described as follows:

Personnel: effect of failure of the item on personnel (worst case always assumed)
Rating 1 = Negligible (no or little damage – bruises / cuts)
Rating 2 = Marginal (minor injuries – treatable, on board)

Rating 3 = Critical (major injuries – requires professional attention)
Rating 4 = Catastrophic (death/permanent disablement)
Equipment: effect of failure on machinery or system in terms of down time and cost of repair:
Rating 1 = Negligible (no or little attention needed – cleaning up / drying / resetting)
Rating 2 = Marginal (minor repair – few hrs lost)
Rating 3 = Critical (major repair – few days lost)
Rating 4 = Catastrophic (destruction of equipment – total system shutdown / ship lost)
Operation: effect of failure on LNG carrier operation in terms of down time:
Rating 1 = No effect (no or little effect)
Rating 2 = Marginal effect (operation affected for a few hours)
Rating 3 = Critical effect (operation affected for a few days)
Rating 4 = Catastrophic effect (operation affected for a few months)
Environment: effect of failure of the item on the environment
Rating 1 = No effect (no or little effect)
Rating 2 = Marginal effect (can be controlled by ship staff)
Rating 3 = Critical effect (requires shore assistance)
Rating 4 = Catastrophic effect (permanent damage to the environment)

Upon assigning a score for each group, a table is generated as shown in Table 1. From this table, a "Total Score" is calculated by summing the score of each individual group for an event. This total score will later be used to assign the membership function for that event using fuzzy rules.

Table 1. Event score.

	Personnel	Equipment	Operation	Environment	Total Score Σx_{ij}
Failure y_1	x_{11}	x_{21}	x_{31}	x_{41}	Σx_{i1}
Failure y_2	x_{12}	x_{22}	x_{32}	x_{42}	Σx_{i2}
Failure y_3	x_{13}	x_{23}	x_{33}	x_{43}	Σx_{i3}
...	...	...	...	...	...
Failure y_n	x_{1n}	x_{2n}	x_{3n}	x_{4n}	Σx_{in}

For example, the fuzzy rules determining the membership function of each event can be divided into 4 categories or hazard classes: HC1, HC2, HC3 and HC4. The maximum score of an event is used to assign that particular event to the appropriate hazard class. Therefore, if an event has exemplary score of {2,2,1,1} for each group respectively, it would be assigned to HC2 (the maximum score for that event is 2 for the "Personnel" and "Environment" categories). The membership functions for each hazard category should be generated based on available historical data, experience and complemented by expert knowledge. If an event has a score of {1,1,1,1}, which means that the effect of the failure is negligible on all categories, then the total effect of that failure on the system and environment should be negligible as well. That is why:

1) If $\sum_{j=1}^{n} x_{ij} = 4$,

then the membership value of "Negligible" fuzzy set: $\mu_N(x) = 1.0$, (N-Negligible) (rule HC1₁)

The minimum score possible in the HC2 category is 5, i.e. {2,1,1,1} or any variation of this score. The maximum possible score is 8, i.e. {2,2,2,2}, therefore the range of membership function between these two extremities have to be assigned so as to ensure a smooth transition between limits and overlapping of functions. Hence, according to Figure 3 (given as example):

2) If $\max(x_{ij}) = 2$, and

$\sum_{j=1}^{n} x_{ij} = 5$ then $\mu_N(x) = 0.75$, $\mu_M(x) = 0.5$, (M-Marginal) (HC2₁)

$\sum_{j=1}^{n} x_{ij} = 6$ then $\mu_M(x) = 1.0$, $\mu_C(x) = 0.2$, (C-Critical) (HC2₂)

$\sum_{j=1}^{n} x_{ij} = 7$ then $\mu_M(x) = 0.7$, $\mu_C(x) = 0.85$, (HC2₃)

$\sum_{j=1}^{n} x_{ij} = 8$ then $\mu_M(x) = 0.1$, $\mu_C(x) = 1.0$,

$\mu_{Ca}(x) = 0.2$, (Ca-Catastrophic) (HC2₄)

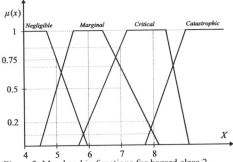

Figure 3. Membership functions for hazard class 2

The minimum score possible in the HC3 category is 6, i.e. {3,1,1,1} or any variation of this score. The

maximum possible score is 12, i.e. {3,3,3,3}. When assigning the linguistic membership function for HC3, it is important to compare the values with those of the HC2 to ensure that it does not contradict the rules generated for that hazard class. For the same total score in HC2 and HC3, the linguistic membership function for HC3 should logically reflect a more severe consequence. For example, for a total score of 8 for HC2 and HC3, which would have a combination of {2,2,2,2} and {3,2,2,1} respectively, using expert judgement, one would say that although both classes have the same total score, a total score of 8 for HC3 would entail a more severe consequence. Hence the membership function for HC3 and a total score of 8 can be 0.5 Critical, 0.5 Catastrophic while the membership function for HC2 with the same total score of 8 is 0.1 Marginal, 1.0 Critical, 0.2 Catastrophic. Using this method, the rules for HC3 can be generated for the other values of its total scores as are reflected below:

3) If $\max(x_{ij}) = 3$, and

$\sum_{j=1}^{n} x_{ij} = 6$ then $\mu_M(x) = 0.5$, $\mu_C(x) = 1.0$, (HC3₁)

$\sum_{j=1}^{n} x_{ij} = 7$ then $\mu_M(x) = 0.1$, $\mu_C(x) = 0.8$,

$\mu_{Ca}(x) = 0.2$, (HC3₂)

$\sum_{j=1}^{n} x_{ij} = 8$ then $\mu_C(x) = 0.5$, $\mu_{Ca}(x) = 0.5$, (HC3₃)

$\sum_{j=1}^{n} x_{ij} = 9$ then $\mu_C(x) = 0.2$, $\mu_{Ca}(x) = 0.8$, (HC3₄)

$\sum_{j=1}^{n} x_{ij} \geq 10$ then $\mu_{Ca}(x) = 1.0$, (HC3₅)

4) If $\max(x_{ij}) = 4$, and $\sum_{j=1}^{n} x_{ij} \geq 7$ then

$\mu_{Ca}(x) = 1.0$, (HC4₁)

Grouping each event into a hazard class allows direct comparison with other events and enables the effects of a failure to be compared based on its linguistic terms assigned to it. For example, if an event A has a score of {3,1,1,2} with a total of 7 and event B has a score of {1,2,2,2} which also gives a total of 7, from experience and expert judgements, it can be said that event A is in most cases more serious in nature. Hence, it should be assigned a linguistic term which must be "more severe" compared to event B. Therefore, the membership functions for events A and B will be obtained from Rules (HC3₂) and (HC2₃), respectively. As each event has been assigned occurrence likelihood and possible consequences, the next step would be to analyse these two parameters and provide a risk ranking number for each event.

4 RISK ASSESSMENT

4.1 *Fuzzy risk*

The risk associated with an event increases as either its severity of the consequences or its occurrence probability increases. Judgement of the severity of possible consequences is, by its very nature, highly subjective. Using a priority table (Table. 2), the "riskiness" of an event in linguistic terms can be obtained. The interpretation of risk ranking is given as below:

Extreme – needs immediate corrective action.
Important – review and corrective action to be carried out.
Moderate – review to be carried out and corrective action implemented if found to be cost effective.
Low – review subject to availability of revenue and time.

Table 2. Risk assessment from a probability and consequence table .

		Severity of consequences			
		Negligi-ble	Marginal	Critical	Cata-strophic
Probability of occurrence	Remote	RN	RM	RC	RCa
	Low	LN	LM	LC	LCa
	Moderate	MN	MM	MC	MCa
	High	HN	HM	HC	HCa
	Very High	VHN	VHM	VHC	VHCa
		Risk			
		Low	Moderate	Im-portant	Extreme

Fuzzy set approach may provide alternative way of assessing risk. The analysis uses linguistic variables to describe severity and probability of occurrence of the failure as in Table. 2 but these parameters are "fuzzified" to determine their degree of membership in each input class using the membership functions developed (Figure 4). The resulting fuzzy inputs are evaluated using the linguistic rule base to yield a classification of the risk of the failure and an associated degree of membership in each class. This fuzzy conclusion is then defuzzified to give a single crisp priority (or crisp risk ranking) for the technical failure.

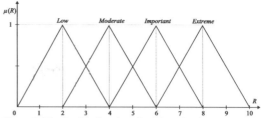
Figure 4. Membership functions for risk evaluation

Rules are evaluated using min-max inference to calculate a numerical conclusion to the linguistic rule based on their input value (Ying 2000, Zalewski 2009). This way the value of obtaining the truth by a rule is determined from the conjunction (i.e. minimum degree of membership of the rule antecedents). Thus this value is taken to be the smallest degree of truth of the rule antecedents. It is then applied to all consequences of the rule. If any fuzzy output is a consequent of more than one rule, that output is set to the highest (maximum) truth-value of all the rules that include it as a consequent. The result of the rule evaluation is a set of fuzzy conclusions that reflect the effects of all the rules whose truth-values are greater than zero. Consider the risk assessment table (Table 2) and membership functions for risk evaluation (Figure 4) where the probability of occurrence is "High", the severity is "Marginal" and their associated degrees of belief are 0.6 and 1.0, respectively. Thus the conclusion Risk = "Important" has a membership value of min (0.6,1.0) = 0.6. To establish how risky the hazard is, this fuzzy conclusion has to be defuzzified to obtain a single crisp result. The defuzzification process creates a single assessment from the fuzzy conclusion set expressing the risk associated with the event, so that corrective actions can be prioritised. Several defuzzification techniques have been developed (Ying 2000). The common technique is centroid or weighted mean of maximum method. This technique averages the points of maximum possibility of each fuzzy conclusion, weighted by their degrees of truth. Hence, if the conclusion from the risk evaluation phase is, for example, 0.6 "Low", 0.2 "Low" and 0.3 "Moderate", the maximum value for each linguistic term is taken. This reduces the conclusion to 0.6 "Low" and 0.3 "Moderate" to be defuzzified.

4.2 *Fuzzy risk of ship grounding*

The following example demonstrates how risk of LNG carrier grounding, while approaching Świnoujście harbour in Poland, can be obtained.

Going back to the equations (4) and (1): $P_G(G|0)$ has to be evaluated by means of statistically viable simulation trials of seaworthy vessel, $P_G(G|T)$ has to be evaluated by means of statistically viable simulation trials of defective vessel at chosen, worst case, fault origin position. The examples of such calculations are presented in (Zalewski 2007, Ślączka 2011). The resultant numeric values of component grounding probabilities at specific waterway's position can be derived by fitting yearly ships' path distributions, so their values will be ratio of yearly occurrences at dangerous isobaths to total ship's occurrences. Suppose for worst case inside harbour heads the probabilities are:

$$P_G(G|0) = 3.6 \times 10^{-4}$$

$$P_G(G|T) = 2.7344 \times 10^{-1}$$

P_T can be derived either as crisp value from historical failure distributions or as fuzzy one. Considering fuzzy probability of technical failure (P_T) - its linguistic values have to be assigned to numeric intervals by expert knowledge / experience.

IMO report of LNG carriers incidents (IMO 2007) presents the results of hazard identification (HAZID) conducted as a one-day workshop with participants from various sectors within the LNG industry, i.e. ship owner/operator, shipyard, ship design office/maritime engineering consultancy, equipment manufacturer, classification society and research centre/university. The results from the hazard identification were recorded in a risk register, which contains a total of 120 hazards within 17 different operational categories. The top ranked hazards according to the outcome is presented in the Table 3. Each hazard is associated with a risk index based on qualitative judgement by the HAZID participants.

Table 3. Results from hazard identification: Top-ranked hazards (IMO 2007)

No.	Hazard	Risk Index
1	Faults in navigation equipment (in coastal waters)	7.0
2	Crew falls or slips onboard	7.0
3	Shortage of crew when LNG trade is increasing	6.8
4	Rudder failure (in coastal waters)	6.8
5	Rudder failure (in manoeuvring)	6.8
6	Severe weather causing vessel to ground/collide (in transit)	6.6
7	Steering and propulsion failure (in manoeuvring)	6.6
8	Severe weather causing vessel to ground/collide (in manoeuvring)	6.6
9	Faults in navigation equipment (in manoeuvring)	6.6
10	Steering and propulsion failure (in coastal waters)	6.6
11	Collision with other ships or facilities (in port)	6.6
12	Terrorist attacks/intentional accidents	6.5

Practically Table 3 shows what faults have to be included into simulation studies to obtain realistic risk assessment of ship's grounding including faulty conditions. Suppose the event "Faults in navigation equipment" during LNG carrier approach operation has the probability of occurrence P_T in fuzzy terms "Moderate" ("High" of membership value 0.4, "Moderate" of membership value 0.9, "Low" of membership value 0.5). It can be obtained by analysing the event fault tree of individual subsystems failures like: gyrocompass, GPS, radar/ARPA, AIS, ECDIS. If fuzzy terms underneath "Low", "Moderate" and "High" probabilities have triangular functions with support intervals set by expertise as:

"Low" $10^{-6} \div 10^{-5}$
"Moderate" $10^{-5} \div 10^{-4}$
"High" $10^{-4} \div 10^{-3}$

then the final support value or defuzzified value of P_T can be calculated by the weighted mean of maximum method. Taking the maximum value for each term of the P_T, that is, $(5.5 \times 10^{-6}, 0.5)$, $(5.5 \times 10^{-5}, 0.9)$, $(5.5 \times 10^{-4}, 0.4)$ the weighted mean is calculated as follows:

$$P_T = \frac{0.5 \times 5.5 \times 10^{-6} + 0.9 \times 5.5 \times 10^{-5} + 0.4 \times 5.5 \times 10^{-4}}{0.5 + 0.9 + 0.4} \approx (10)$$

$$\approx 1.51 \times 10^{-4}$$

Inputting to (4):
$P_{Ac} \approx 4.01 \times 10^{-4}$
or in fuzzy terms: 0.8 "Remote" (it is far below acceptable grounding probability in non-tidal areas of 7×10^{-3} (Gucma 2001))

The severity of consequences of this event in case of ship's grounding is 1.0 "Critical" ("Critical" of membership value 1.0) and 0.3 "Catastrophic". Therefore, the following terms of risk are generated (Table 2) by "min" rule:
0.8 "Remote", 1.0 "Critical" = RC = 0.8 "Moderate"
0.8 "Remote", 0.3 "Catastrophic" = RCa = 0.3 "Moderate"
and by "max" rule:
R="Moderate" with membership 0.8,
or in crisp terms from the Figure 4:
R=4.2

5 CONCLUSIONS

Based on the research presented herein, the following conclusions can be drawn:
– Presented method quantifies risk value by linguistic term and its corresponding fuzzy support value. Though scale of support values is arbitrary, it can be modified freely for further analytical calculation.
– Presented method can be easily applied into assessment of simulation scenarios during LNG carrier navigators / operators' training or simulation based safety studies.

REFERENCES

IMO 2002. *Guidelines for Formal Safety Assessment (FSA) for use in the IMO rule-making process*, IMO MSC/Circ.1023-MEPC/Circ.392.

IMO 2007. *FSA - Liquefied Natural Gas (LNG) carriers submitted by Denmark*, IMO MSC83/21/1.

Gucma S. 2001. *Inżynieria ruchu morskiego,* in Polish, Okrętownictwo i Żegluga. Gdańsk.

Krystek R. (ed.) 2009. *Zintegrowany system bezpieczeństwa transportu tom. 2*, in Polish, WKiŁ, Warszawa.

Pillay A. (ed.) 2003. *Technology and Safety of Marine Systems*, Elsevier Ocean Engineering Series.

Ślączka W. 2011. *Estimation of the consequences of LNG vessel tank leakage in the port of Świnoujście*. Scientific Journals Maritime University of Szczecin, 2011, 25(97) pp. 70–76.

Ying H. 2000. *Fuzzy Control and Modelling, Analytical Foundations and Applications.* IEEE Press, New York.

Zalewski P., Tomczak A. 2007: Analysis of navigational safety by means of simulation studies in the Marine Traffic Engineering Centre in Szczecin, MTE Conference Świnoujście 2007.

Zalewski P. 2009. *Fuzzy Fast Time Simulation Model of Ship's Manoeuvring*, subchapter in Marine Navigation and Safety of Sea Transportation" edited by Weintrit, CRC Press/Balkema, Taylor & Francis Group, London.

Safety, Reliability and Risk Assessment
Advances in Marine Navigation – Marine Navigation and Safety of Sea Transportation – Weintrit (ed.)

Generic Competencies for Resilient Systems

S. Möckel, M. Brenker & S. Strohschneider
Department for Intercultural Communication and Cultural Studies (IWK), Friedrich-Schiller-University Jena, Germany

ABSTRACT: It is popular wisdom that humans are fallible and there are numerous publications that stress the importance of "human error" in the chain of events that lead to a maritime accident. However, humans are often also the last defense that prevents maritime casualties from happening. They should, therefore, rather be considered as a safety contributing factor than the main cause for accidents. It is the people at the sharp end who have the last chance to bring back a malfunctioning system into a well-defined state. Backed up by observations published in official accidents reports we claim that educating seamen with generic competencies concerning information management, communication and coordination, problem-solving as well as effect control would enhance safety onboard particularly in complex critical situations.

1 INTRODUCTION

Safety is among the most prominent topics in the maritime domain. Often, it is instantly associated with technological innovation and the replacement of traditional nautical instruments. This development is supplemented by International Conventions such as International Regulations for Preventing Collisions at Sea (COLREGS), the Standards of Training, Certification and Watchkeeping (STCW) and standards for safe management and operation of ships (ISM Code) which have been adopted by regulating bodies.

While improving technology and regulatory respective standardization efforts are treated as pillars on which maritime safety rests, the seafaring personnel is often considered as the error-prone and safety-critical element within the world of shipping – with human error being the main cause of many maritime casualties (Allianz 2012, 2013 Hetherington 2006, Strohschneider 2010).

In a way, this critical view of the human factor is limiting the approaches taken towards increasing safety. This paper aims at challenging the traditional perspective on the human element in the maritime domain: Acknowledging the potentials while being aware of its fallibilities and thus making the human factor the third pillar in the concept of maritime safety.

2 TOWARDS A PROACTIVE CONCEPT OF SAFETY

Concerning our claim we want to draw attention to the understanding of safety advocated by the International Maritime Organization (IMO) which has recently shifted from a purely reactive approach towards a more proactive concept of safety and security at sea (Carbone 2005, Brenker & Strohschneider 2012). Celebrating 100 years of Safety of Life at Sea (SOLAS) we should remind ourselves of that very event which triggered the whole development: The maiden voyage of the Titanic which sank after colliding with an iceberg and caused the loss of 1,517 lives. The investigations by members of the US Senate and the British Parliament revealed tremendous safety flaws. For instance, the ratio of tonnage and the number of required rescue boats did not take into account the actual number of passengers at that point of time. Thus, the Titanic had only 20 rescue boats with a capacity for a total of 1,178 persons whereas 2,200 persons were onboard (Frey et al. 2010).

Although SOLAS was a tremendous improvement for safety at sea at that time it was by all means a reactive approach: It considered only those factors that had been causal for the sinking of the Titanic. Regulations were tailored to prevent similar accidents. Studying the past thoroughly is certainly not futile but tailoring recommendations in order to prevent what has already happened can be considered a reactive understanding of safety, one that has shaped the thinking of a majority of safety-related industries until the 1980ies (Reason 1990, Brenker & Strohschneider 2012).

However, in the past two decades, starting around 1990, the IMO adopted a more proactive approach towards safety which led to the addition of SOLAS Chapter IX in 1994 and a revised version of the Standards of Training, Certification and Watchkeeping (STCW) in 1995. Following these developments, efforts towards the development of a better safety-culture have been undertaken in the whole shipping industry: The recent adoption of the Manila Amendments to the STCW emphasizes, for instance, concepts such as "marine environmental awareness", as key concepts of proactive behavior which is trained in Maritime Resource Management courses (see also Brenker & Strohschneider 2012).

Still, one could argue that instead of educating the human element in proactive behaviors, training in the use of checklists, handbooks, and in standardized operating procedures (SOP) are used to eliminate humans' supposedly negative impact. Even the implementation of the Manila Amendments in day-to-day operations relies firmly on the use of SOPs; safety audits, for instance, still follow checklists.

3 ACCIDENTS AS NON-ROUTINE SITUATIONS

According to the IMO Code, safety improvements shall, among other sources, be based on the analysis of accidents. Organizations such as the German Federal Bureau for Maritime Casualty Investigation (BSU) are institutions established to examine causes and factors of maritime accidents and derive safety recommendation for the future (BSU 2013).

In order to use official accident reports for scientific purposes, it is important to ponder what accidents reports reveal and about what they in fact remain silent.

These reports refer to events on the tip of the accident pyramid (Grech et al. 2008). Based on accident reports there are hardly any conclusions to be drawn about actions or factors that actually prevented accidents. Events which might provide insight into this issue are incidents, near misses and unsafe acts which occur more frequently and do not necessarily result in accidents (see Fig. 1). Since the incident report systems in place today are handled rather sloppily in practice, there is basically no data accessible which allows us to learn about those factors and actions.

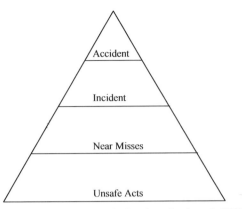

Figure 1. Accident pyramid taken from Grech et al. (2008:17) visualizing the different frequencies with which unsafe acts, near misses, incidents and accidents occur

Coming back to the discussion of accident investigations, there is certainly a lot to be learned from accidents: Concrete examples of things that can go wrong and how people actually behaved in critical situations. However, the official reports represent a rather restricted access to safety at sea, since they can only provide insights into failures of safety measures. With their rather reactive and practically orientated scope, accident investigation reports represent examples of highly non-routine situations that sometimes illustrate the limits of existing safety practices. This makes them a valuable and comprehensive source for safety research.

Accident situations distinguish themselves from routinized standard operations on board a vessel by a set of specific characteristics: Accidents which are categorized as (very) serious maritime casualties pose severe threats to human life, the integrity of the vessel, as well as to the ecological and economic environment. Hence, the time shortly before, during and after the occurrence of an accident can be regarded as highly non-routine. These kind of critical situations can be described as complex: unsafe, uncertain, non-transparent and highly dynamic with crucial decisions to be taken based on conflicting, erroneous or even lacking information (Borodzicz 2004).

4 IRONIES OF RISK MITIGATION AND MANAGEMENT

Analyzing accident reports from the BSU (2013) published between 2003 and 2012 we found evidence that there might be a need for safety recommendations beyond the pillars of regulatory issues (working procedures and standardization) and technology improvements. Some observations made by accident investigators show that the human element involved – through its knowledge, skills or behavior – could have made a difference for the better in the course of events.

Following Bainbridge's (1983) ideas about "the ironies of automation" in which the author discusses the unintended consequences of automation (an expansion rather than elimination of operator related problems), we present selected casualties whose causes and aggravating factors suggest that several procedures and technologies, implemented for safety had, at least in these cases, a rather detrimental effect. In Table 1 we group those accidents according to six ironies which we will explain in the following paragraphs.

Table 1. Selected BSU (2013) reports which exemplify six ironies of risk mitigation and management

No.	Observation / Irony
	Unlikely and unexpected events
176/05	Installation of a wrong shut-off valve caused a fire on a container ship
262/03	Unexpected rupture of a fairlead shackle led to hospitalization of three crew members
637/06	Death of a seaman and three injured after wave
07/10	Unexpected weather conditions led to foundering
	Non-compliant behavior
09/06	Collision as a result of multiple non-compliant behaviors concerning crossing the shipping channel and right of way
181/04	Assuming that bow thrusters are shut off during diving sessions a diver was mortally injured
455/05	Omitting continuous positioning resulted in touching the sea bottom
	Safety flaws and ambiguity
155/04	In the course of the investigation it could not be determined who broke the right of way, resulting in a collision
	Diffuse status of information
167/08	Stranding on an uncharted reef
198/02	Differing convoy lists sent to a tanker are causal for a collision in the Suez Canal
119/05 & 156/03	Contradictory information from pilot and a crew member on lookout about an object in the shipping channel led to a collision
	Inadequate communication
510/09	Communication problems between the pilot and helmsman led to collision
107/08	Communication problems between the captain and the pilot led to a collision
115/06	Discrepancies in arranging the maneuver led to a collision
	Technology brings procedural change
166/05	Maloperation of the autopilot and poor observation of the autopilot's effects led to the death of a crew member
19/03	Restricted availability of the radar due to weather conditions led to a collision. Recommendation about regular training on the usage of navigational equipment.

Unlikely and unexpected events: In investigation reports one repeatedly comes across the phrase that "events took a sudden and unexpected course". The crews had no handbook available nor was there an SOP in place that could have helped to restore safety in this particular situation. Hazards of seafaring are legendary. They are caused by sudden weather changes, phenomena such as single (freak) waves, or the unexpected malfunctioning of instruments or machinery. Whenever we think we have ways and means to deal effectively with every course of events, even the unlikely ones, there are always exceptions no one has ever thought of (Taleb 2004, 2010). SOPs and well-rehearsed emergency drills cannot ever be comprehensive measures to attain safety in any and all situations – we need more ways to deal with uncertainty.

Non-compliant behavior: The rationale behind establishing rules is the firm belief that everyone adheres to them. Adherence to safety rules is an arduous task and people who choose to neglect them in favor of focusing on other aspects of their work are often even rewarded – as long as nothing goes wrong (Dekker 2005).

Safety flaws and ambiguity: In spite of thoroughly devised and adhered to rules and regulations, there are safety flaws where the application of a rule and procedures is ambiguous. We discovered a report where even in the aftermath of an investigation the correct application of rules and procedures could not be determined. It might not be the rules and regulations that are obscure, but complex critical situations definitely are.

Diffuse state of information: It is a characteristic feature of emergency situations that there is a diffuse state of information. Investigations refer to the unavailability of critical information as factors contributing to an accident and therefore propose the integration of displays that would make them available on existing bridges. Ironically, it is a common observation that there already is an overload of information on bridges, which holds especially true in critical situations (Strohschneider et al. 2006).

Inadequate communication: One could also argue that inadequate communication is the result of the existence of rules, well-trained SOP and automation. An increasing availability of data on the bridge in combination with one-man watch schedules reduces the need for interpersonal communication, so that this essential skill withers away (Strohschneider 2010, Dekker et al. 2008). However, particularly in critical and piloting situations effective communication between the bridge team and the assisting pilots would be an essential tool which is backed up by several reports listing inadequate communication as one cause for an accident.

Technology brings procedural change: Technology comes along with procedural change which is the key message of Bainbridge's (1983) observations. Newly introduced (technical) instruments put different requirements on the operator ranging from purely operational skills, to the integration into existing procedures and, finally, the management of critical situations when technology fails. The introduction of ECDIS as a mandatory navigational instrument, for instance, has sparked a debate about the socio-technical error-proneness of the technology and about the navigators' needs for special training (Tang 2009, Jie & Xian-Zhou 2008, Allianz 2013). Sifting through accident reports we found examples of procedural operator errors causing accidents.

Only in some cases the ironies mentioned may have been the major accident causes. Yet they can be definitely regarded as contributing factors to a fatal course of events. They exemplify that measures intended to raise the safety level might under certain circumstances have unintended, and even contradictory, side effects.

5 GENERIC COMPETENCIES FOR RESILIENT SYSTEMS

Technology and regulations, intended to avoid critical situations will, on their own, not be enough to achieve the best possible levels of safety. "[T]he very rules, procedures, and techniques used to bring about excellence in emergency situations may actually contribute to failure in crisis" (Borodzicz 2004:416). A point of view that has been adopted in recent years in the aviation domain as a result of research in high-risk and high-reliability environments: The human element has to be trusted (and supported) in dealing with critical situations instead of being eliminated from the control loop (Dekker et al. 2008).

Reason (1990) distinguishes between people "at the sharp end" who are located at the place in time of the accident (i.e. the crew of seafarers) and those people "at the blunt end" who are indirectly involved into the happenings as, e.g., industrial engineers, agents, policy makers, or designers (Celik et al. 2007). In terms of proactive risk management onboard, the "generic competencies" could be beneficial in mastering complex critical situations and allow the seafarer to mitigate or to manage them successfully. We claim that besides occupational skills and knowledge there is also the need for a set of domain-independent generic competencies that help seafarers at the sharp end to handle critical situations. These will be elaborated in the following paragraphs:

In critical situation seafarers face information overload as well as erroneous, contradicting, incomplete or even lacking information (cf. Tab. 1: diffuse state of information). The seafarer has to learn to cope with these circumstances in a quickly developing situation of stress and threat. The ability to develop strategies to handle the information available and to analyze in order to make valid decisions is called *Information Management* (Bergström et al. 2008, Dörner 1996, Strohschneider 2010).

In routine situations and some particular critical situations (such as man-over-board, abandon-ship, or fire drills) responsibilities and functions onboard are clearly structured. Yet, in a critical situation these structures might need to be adjusted according to the given circumstances. *Communication and Coordination* (cf. Tab. 1: inadequate communication) are indispensable competencies to articulate causal coherences and adapt to unfolding events (Bergström et al. 2008, Strohschneider 2010).

Non-routine situations are characterized by uncertainty as well as their dynamic character (cf. Tab. 1: unlikely and unexpected events, safety flaws and ambiguity). Therefore, decision-making has to take into account all available and relevant information while still being aware of current developments. The process of continuously structuring decision-making and the implementation of decisions is described by *Decision and Implementation*. This competency helps to make to decisions based on what is actually happening and to develop alternatives for action (Bergström et al. 2008, Dörner 1996, Strohschneider 2010).

In rapidly progressing situations it is vital to perform *Effect Control* (cf. Tab. 1: non-compliant behavior, technology brings procedural change). This is the process of checking whether the intended effects of actions are achieved (or not) and whether the situation develops according to or in contrast to the expectations (Bergström et al. 2008, Dörner 1996, Strohschneider 2010)

This set of generic competencies supplements requirements like "situation awareness" or "shared mental models" that are often referred to in the human factors literature as being critical for safe voyages (Stanton et al 2001, Stout et al. 1999). It is comparable to Dörner's (1996) model of decision and problem solving competencies which describes skills that help in transferring knowledge and analogies from one context to another to allow for flexible problem solving.

Predefined and well-rehearsed SOPs, can only prepare for expected critical situations and might fail under (slightly) different conditions, such as similar but yet different scenarios or in the absence of key players. In these situations, a more flexible approach seems promising: "It was found that in every single case of a successfully managed crisis event, the positive outcome could be directly linked to creative or flexible rule breaking by key decision makers in the response" (Borodzicz 2004:418).

6 CONCLUSION

Our call for educating seafarers on generic competencies is not an appeal to ban emergency drills or SOP. Rather, it is an appeal to question them. There is a place and time for each SOP and each regulation – but also for generic competencies. Seamen should trust their own knowledge and skills in decision making and be able to abandon rules and routines if they are detrimental to the safety of crew, ship, or environment. We argue that the human element has unique abilities in dealing with critical and dynamic situations and thus can contribute to the system's recovery from non-routine or critical situations. These abilities do not come out of nowhere, they have to be trained and further developed.

Therefore, we make the case for the training of generic competences, a set of competences that reach beyond occupationally anchored skills and facilitate the handling of new and uncertain situations as an element of Maritime and Bridge Resource Management courses to prepare seafarers for the unexpected.

7 CHALLENGES IN THE MARITIME DOMAIN

Educating seamen on generic competencies confronts us with many challenges. They range from educational and didactic questions to challenges which distinguish the maritime domain from many workplaces ashore. Two questions seem to be of special importance:

1 How can generic competencies be taught in an effective and sustained way? This is a current research question in various domains (Bergström et al. 2009, Heijke et al. 2003 Strohschneider & Gerdes 2004).
2 Who are the key players to be educated? Bearing in mind that crews are affected by high fluctuation (Carbone 2005) and have to work across language barriers (Kahveci et al. 2002, Sampson & Zhao 2003) this becomes a major issue. How can we assure that crews have collectively acquired adequate generic competencies so that the level of safety onboard is actually enhanced?

We do not claim that we already have answers to these questions. However, we argue that trusting the human element at the sharp end and acknowledging its contribution to successful mastering of critical situations is a proactive measure for safety management. It depends on the conception of the human element whether a flexible handling of critical situations in order to return to a routine state is judged as rule breaking or a paradigm shift (Borodzicz 2004) in maritime safety.

ACKNOWLEDGEMENTS

The paper is a product of the MarNet Project which is supported by the German Federal Ministry of Economics and Technology (BMWi). We gratefully acknowledge their support and would also like to thank Leoni Liebetrau, Philine Meyreiß and Michael Babinszki for their preparatory work.

REFERENCES

Allianz Global Corporate & Specialty AG 2012. Safety and shipping 1912-2012.

Allianz Global Corporate & Specialty AG. 2013. Safety and shipping review 2013.

Bainbridge, L. 1983. Ironies of automation. *Automatica 19*: 775-779.

Bergström, J., Dahlström, N., van Winsen, R., Lützhoft, M., Dekker, S. & Nyce, J. 2009. Rule- and role retreat: An empirical study of procedures and resilience. *Journal of Maritime Research 6(1)*: 75-90.

Bergström, J.; Petersen, K. & Dahlström, N. 2008. Securing organizational resilience in escalating situations: Development of skills for crisis and disaster management. In: Hollnagel, E.; Pieri, F. & Rigaud, E. (eds.) *Proceedings of the Third Resilience Engineering Symposium, Antibes Juan-les-Pins, 28-30 October*. Paris: Ecole de mines de Paris: 11-17.

Borodzicz, E. P. 2004. The missing ingredient is the value of flexibility. *Simulation & Gaming* 35(3): 414-426.

Brenker, M. & Strohschneider, S. 2012. The MarNet Project: Assessing seafarers' demands for IMO's Safe Return to Port. In Deutsche Gesellschaft für Ortung und Navigation (eds.), *International Symposium Information on Ships (ISIS) 2012* (CD-ROM). Hamburg: Deutsche Gesellschaft für Ortung und Navigation e.V. (DGON).

BSU: Federal Bureau of Maritime Casualty Investigation 2013. Website of the Federal Bureau of Maritime Casualty Investigation, http://www.bsu-bund.de/EN accessed 03. January 2013.

Carbone, V. 2005. Developments in the labor market. In Leggate, H., McConville, J. & Morvillo, A. (eds.), *International Maritime Transport: Perspectives* New York: Routledge: 61-74

Celik M. & Er I.D. 2007. Identifying the potential roles of design-based failures on Human Errors in shipboard operations. *TransNav - International Journal on Marine Navigation and Safety of Sea Transportation 1(3)*: 339-343, 2007.

Dekker, S. 2005. Ten Questions about human error: A new view of human errors and system safety. Hillsdale: Erlbaum.

Dekker, S., Dahlström, N., van Winsen, R. & Nyce, J. 2008. Crew resilience and simulator training in aviation. In Hollnagel, E., Woods, D. D. & Leveson, N. (eds.). *Resilience Engineering Perspectives*, Aldershot: Ashgate: 119-126.

Dörner, D. 1996. The logic of failure: Recognizing and avoiding error in complex situations. New York: Metropolitan Books.

Frey, B. S., Savage, D. A., & Torgler, B. 2010. Interaction of natural survival instincts and internalized social norms exploring the Titanic and Lusitania disasters. In *Proceedings of the National Academy of Sciences* 107(11): 4862-4865.

Grech M. R., Horberry T. J. & Koester, T. 2008. Human Factors in the Maritime Domain. Boca Raton: CRC Press.

Heijke, H., Meng, C. & Ris, C. 2003. Fitting to the job: The role of generic and vocational competencies in adjustment and performance. *Labour Economics* 10: 215–229.

Hetherington, C., Flin, R. & Mearns, K. 2006. Safety in shipping: The human element. *Journal of Safety Research 37:* 401–411.

Jie, W. & Xian-Zhong, H. 2008. The error chain in using Electronic Chart Display and Information Systems. In *Proceedings of the IEEE International Conference on ProSystems, Man and Cybernetics (SMC), Singapore*, 12-15 October: 1895-1899.

Kahveci, E., Lane, T. & Sampson, H. 2002. Transnational seafarer communities. Cardiff: Seafarers International Research Centre, Cardiff University.

Reason, J. 1990. Human Error. Cambridge: Cambridge University Press.

Sampson, H. & Zhao, M. 2003. Multilingual crews: Communication and the operation of ships. *World Englishes 22(1):* 31-43.

Stanton, N.A., Chambers, P.R.G. & Piggott, J. 2001. Situational awareness and safety. *Safety Science 39(3):* 189-204.

Stout, R. J., Cannon-Bowers, J. A., Salas, E. & Milanovich, D. M. 1999. Planning, Shared Mental Models, and coordinated performance: An empirical link is established. *Human Factors 41(1):* 61-71.

Strohschneider, S. 2010. Technisierungsstrategien und der Human Factor. In Zoche, P., Kaufmann, S. & Haverkamp, R. (eds.) *Zivile Sicherheit: Gesellschaftliche Dimensionen gegenwärtiger Sicherheitspolitiken.* Bielefeld: transcript Verlag: 161-177.

Strohschneider, S., Meck, U. & Brüggemann, U. 2006. Human factors in ship bridge design: Some insights from the DGON-BRIDGE-Project. Hamburg: Deutsche Gesellschaft für Ortung und Navigation e.V. (DGON).

Strohschneider, S. & Gerdes, J. 2004. MS ANTWERPEN: Emergency management training for low risk environments. *Simulation & Gaming 35:* 394-413.

Taleb, N. N. 2004. Fooled by randomness: The hidden role of chance in life and in the markets. London: Penguin Books.

Taleb, N. N. 2010. The black swan: the impact of the highly improbable. New York: Random House Trade Paperbacks.

Tang, L. 2009. Training and technology: Potential issues for shipping. In SIRC Symposium, 8-9 July, Cardiff: Seafarers International Research Centre, Cardiff University.

AUTHOR INDEX